Technological Advances in STEEL DESIGNER

Technological Advances in STEEL DESIGNER

Naseer Khan

RANDOM
RANDOM PUBLICATIONS
NEW DELHI - 110 002 (INDIA)

Technological Advances in Steel Designer

ISBN 978-93-51114-58-1

Published in 2014 in India by
RANDOM PUBLICATIONS
4376-A/4B, Gali Murari Lal, Ansari Road
New Delhi-110 002
Phone: +9111-43580356, 23289044
E-mail: randomexports@gmail.com; sales@randompublications.com; info@randompublications.com

Reprint 2021

Type Setting by: Friends Media, Delhi-110089
Printed at : Thomson Press (India) Ltd

Preface

Steel is an alloy of iron, with carbon, which may contribute up to 2.1% of its weight. Carbon, other elements, and inclusions within iron act as hardening agents that prevent the movement of dislocations that naturally exist in the iron atom crystal lattices. Varying the amount of alloying elements, their form in the steel either as solute elements, or as precipitated phases, retards the movement of those dislocations that make iron so ductile and so weak, and so it controls qualities such as the hardness, ductility, and tensile strength of the resulting steel. Steel can be made stronger than pure iron, but only by trading away ductility, of which iron has an excess. Structural steel is steel construction material, a profile, formed with a specific shape or cross section and certain standards of chemical composition and mechanical properties. Structural steel shape, size, composition, strength, storage, etc., is regulated in most industrialized countries.

Steel design, or more specifically, structural steel design, is an area of knowledge of structural engineering used to design steel structures. The structures can range from schools to homes to bridges. In structural engineering, a structure is a body or combination of pieces of rigid bodies in space to form a fitness system for supporting loads. Structures such as buildings, bridges, aircraft and ships are all examples under steel structure. The effects of loads on structures are determined through structural analysis. Steel structure is steel construction material, a profile, formed with a specific shape or cross section and certain standards of chemical composition and mechanical properties. There are currently two common methods of steel design: The first (and older) method is the Allowable Strength Design (ASD) method. The second (newer) is the Load and Resistance Factor Design (LRFD) method. The design process encompasses the architectural design, the development of the structural concept , the analysis of the steel structure and the verification of members. Steel solutions are lighter than their concrete equivalents, with the opportunity to provide more column-free flexible floor space, less foundations and a fast, safe construction programme. For the designer, a steel solution means reliable materials, known material and section properties, precise off-site manufacture and extensive support including software, design guides and easy to use resistance tables. Steel is ideally suited for design. Material properties are known and member properties are accurate, meaning that analysis is precise. Design rules are clear mature, without undue conservatism, having been developed over many decades. There is a wealth of support resources, including software, to facilitate efficient design. The choice and design of the primary structure is a fundament part of the concept design of buildings, and ideally should be integrated with the development of the architectural design. Meeting client, planning and Building Regulation requirements are paramount, but there will be a range of structural forms that meet these requirements, each with its own advantages. The merits of different structural forms should be reviewed against the requirements for the structure. The principal structural elements of a typical multi-storey building comprise floors, beams and columns. A wide variety of alternative forms and arrangements can be used in multi-storey steel framed structures to deliver the benefits.

This comprehensive book covers the fundamentals of structural steel design with an emphasis on the design of members and their connections, rather than the integrated design of buildings. The book is designed so that instructors can easily teach LRFD, ASD, or both, time-permitting. The application of fundamental principles is encouraged for design procedures as well as for practical design, but a theoretical approach is also provided to enhance student development.

I thank all members of my team who have helped in the preparation of the book. My special thanks go to "Random Publications" who have published the book.

— *Naseer Khan*

Contents

1

Introduction

Steel

Steel is an alloy of iron, with carbon being the primary alloying element. The carbon content of steel is between 0.002% and 2.1% by weight. Too little carbon content leaves (pure) iron quite soft, ductile, and weak. Carbon contents higher than those of steel make an alloy commonly called pig iron that is brittle and not malleable. Additional elements may be present in steel: manganese, phosphorus, sulfur, silicon, and traces of oxygen, nitrogen, and aluminium.

Alloy steel is steel to which additional alloying elements have been intentionally added to modify the characteristics of steel. Common alloying elements include: manganese, nickel, chromium, molybdenum, boron, titanium, vanadium, and niobium.

Overview

Carbon, other elements, and inclusions within iron act as hardening agents that prevent the movement of dislocations that naturally exist in the iron atom crystal lattices. Varying the amount of alloying elements, their form in the steel either as solute elements, or a precipitated phases, retards the movement of those dislocations that make iron so ductile and so weak, and so it controls qualities such as the hardness, ductility, and tensile strength of the resulting steel. Steel can be made stronger than pure iron, but only by trading it for ductility, of which iron has an excess.

Alloys with a higher than 2.1% carbon content, depending on other element content and possibly on processing, are known as cast iron. Cast iron is not malleable even when hot, but it can be formed by casting as it has a lower melting point than steel and good castability properties. Steel is also distinguishable from wrought iron (now largely obsolete), which may contain a small amount of carbon but large amounts of slag. Note that the percentages of carbon and other elements quoted are on a weight basis.

Although steel had been produced in bloomery furnaces for thousands of years, steel's use expanded extensively after more efficient production methods were devised in the 17th century for blister steel and then crucible steel. With the invention of the Bessemer process in the mid-19th century, a new era of mass-produced steel began. This was followed by Siemens-Martin process and then Gilchrist-Thomas process that refined the quality of steel. With their introductions, mild steel replaced wrought iron.

Further refinements in the process, such as basic oxygen steelmaking (BOS), further lowered the cost of production, while increasing the quality of the metal and largely replaced earlier methods. Today, steel is one of the most common materials in the world, with more than 1.3 billion tons produced annually. It is a major component in buildings, infrastructure, tools, ships, automobiles, machines, appliances, and weapons. Modern steel is generally identified by various grades defined by assorted standards organisations.

Material Properties

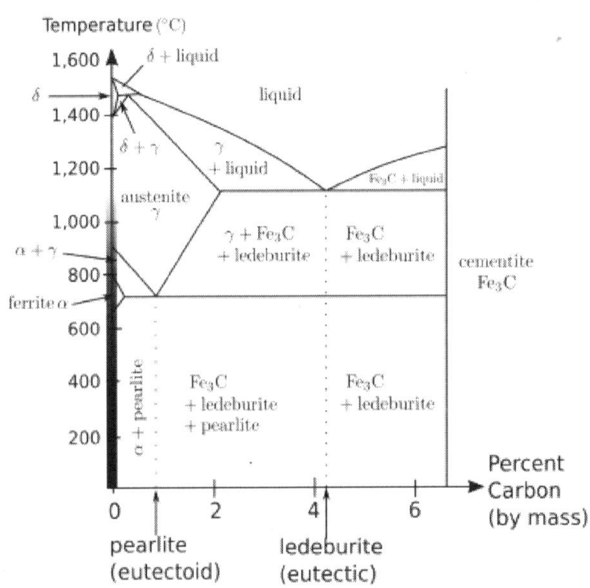

Figure: *Iron-carbon phase diagram, showing the conditions necessary to form different phases*

Iron is found in the Earth's crust only in the form of an ore, usually an iron oxide, such as magnetite, hematite etc. Iron is extracted from iron ore by removing the oxygen by combining it with a preferred chemical partner such as carbon that is lost to the atmosphere as carbon dioxide. This process, known as smelting, was first applied to metals with lower melting points, such as tin, which melts at approximately 250 °C (482 °F) and copper, which melts at approximately 1,100 °C (2,010 °F). In comparison, cast iron melts at approximately 1,375 °C (2,507 °F). Small quantities of iron were smelted in ancient times, in the solid state, by heating the ore buried in a charcoal fire and welding the metal together with a hammer, squeezing out the impurities. With care, the carbon content could be controlled by moving it around in the fire.

All of these temperatures could be reached with ancient methods that have been used since the Bronze Age. Since the oxidation rate of iron increases rapidly beyond 800 °C (1,470 °F), it is important that smelting take place in a low-oxygen environment. Unlike copper and tin, liquid or solid iron dissolves carbon quite readily. Smelting results in an alloy (pig iron) that contains too much carbon to be called steel. The excess carbon and other impurities are removed in a subsequent step.

Other materials are often added to the iron/carbon mixture to produce steel with desired properties. Nickel and manganese in steel add to its tensile strength and make the austenite form of the iron-carbon solution more stable, chromium increases hardness and melting temperature, and vanadium also increases hardness while making it less prone to metal fatigue.

To inhibit corrosion, at least 11% chromium is added to steel so that a hard oxide forms on the metal surface; this is known as stainless steel. Tungsten interferes with the formation of cementite, allowing martensite to preferentially form at slower quench rates, resulting in high speed steel. On the other hand, sulfur, nitrogen, and phosphorus make steel more brittle, so these commonly found elements must be removed from the steel melt during processing.

The density of steel varies based on the alloying constituents but usually ranges between 7,750 and 8,050 kg/m^3 (484 and 503 lb/cu ft), or 7.75 and 8.05 g/cm^3 (4.48 and 4.65 oz/cu in).

Even in a narrow range of concentrations of mixtures of carbon and iron that make a steel, a number of different metallurgical structures, with very different properties can form. Understanding such properties is essential to making quality steel. At room temperature, the most stable form of iron is the body-centred cubic (BCC) structure called ferrite or á-iron. It is a fairly soft metal that can dissolve only a small concentration of carbon, no more than 0.021 wt% at 723 °C (1,333 °F), and only 0.005% at 0 °C (32 °F). At 910°C pure iron transforms into a face-centred cubic (FCC) structure, called austenite or ã-iron. The FCC structure of austenite can dissolve considerably more carbon, as much as 2.1% (38 times that of ferrite) carbon at 1,148 °C (2,098 °F), which reflects the upper carbon content of steel, beyond which is cast iron.

When steels with less than 0.8% carbon, known as a hypoeutectoid steel, are cooled, the austenitic phase (FFC) of the mixture attempts to revert to the ferrite phase (BCC), resulting in an excess of carbon. One way for carbon to leave the austenite is for it to precipitate out of solution as cementite, leaving behind iron that is low enough in carbon to take the form of ferrite, resulting in a ferrite matrix with cementite inclusions. Cementite is a hard and brittle intermetallic compound with the chemical formula of Fe_3C. At the eutectoid, 0.8% carbon, the cooled structure takes the form of pearlite, named for its resemblance to mother of pearl. It is a lamellar structure of ferrite and cementite. For steels that have more than 0.8% carbon, the cooled structure takes the form of pearlite and cementite.

Perhaps the most important polymorphic form of steel is martensite, a metastable phase that is significantly stronger than other steel phases. When the steel is in an austenitic phase and then quenched rapidly, it forms into martensite, as the atoms "freeze" in place when the cell structure changes from FCC to BCC. Depending on the carbon content, the martensitic phase takes different forms. Below approximately 0.2% carbon, it takes an α ferrite BCC crystal form, but at higher carbon content it takes a body-centred tetragonal (BCT) structure. There is no thermal activation energy for the transformation from austenite to martensite. Moreover, there is no compositional change so the atoms generally retain their same neighbours.

Martensite has a lower density than does austenite, so that the transformation between them results in a change of volume. In this case, expansion occurs. Internal stresses from this expansion generally take the form of compression on the crystals of martensite and tension on the remaining ferrite, with a fair amount of shear on both constituents. If quenching is done improperly, the internal stresses can cause a part to shatter as it cools. At the very least, they cause internal work hardening and other microscopic imperfections. It is common for quench cracks to form when steel is water quenched, although they may not always be visible.

Carbon Steel

Carbon steel is steel in which the main interstitial alloying constituent is carbon in the range of 0.12–2.0%. The American Iron and Steel Institute (AISI) defines carbon steel as the following: "Steel is considered to be carbon steel when no minimum content is specified or required for chromium, cobalt, molybdenum, nickel, niobium, titanium, tungsten, vanadium or zirconium, or any other element to be added to obtain a desired alloying effect; when the specified minimum for copper does not exceed 0.40 percent; or when the maximum content specified for any of the following elements does not exceed the percentages noted: manganese 1.65, silicon 0.60, copper 0.60."

The term "carbon steel" may also be used in reference to steel which is not stainless steel; in this use carbon steel may include alloy steels.

As the carbon percentage content rises, steel has the ability to become harder and stronger through heat treating; however it becomes less ductile. Regardless of the heat treatment, a higher carbon content reduces weldability. In carbon steels, the higher carbon content lowers the melting point.

Types

Carbon steel is broken down into four classes based on carbon content:

Mild and Low-Carbon Steel

Mild steel, also called as plain-carbon steel, is the most common form of steel because its price is relatively low while it provides material properties that are acceptable for many applications, more so than iron. Low-carbon steel contains approximately 0.05–0.3% carbon making it malleable and ductile. Mild steel has a relatively low tensile strength, but it is cheap and malleable; surface hardness can be increased through carburizing.

It is often used when large quantities of steel are needed, for example as structural steel. The density of mild steel is approximately 7.85 g/cm^3 (7850 kg/m^3 or 0.284 lb/in^3) and the Young's modulus is 210 GPa (30,000,000 psi).

Low-carbon steels suffer from *yield-point runout* where the material has two yield points. The first yield point (or upper yield point) is higher than the second and the yield drops dramatically after the upper yield point. If a low-carbon steel is only stressed to some point between the upper and lower yield point then the surface may develop Lüder bands. Low-carbon steels contain less carbon than other steels and are easier to cold-form, making them easier to handle.

Higher Carbon Steels

Carbon steels which can successfully undergo heat-treatment have a carbon content in the range of 0.30–1.70% by weight. Trace impurities of various other elements can have a significant effect on the quality of the resulting steel. Trace amounts of sulfur in particular make the steel red-short, that is, brittle and crumbly at working temperatures. Low-alloy carbon steel, such as A36 grade, contains about 0.05% sulfur and melts around 1,426–1,538 °C (2,599–2,800 °F). Manganese is often added to improve the hardenability of low-carbon steels. These additions turn the material into a low-alloy steel by some definitions, but AISI's definition of carbon steel allows up to 1.65% manganese by weight.

Low Carbon Steel

<0.3% carbon content.

Medium Carbon Steel

Approximately 0.30–0.59% carbon content. Balances ductility and strength and has good wear resistance; used for large parts, forging and automotive components.

High-Carbon Steel

Approximately 0.6–0.99% carbon content. Very strong, used for springs and high-strength wires.

Ultra-High-Carbon Steel

Approximately 1.0–2.0% carbon content. Steels that can be tempered to great hardness. Used for special purposes like (non-industrial-purpose) knives, axles or punches. Most steels with more than 1.2% carbon content are made using powder metallurgy. Note that steel with a carbon content above 2.14% is considered cast iron.

Heat Treatment

The purpose of heat treating carbon steel is to change the mechanical properties of steel, usually ductility, hardness, yield strength, or impact resistance. Note that the electrical and thermal conductivity are only slightly altered. As with most strengthening techniques for steel, Young's modulus (elasticity) is unaffected. All treatments of steel trade ductility for increased strength and vice versa. Iron has a higher solubility for carbon in the austenite phase; therefore all heat treatments, except spheroidizing and process annealing, start by heating the steel to a temperature at which the austenitic phase can exist. The steel is then quenched (heat drawn out) at a high rate causing cementite to precipitate and finally the remaining pure iron to solidify.

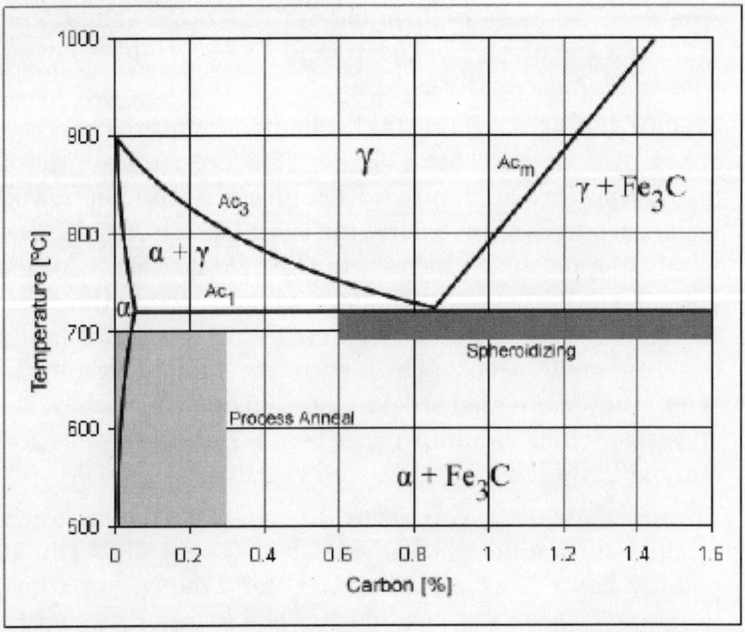

Figure: *Iron-carbon phase diagram, showing the temperature and carbon ranges for certain types of heat treatments.*

The rate at which the steel is cooled through the eutectoid temperature affects the rate at which carbon diffuses out of austenite and forms cementite. Generally speaking, cooling swiftly will leave iron carbide finely dispersed and produce a fine grained pearlite (until the martensite critical temperature is reached) and cooling slowly will give a coarser pearlite. Cooling a hypoeutectoid steel (less than 0.77 wt% C) results in a lamellar-pearlitic structure of iron carbide layers with α-ferrite (pure iron) between.

If it is hypereutectoid steel (more than 0.77 wt% C) then the structure is full pearlite with small grains (larger than the pearlite lamella) of cementite scattered throughout. The relative amounts of constituents are found using the lever rule.

The following is a list of the types of heat treatments possible:

- Spheroidizing: Spheroidite forms when carbon steel is heated to approximately 700 °C for over 30 hours. Spheroidite can form at lower temperatures but the time needed drastically increases, as this is a diffusion-controlled process. The result is a structure of rods or spheres of cementite within primary structure (ferrite or pearlite, depending on which side of the eutectoid you are on). The purpose is to soften higher carbon steels and allow more formability. This is the softest and most ductile form of steel. The image to the right shows where spheroidizing usually occurs.
- Full annealing: Carbon steel is heated to approximately 40 °C above Ac3 or Ac1 for 1 hour; this ensures all the ferrite transforms into austenite (although cementite might still exist if the carbon content is greater than the eutectoid). The steel must then be cooled slowly, in the realm of 20 °C (68 °F) per hour. Usually it is just furnace cooled, where the furnace is turned off with the steel still inside. This results in a coarse pearlitic structure, which means the "bands" of pearlite are thick. Fully annealed steel is soft and ductile, with no internal stresses, which is often necessary for cost-effective forming. Only spheroidized steel is softer and more ductile.
- Process annealing: A process used to relieve stress in a cold-worked carbon steel with less than 0.3 wt% C. The steel is usually heated up to 550–650 °C for 1 hour, but sometimes temperatures as high as 700 °C. The image rightward shows the area where process annealing occurs.
- Isothermal annealing: It is a process in which hypoeutectoid steel is heated above the upper critical temperature and this temperature is maintained for a time and then the temperature is brought down below lower critical temperature and is again maintained. Then finally it is cooled at room temperature. This method rids any temperature gradient.
- Normalizing: Carbon steel is heated to approximately 55 °C above Ac3 or Acm for 1 hour; this ensures the steel completely transforms to austenite. The steel is then air-cooled, which is a cooling rate of approximately 38 °C (100 °F) per minute. This results in a fine pearlitic structure, and a more-uniform structure. Normalized steel has a higher strength than annealed steel; it has a relatively high strength and ductility.

- Quenching: Carbon steel with at least 0.4 wt% C is heated to normalizing temperatures and then rapidly cooled (quenched) in water, brine, or oil to the critical temperature. The critical temperature is dependent on the carbon content, but as a general rule is lower as the carbon content increases. This results in a martensitic structure; a form of steel that possesses a super-saturated carbon content in a deformed body-centred cubic (BCC) crystalline structure, properly termed body-centred tetragonal (BCT), with much internal stress. Thus quenched steel is extremely hard but brittle, usually too brittle for practical purposes. These internal stresses cause stress cracks on the surface. Quenched steel is approximately three to four (with more carbon) fold harder than normalized steel.
- Martempering (Marquenching): Martempering is not actually a tempering procedure, hence the term "marquenching". It is a form of isothermal heat treatment applied after an initial quench of typically in a molten salt bath at a temperature right above the "martensite start temperature". At this temperature, residual stresses within the material are relieved and some bainite may be formed from the retained austenite which did not have time to transform into anything else. In industry, this is a process used to control the ductility and hardness of a material. With longer marquenching, the ductility increases with a minimal loss in strength; the steel is held in this solution until the inner and outer temperatures equalize. Then the steel is cooled at a moderate speed to keep the temperature gradient minimal. Not only does this process reduce internal stresses and stress cracks, but it also increases the impact resistance.
- Quench and tempering: This is the most common heat treatment encountered, because the final properties can be precisely determined by the temperature and time of the tempering. Tempering involves reheating quenched steel to a temperature below the eutectoid temperature then cooling. The elevated temperature allows very small amounts of spheroidite to form, which restores ductility, but reduces hardness. Actual temperatures and times are carefully chosen for each composition.
- Austempering: The austempering process is the same as martempering, except the steel is held in the molten salt bath through the bainite transformation temperatures, and then

moderately cooled. The resulting bainite steel has a greater ductility, higher impact resistance, and less distortion. The disadvantage of austempering is it can only be used on a few steels, and it requires a special salt bath.

Case Hardening

Case hardening processes harden only the exterior of the steel part, creating a hard, wear resistant skin (the "case") but preserving a tough and ductile interior. Carbon steels are not very hardenable; therefore wide pieces cannot be through-hardened. Alloy steels have a better hardenability, so they can through-harden and do not require case hardening. This property of carbon steel can be beneficial, because it gives the surface good wear characteristics but leaves the core tough.

Forging Temperature of Steel

Steel Type	*Maximum forging temperature (°F / °C)*	*Burning temperature (°F / °C)*
1.5% carbon	1920 / 1049	2080 / 1138
1.1% carbon	1980 / 1082	2140 / 1171
0.9% carbon	2050 / 1121	2230 / 1221
0.5% carbon	2280 / 1249	2460 / 1349
0.2% carbon	2410 / 1321	2680 / 1471
3.0% nickel steel	2280 / 1249	2500 / 1371
3.0% nickel–chromium steel	2280 / 1249	2500 / 1371
5.0% nickel (case-hardening) steel	2320 / 1271	2640 / 1449
Chromium–vanadium steel	2280 / 1249	2460 / 1349
High-speed steel	2370 / 1299	2520 / 1382
Stainless steel	2340 / 1282	2520 / 1382
Austenitic chromium–nickel steel	2370 / 1299	2590 / 1421
Silico-manganese spring steel	2280 / 1249	2460 / 1349

History of Steelmaking

Ancient Steel: Steel was known in antiquity, and may have been produced by managing bloomeries, or iron-smelting facilities, in which the bloom contained carbon.

The earliest known production of steel is a piece of ironware excavated from an archaeological site in Anatolia (Kaman-Kalehoyuk) and is about 4,000 years old. Other ancient steel comes from East

Africa, dating back to 1400 BC. In the 4th century BC steel weapons like the Falcata were produced in the Iberian Peninsula, while Noric steel was used by the Roman military.

Steel was produced in large quantities in Sparta around 650BC.

The Chinese of the Warring States (403–221 BC) had quench-hardened steel, while Chinese of the Han Dynasty (202 BC – 220 AD) created steel by melting together wrought iron with cast iron, gaining an ultimate product of a carbon-intermediate steel by the 1st century AD. The Haya people of East Africa invented a type of furnace they used to make carbon steel at 1,802 °C (3,276 °F) nearly 2,000 years ago.

Wootz Steel and Damascus Steel

Evidence of the earliest production of high carbon steel in the Indian Subcontinent was found in Samanalawewa area in Sri Lanka. Wootz steel was produced in India by about 300 BC. However, the steel was an old technology in India when King Porus presented a Steel sword to the Emperor Alexander in 326 BC. The steel technology obviously existed before 326 BC as steel was being exported to the Arab World at that time. Since the technology was acquired from the Tamilians from South India, the origin of steel technology in India can be conservatively estimated at 400-500 BC.

Along with their original methods of forging steel, the Chinese had also adopted the production methods of creating Wootz steel, an idea imported into China from India by the 5th century AD. In Sri Lanka, this early steel-making method employed a unique wind furnace, driven by the monsoon winds, capable of producing high-carbon steel.

Also known as Damascus steel, wootz is famous for its durability and ability to hold an edge. It was originally created from a number of different materials including various trace elements.

It was essentially a complicated alloy with iron as its main component. Recent studies have suggested that carbon nanotubes were included in its structure, which might explain some of its legendary qualities, though given the technology available at that time, they were produced by chance rather than by design. Natural wind was used where the soil containing iron was heated by the use of wood.

The ancient Sinhalese managed to extract a ton of steel for every 2 tons of soil, a remarkable feat at the time. One such furnace was found in Samanalawewa and archaeologists were able to produce steel as the ancients did.

Crucible steel, formed by slowly heating and cooling pure iron and carbon (typically in the form of charcoal) in a crucible, was produced in Merv by the 9th to 10th century AD. In the 11th century, there is evidence of the production of steel in Song China using two techniques: a "berganesque" method that produced inferior, inhomogeneous steel and a precursor to the modern Bessemer process that used partial decarbonization via repeated forging under a cold blast.

Modern Steelmaking

Since the 17th century the first step in European steel production has been the smelting of iron ore into pig iron in a blast furnace. Originally using charcoal, modern methods use coke, which has proven more economical.

Processes Starting from Bar Iron

In these processes pig iron was "fined" in a finery forge to produce bar iron (wrought iron), which was then used in steel-making.

The production of steel by the cementation process was described in a treatise published in Prague in 1574 and was in use in Nuremberg from 1601. A similar process for case hardening armour and files was described in a book published in Naples in 1589. The process was introduced to England in about 1614 and used to produce such steel by Sir Basil Brooke at Coalbrookdale during the 1610s.

The raw material for this process were bars of wrought iron. During the 17th century it was realised that the best steel came from oregrounds iron of a region north of Stockholm, Sweden. This was still the usual raw material source in the 19th century, almost as long as the process was used.

Crucible steel is steel that has been melted in a crucible rather than having been forged, with the result that it is more homogeneous. Most previous furnaces could not reach high enough temperatures to melt the steel. The early modern crucible steel industry resulted from the invention of Benjamin Huntsman in the 1740s. Blister steel (made as above) was melted in a crucible or in a furnace, and cast (usually) into ingots.

Processes Starting from Pig Iron

The modern era in steelmaking began with the introduction of Henry Bessemer's Bessemer process in 1855, the raw material for which was pig iron. His method let him produce steel in large quantities cheaply, thus mild steel came to be used for most purposes for which wrought iron was formerly used. The Gilchrist-Thomas process (or basic Bessemer process*) was an improvement to the Bessemer process, made by lining the converter with a basic material to remove phosphorus.*

Another 19th-century steelmaking process was the Siemens-Martin process, which complemented the Bessemer process. It consisted of co-melting bar iron (or steel scrap) with pig iron.

These methods of steel production were rendered obsolete by the Linz-Donawitz process of basic oxygen steelmaking (BOS), developed in the 1950s, and other oxygen steel making methods. Basic oxygen steelmaking is superior to previous steelmaking methods because the oxygen pumped into the furnace limits impurities that previously had entered from the air used. Today, electric arc furnaces (EAF) are a common method of reprocessing scrap metal to create new steel. They can also be used for converting pig iron to steel, but they use a lot of electricity (about 440 kWh per metric ton), and are thus generally only economical when there is a plentiful supply of cheap electricity.

Steel Industry

It is common today to talk about "the iron and steel industry" as if it were a single entity, but historically they were separate products. The steel industry is often considered an indicator of economic progress, because of the critical role played by steel in infrastructural and overall economic development.

In 1980, there were more than 500,000 U.S. steelworkers. By 2000, the number of steelworkers fell to 224,000.

The economic boom in China and India has caused a massive increase in the demand for steel in recent years. Between 2000 and 2005, world steel demand increased by 6%. Since 2000, several Indian and Chinese steel firms have risen to prominence like Tata Steel (which bought Corus Group in 2007), Shanghai Baosteel Group Corporation and Shagang Group. Arcelor Mittal is however the world's largest steel producer.

In 2005, the British Geological Survey stated China was the top steel producer with about one-third of the world share; Japan, Russia, and the US followed respectively.

In 2008, steel began trading as a commodity on the London Metal Exchange. At the end of 2008, the steel industry faced a sharp downturn that led to many cut-backs.

The world steel industry peaked in 2007. That year, ThyssenKrupp spent $12 billion to build the two most modern mills in the world, in

Calvert, Alabama and Sepetiba, Rio de Janeiro, Brazil. The worldwide Great Recession starting in 2008, however, sharply lowered demand and new construction, and so prices fell. ThyssenKrupp lost $11 billion on its two new plants, which sold steel below the cost of production. Finally in 2013, ThyssenKrupp offered the plants for sale at under $4 billion.

Microbial Degradation

Bacteria such as those consuming the hull of the Titanic can separate the iron from carbon in steel and consume it, eventually leaving only rust.

Uses

Iron and steel are used widely in the construction of roads, railways, other infrastructure, appliances, and buildings. Most large modern structures, such as stadiums and skyscrapers, bridges, and airports, are supported by a steel skeleton. Even those with a concrete structure employ steel for reinforcing. In addition, it sees widespread use in major appliances and cars. Despite growth in usage of aluminium, it is still the main material for car bodies. Steel is used in a variety of other construction materials, such as bolts, nails, and screws.

Other common applications include shipbuilding, pipelines, mining, offshore construction, aerospace, white goods (e.g. washing machines), heavy equipment such as bulldozers, office furniture, steel wool, tools, and armour in the form of personal vests or vehicle armour (better known as rolled homogeneous armour in this role).

Historical

Before the introduction of the Bessemer process and other modern production techniques, steel was expensive and was only used where no cheaper alternative existed, particularly for the cutting edge of knives, razors, swords, and other items where a hard, sharp edge was needed. It was also used for springs, including those used in clocks and watches.

With the advent of speedier and thriftier production methods, steel has become easier to obtain and much cheaper. It has replaced wrought iron for a multitude of purposes. However, the availability of plastics in the latter part of the 20th century allowed these materials to replace steel in some applications due to their lower fabrication cost and weight. Carbon fiber is replacing steel in some cost insensitive applications such as aircraft, sports equipment and high end automobiles.

Long Steel

- As reinforcing bars and mesh in reinforced concrete
- Railroad tracks
- Structural steel in modern buildings and bridges
- Wires
- Input to reforging applications

Flat Carbon Steel

- Major appliances
- Magnetic cores
- The inside and outside body of automobiles, trains, and ships.

Stainless Steel

In metallurgy, stainless steel, also known as inox steel or inox from French "*inoxydable*", is a steel alloy with a minimum of 10.5% chromium content by mass.

Stainless steel does not readily corrode, rust or stain with water as ordinary steel does, but despite the name it is not fully stain-proof, most notably under low-oxygen, high-salinity, or poor-circulation environments. There are different grades and surface finishes of stainless steel to suit the environment the alloy must endure. Stainless steel is used where both the properties of steel and resistance to corrosion are required.

Stainless steel differs from carbon steel by the amount of chromium present. Unprotected carbon steel rusts readily when exposed to air and moisture. This iron oxide film (the rust) is active and accelerates corrosion by forming more iron oxide, and due to the greater volume of the iron oxide this tends to flake and fall away. Stainless steels contain sufficient chromium to form a passive film of chromium oxide, which prevents further surface corrosion by blocking oxygen diffusion to the steel surface and blocks corrosion from spreading into the metal's internal structure, and due to the similar size of the steel and oxide ions they bond very strongly and remain attached to the surface.

Passivation only occurs if the proportion of chromium is high enough and oxygen is present.

History

A few corrosion-resistant iron artifacts survive from antiquity. A famous example is the Iron Pillar of Delhi, erected by order of Kumara Gupta I around AD 400. Unlike stainless steel, however, these artifacts owe their durability not to chromium but to their high phosphorus content, which, together with favourable local weather conditions, promotes the formation of a solid protective passivation layer of iron oxides and phosphates, rather than the non-protective cracked rust layer that develops on most ironwork.

The corrosion resistance of iron-chromium alloys was first recognised in 1821 by French metallurgist Pierre Berthier, who noted their resistance against attack by some acids and suggested their use in cutlery. Metallurgists of the 19th century were unable to produce the combination of low carbon and high chromium found in most modern stainless steels, and the high-chromium alloys they could produce were too brittle to be practical.

In the late 1890s Hans Goldschmidt of Germany developed an aluminothermic (thermite) process for producing carbon-free chromium. Between 1904 and 1911 several researchers, particularly Leon Guillet of France, prepared alloys that would today be considered stainless steel.

Friedrich Krupp Germaniawerft built the 366-ton sailing yacht *Germania* featuring a chrome-nickel steel hull in Germany in 1908. In 1911, Philip Monnartz reported on the relationship between chromium content and corrosion resistance. On October 17, 1912, Krupp engineers Benno Strauss and Eduard Maurer patented austenitic stainless steel as Thyssen Krupp Nirosta.

Similar developments were taking place contemporaneously in the United States, where Christian Dantsizen and Frederick Becket were industrializing ferritic stainless steel. In 1912, Elwood Haynes applied for a US patent on a martensitic stainless steel alloy, which was not granted until 1919.

Also in 1912, Harry Brearley of the Brown-Firth research laboratory in Sheffield, England, while seeking a corrosion-resistant alloy for gun barrels, discovered and subsequently industrialized a martensitic stainless steel alloy. The discovery was announced two years later in a January 1915 newspaper article in *The New York Times*. The metal was later marketed under the 'Staybrite' brand by Firth Vickers in England and was used for the new entrance canopy for the Savoy Hotel in London in 1929. Brearley applied for a US patent during 1915 only to find that Haynes had already registered a patent. Brearley and Haynes pooled their funding and with a group of investors formed the American Stainless Steel Corporation, with headquarters in Pittsburgh, Pennsylvania.

In the beginning stainless steel was sold in the US under different brand names like 'Allegheny metal' and 'Nirosta steel'. Even within the metallurgy industry the eventual name remained unsettled; in 1921 one trade journal was calling it "unstainable steel". In 1929, before the Great Depression hit, over 25,000 tons of stainless steel were manufactured and sold in the US.

Properties

Oxidation: High oxidation resistance in air at ambient temperature is normally achieved with additions of a minimum of 13% (by weight) chromium, and up to 26% is used for harsh environments. The chromium forms a passivation layer of chromium(III) oxide (Cr_2O_3) when exposed to oxygen. The layer is too thin to be visible, and the metal remains lustrous and smooth. The layer is impervious to water and air, protecting the metal beneath, and this layer quickly reforms when the surface is scratched. This phenomenon is called passivation and is seen in other metals, such as aluminium and titanium. Corrosion resistance can be adversely affected if the component is used in a non-oxygenated environment, a typical example being underwater keel bolts buried in timber.

When stainless steel parts such as nuts and bolts are forced together, the oxide layer can be scraped off, allowing the parts to weld together. When forcibly disassembled, the welded material may be torn and pitted, an effect known as galling. This destructive galling can be avoided by the use of dissimilar materials for the parts forced together, for example bronze and stainless steel, or even different types of stainless steels (martensitic against austenitic). However, two different alloys electrically connected in a humid environment may act as Voltaic pile and corrode faster. Nitronic alloys made by selective alloying with manganese and nitrogen may have a reduced tendency to gall. Additionally, threaded joints may be lubricated to prevent galling.

Acids

Stainless steel is generally highly resistant to attack from acids, but this quality depends on the kind and concentration of the acid, the surrounding temperature, and the type of steel. Type 304 is resistant to sulfuric acid at room temperature, even in high concentrations, but type 316 and 317 are only resistant at low concentrations. All types of stainless steel resist attack from phosphoric acid, 316 and 317 more so than 304; and Types 304L and 430 have been successfully used with nitric acid. Hydrochloric acid will damage any kind of stainless steel, and should be avoided.

Bases

The 300 series of stainless steel grades is unaffected by any of the weak bases such as ammonium hydroxide, even in high concentrations and at high temperatures. The same grades of stainless exposed to stronger bases such as sodium hydroxide at high concentrations and high temperatures will likely experience some etching and cracking, especially with solutions containing chlorides.

Organics

Types 316 and 317 are both useful for storing and handling acetic acid, especially in solutions where it is combined with formic acid and when aeration is not present (oxygen helps protect stainless steel under such conditions), though 317 provides the greatest level of resistance to corrosion. Type 304 is also commonly used with formic acid though it will tend to discolour the solution.

All grades resist damage from aldehydes and amines, though in the latter case grade 316 is preferable to 304; cellulose acetate will damage 304 unless the temperature is kept low. Fats and fatty acids only affect grade 304 at temperatures above 150 °C (302 °F), and grade 316 above 260 °C (500 °F), while 317 is unaffected at all temperatures. Type 316L is required for processing of urea.

Electricity and Magnetism

Similarly to steel, stainless steel is a relatively poor conductor of electricity, with a few percent of the electrical conductivity of copper. Ferritic and martensitic stainless steels are magnetic. Austenitic stainless steels are non-magnetic.

Applications

Stainless steel's resistance to corrosion and staining, low maintenance and familiar lustre make it an ideal material for many applications. There are over 150 grades of stainless steel, of which fifteen are most commonly used.

The alloy is milled into coils, sheets, plates, bars, wire, and tubing to be used in cookware, cutlery, household hardware, surgical instruments, major appliances, industrial equipment (for example, in sugar refineries) and as an automotive and aerospace structural alloy and construction material in large buildings. Storage tanks and tankers used to transport orange juice and other food are often made of stainless steel, because of its corrosion resistance.

This also influences its use in commercial kitchens and food processing plants, as it can be steam-cleaned and sterilized and does not need paint or other surface finishes. Stainless steel is used for jewellery and watches with 316L being the type commonly used for such applications. It can be re-finished by any jeweller and will not oxidize or turn black.

Some firearms incorporate stainless steel components as an alternative to blued or parkerized steel. Some handgun models, such as the Smith & Wesson Model 60 and the Colt M1911 pistol, can be made entirely from stainless steel. This gives a high-luster finish similar in appearance to nickel plating. Unlike plating, the finish is not subject to flaking, peeling, wear-off from rubbing (as when repeatedly removed from a holster), or rust when scratched. Some automotive manufacturers use stainless steel as decorative highlights in their vehicles.

SAE Steel Grades

SAE International, as a standards organisation, maintains several alloy numbering systems, one of which, for steel grades, is the SAE steel grades system. In the 1930s and 1940s the American Iron and Steel Institute (AISI) and SAE were both involved in efforts to standardize such a numbering system for steels. These efforts were similar and overlapped significantly. For several decades the systems were united into a joint system designated the AISI/SAE steel grades. In 1995 the AISI turned over future maintenance of the system to SAE because the AISI never wrote any of the specifications.

Today steel quotes and certifications commonly make reference to both SAE and AISI, not always with precise differentiation. For example, in the alloy/grade field, a cert might say "4140", "AISI 4140", or "SAE 4140", and in most light-industrial applications any of the above is accepted as adequate, and considered equivalent, for the job at hand, as long as the specific specification called out by the designer (for example, "4140 bar per ASTM-A108" or "4140 bar per AMS 6349") is certified to on the certificate. The alloy number is simply a general classifier, whereas it is the specification itself that narrows down the steel to a very specific standard.

The SAE steel grade system's correspondence to other alloy numbering systems, such as the ASTM-SAE unified numbering system (UNS), can be seen in cross-referencing tables (including the ones given below). The AISI system used a letter prefix to denote the steelmaking process. The prefix "C" denoted open-hearth furnace, electric arc furnace or basic oxygen furnace, while "E" denotes electric arc furnace steel. A letter "L" within the grade name indicates lead as an added ingredient.

Carbon and Alloy Steel

Carbon steels and alloy steels are designated by a four digit number, where the first digit indicates the main alloying element(s), the second digit indicates the secondary alloying element(s), and the last two digits indicate the amount of carbon, in hundredths of a percent (basis points) by weight. For example, a 1060 steel is a plain-carbon steel containing 0.60 wt% C.

An "H" suffix can be added to any designation to denote hardenability is a major requirement. The chemical requirements are loosened but hardness values defined for various distances on a Jominy test.

Major Classifications of Steel

SAE Designation	*Type*
1xxx	Carbon steels
2xxx	Nickel steels
3xxx	Nickel-chromium steels
4xxx	Molybdenum steels
5xxx	Chromium steels
6xxx	Chromium-vanadium steels
7xxx	Tungsten steels
8xxx	Nickel-chromium-molybdenum steels
9xxx	Silicon-manganese steels

Carbon and Alloy Steel Grades

SAE Designation	*Type*
	Carbon Steels
10xx	Plain carbon (Mn 1.00% max)
11xx	Resulfurized
12xx	Resulfurized and rephosphorized
15xx	Plain carbon (Mn 1.00% to 1.65%)
	Manganese Steels
13xx	Mn 1.75%
	Nickel Steels
23xx	Ni 3.50%
25xx	Ni 5.00%

	Nickel-Chromium Steels
31xx	Ni 1.25%, Cr 0.65% or 0.80%
32xx	Ni 1.25%, Cr 1.07%
33xx	Ni 3.50%, Cr 1.50% or 1.57%
34xx	Ni 3.00%, Cr 0.77%
	Molybdenum Steels
40xx	Mo 0.20% or 0.25% or 0.25% Mo & 0.042 S
44xx	Mo 0.40% or 0.52%
	Chromium-Molybdenum (Chromoly) Steels
41xx	Cr 0.50% or 0.80% or 0.95%, Mo 0.12% or 0.20% or 0.25% or 0.30%
	Nickel-Chromium-Molybdenum Steels
43xx	Ni 1.82%, Cr 0.50% to 0.80%, Mo 0.25%
43BVxx	Ni 1.82%, Cr 0.50%, Mo 0.12% or 0.35%, V 0.03% min
47xx	Ni 1.05%, Cr 0.45%, Mo 0.20% or 0.35%
81xx	Ni 0.30%, Cr 0.40%, Mo 0.12%
81Bxx	Ni 0.30%, Cr 0.45%, Mo 0.12%, and added boron
86xx	Ni 0.55%, Cr 0.50%, Mo 0.20%
87xx	Ni 0.55%, Cr 0.50%, Mo 0.25%
88xx	Ni 0.55%, Cr 0.50%, Mo 0.35%
93xx	Ni 3.25%, Cr 1.20%, Mo 0.12%
94xx	Ni 0.45%, Cr 0.40%, Mo 0.12%
97xx	Ni 0.55%, Cr 0.20%, Mo 0.20%
98xx	Ni 1.00%, Cr 0.80%, Mo 0.25%
	Nickel-Molybdenum Steels
46xx	Ni 0.85% or 1.82%, Mo 0.20% or 0.25%
48xx	Ni 3.50%, Mo 0.25%
	Chromium Steels
50xx	Cr 0.27% or 0.40% or 0.50% or 0.65%
50xxx	Cr 0.50%, C 1.00% min
50Bxx	Cr 0.28% or 0.50%, and added boron
51xx	Cr 0.80% or 0.87% or 0.92% or 1.00% or 1.05%
51xxx	Cr 1.02%, C 1.00% min
51Bxx	Cr 0.80%, and added boron
52xxx	Cr 1.45%, C 1.00% min

	Chromium-Vanadium Steels
61xx	Cr 0.60% or 0.80% or 0.95%, V 0.10% or 0.15% min
	Tungsten-Chromium Steels
72xx	W 1.75%, Cr 0.75%
	Silicon-Manganese Steels
92xx	Si 1.40% or 2.00%, Mn 0.65% or 0.82% or 0.85%, Cr 0.00% or 0.65%
	High-Strength Low-Alloy Steels
9xx	Various SAE grades
xxBxx	Boron steels
xxLxx	Leaded steels

2

Material and Design Requirements

Steelmaking

Steelmaking is the process for producing steel from iron and ferrous scrap. In steelmaking, impurities such as nitrogen, silicon, phosphorus, and excess carbon are removed from the raw iron, and alloying elements such as manganese, nickel, chromium and vanadium are added to produce different grades of steel. Limiting dissolved gases such as nitrogen and oxygen, and entrained impurities (termed "inclusions") in the steel is also important to ensure the quality of the products cast from the liquid steel. There are two major processes for making steel, namely basic oxygen steelmaking which has liquid pig-iron from the blast furnace and scrap steel as the main feed materials, and electric arc furnace (EAF) steelmaking which uses scrap steel or direct reduced iron (DRI) as the main feed materials. Oxygen steelmaking is fuelled predominantly by the exothermic nature of the reactions inside the vessel where as in EAF steelmaking, electrical energy is used to melt the solid scrap and/or DRI materials. In recent times, EAF steelmaking technology has evolved closer to oxygen steelmaking as more chemical energy is introduced into the process.

History

Steelmaking has played a crucial role in the development of modern technological societies. Cast iron is a hard brittle material that is difficult to work, whereas steel is malleable, relatively easily formed and a versatile material. For much of human history, steel has only been able to be made in small quantities but since the invention of the Bessemer process in the 19th century and subsequent technological developments in injection technology and process control, mass production of steel has become an integral part of the world's economy and a key indicator of technological development. The earliest means of producing steel was in a bloomery.

Early modern methods of producing steel were often labour-intensive and highly skilled arts. See:

- finery forge, in which the German finery process could be managed to produce steel.
- blister steel and crucible steel.

An important aspect of the industrial revolution was the development of large-scale methods of producing forgeable metal (bar iron or steel). The puddling furnace was initially a means of producing wrought iron, but was later applied to steel production.

The real revolution in steelmaking only began at the end of the 1850s when the Bessemer process became the first successful method of steelmaking in quantity, followed by the open-hearth furnace.

Modern Processes

Modern steelmaking processes can be broken into two categories: primary and secondary steelmaking. Primary steelmaking involves converting liquid iron from a blast furnace and steel scrap into steel via basic oxygen steelmaking or melting scrap steel and/or direct reduced iron (DRI) in an electric arc furnace. Secondary steelmaking involves refining of the crude steel before casting and the various operations are normally carried out in ladles. In secondary metallurgy, alloying agents are added, dissolved gases in the steel are lowered, inclusions are removed or altered chemically to ensure that high-quality steel is produced after casting.

Basic Oxygen Steelmaking

Basic oxygen steelmaking (BOS, BOP, BOF, and OSM), also known as Linz-Donawitz-Verfahren steelmaking or the oxygen converter process is a method of primary steelmaking in which carbon-rich molten pig iron is made into steel. Blowing oxygen through molten pig iron lowers the carbon content of the alloy and changes it into low-carbon steel. The process is known as *basic* due to the type of refractories—calcium oxide and magnesium oxide—that line the vessel to withstand the high temperature of the molten metal.

The process was developed in 1948 by Robert Durrer and commercialized in 1952–1953 by Austrian VOEST and ÖAMG. The LD converter, named after the Austrian towns Linz and Donawitz (a district of Leoben) is a refined version of the Bessemer converter where blowing of air is replaced with blowing oxygen. It reduced capital cost of the plants, time of smelting, and increased labour productivity. Between 1920 and 2000, labour requirements in the industry decreased by a factor of 1,000, from more than 3 worker-hours per tonne to just 0.003. The vast majority of steel manufactured in the world is produced using the basic oxygen furnace; in 2000, it accounted for 60% of global steel output. Modern furnaces will take a charge of iron of up to 350 tons and convert it into steel in less than 40 minutes, compared to 10–12 hours in an open hearth furnace.

History

The basic oxygen process developed outside of traditional "big steel" environment. It was developed and refined by a single man, Swiss engineer Robert Durrer, and commercialized by two small steel companies in allied-occupied Austria, which had not yet recovered from the destruction of World War II.

In 1858, Henry Bessemer patented a steelmaking process involving oxygen blowing for decarburizing molten iron (UK Patent No. 2207). For nearly a hundred years commercial quantities of oxygen were not available at all or were too expensive, and the invention remained unused. During World War II German (C. V. Schwartz), Belgian (John Miles) and Swiss (Durrer and Heinrich Heilbrugge) engineers proposed their versions of oxygen-blown steelmaking, but only Durrer and Heilbrugge brought it to mass-scale production.

In 1943, Durrer, formerly a professor at the Berlin Institute of Technology, returned to Switzerland and accepted a seat on the board of Roll AG, the country's largest steel mill. In 1947 he purchased the first small 2.5-ton experimental converter from the U. S., and on April 3, 1948 the new converter produced its first steel. The new process could conveniently process large amounts of scrap metal with only a small proportion of primary metal necessary. In the summer of 1948 Roll AG and two Austrian state-owned companies, VOEST and ÖAMG, agreed to commercialize the Durrer process.

By June 1949, VOEST developed an adaptation of Durrer's process, known as the LD (Linz-Donawitz) process. In December 1949, VOEST and ÖAMG committed to building their first 30-ton oxygen converters. They were put into operation in November 1952 (VOEST in Linz) and May 1953 (ÖAMG, Donawitz) and temporarily became the leading edge of the world's steelmaking, causing a surge in steel-related research. Thirty-four thousand businesspeople and engineers visited the VOEST converter by 1963. The LD process reduced processing time and capital costs per ton of steel, contributing to the competitive advantage of Austrian steel. VOEST eventually acquired the rights to market the new technology. However, errors made by the VOEST and the ÖAMG management in licensing their technology made control over its adoption in Japan impossible and by the end of the 1950s the Austrians lost their competitive edge.

The original LD process consisted in blowing oxygen over the top of the molten iron through the water-cooled nozzle of a vertical lance. In the 1960s steelmakers introduced bottom-blown converters and introduced inert gas blowing for stirring the molten metal and removing the phosphorus impurities.

In the Soviet Union, some experimental production of steel using the process was done in 1934, but industrial use was hampered by lack of efficient technology to produce liquid oxygen. In 1939, the Russian physicist Pyotr Kapitsa perfected the design of the centrifugal turboexpander.

The process was put to use in 1942-1944. Most turboexpanders in industrial use since then have been based on Kapitsa's design and centrifugal turboexpanders have taken over almost 100 percent of the industrial gas liquefaction and in particular the production of liquid oxygen for steelmaking.

The big American steelmakers caught up late with the new technology; the first oxygen converters in the United States were launched at the end of 1954 by McLouth Steel in Trenton, Michigan, which accounted for less than 1 per cent of the national steel market. U.S. Steel and Bethlehem Steel introduced the oxygen process only in 1964. By 1970 half of world's and 80% of Japan's steel output was produced in oxygen converters. In the last quarter of the 20th century basic oxygen converters were gradually replaced by the electric arc furnace. In Japan the share of LD process decreased from 80% in 1970 to 70% in 2000; worldwide share of the basic oxygen process stabilized at 60%.

Process

Basic oxygen steelmaking is a primary steelmaking process for converting the molten pig iron into steel by blowing oxygen through a lance over the molten pig iron inside the converter. The converter used for steel making is called as Basic Oxygen Furnace because of the exothermic heat generated by the oxidation reactions during blowing.

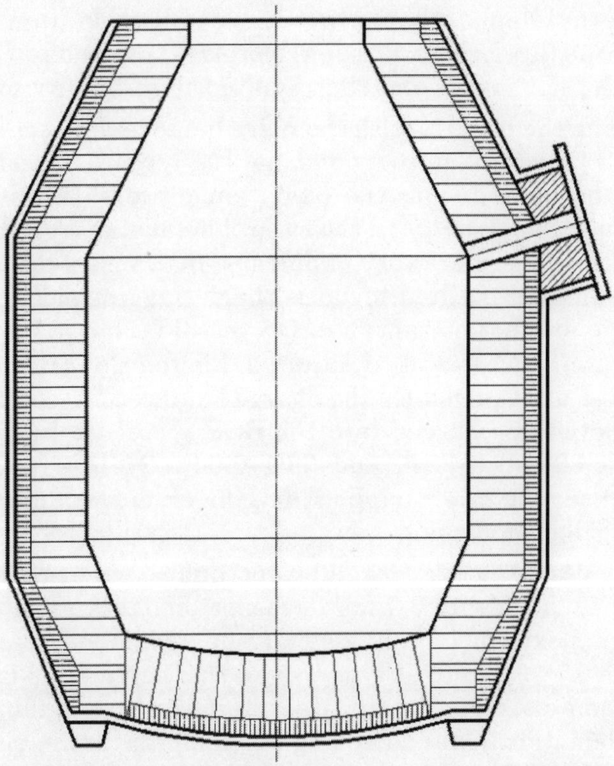

Figure: *Cross-section of a basic oxygen furnace*

The basic oxygen steel-making process is as follows:

1. Molten pig iron (sometimes referred to as "hot metal") from a blast furnace is poured into a large refractory-lined container called a ladle;
2. The metal in the ladle is sent directly for basic oxygen steelmaking or to a pretreatment stage. High purity oxygen at a pressure of 100-150 psi (pounds per inch square) is introduced at supersonic speed onto the surface of the iron bath through a water-cooled lance, which is suspended in the vessel and kept at few feet above the bath. Pretreatment of the blast furnace hot metal is done externally to reduce sulphur, silicon, and phosphorus before charging the hot metal into the converter. In external desulphurising pretreatment, a lance is lowered into the molten iron in the ladle and several hundred kilograms of powdered magnesium are added and the sulphur impurities are reduced to magnesium sulphide in a violent exothermic reaction. The sulfide is then raked off. Similar pretreatments are possible for external desiliconisation and

external dephosphorisation using mill scale (iron oxide) and lime as fluxes. The decision to pretreat depends on the quality of the hot metal and the required final quality of the steel.

3. Filling the furnace with the ingredients is called *charging*. The BOS process is autogenous, i.e. the required thermal energy is produced during the oxidation process. Maintaining the proper *charge balance*, the ratio of hot metal, from melt, to cold scrap, is therefore very important. BOS vessel can be tilted up to 360° and is tilted towards the deslagging side for charging scrap and hot metal. The BOS vessel is charged with steel or iron scrap(25%-30%) if required. Molten iron from the ladle is added as required for the charge balance. A typical chemistry of hotmetal charged into the BOS vessel is: 4% C, 0.2–0.8% Si, 0.08%–0.18% P, and 0.01–0.04% S all of which can be oxidised by the supplied oxygen except sulphur (requires reducing condition).
4. The vessel is then set upright and a water-cooled, copper tipped lance with 3-7 nozzles is lowered down into it and high purity oxygen is delivered at supersonic speeds. The lance "blows" 99% pure oxygen over the hot metal, igniting the carbon dissolved in the steel, to form carbon monoxide and carbon dioxide, and causing the temperature to rise to about 1700°C. This melts the scrap, lowers the carbon content of the molten iron and helps remove unwanted chemical elements. It is this use of pure oxygen instead of air that improves upon the Bessemer process, as the nitrogen (a particularly undesirable element) and other gases in air do not react with the charge.
5. Fluxes (burnt lime or dolomite) are fed into the vessel to form slag, to maintain basicity more than 3 and absorbs impurities during the steelmaking process. During "blowing," churning of metal and fluxes in the vessel forms an emulsion, that facilitates the refining process. Near the end of the blowing cycle, which takes about 20 minutes, the temperature is measured and samples are taken. A typical chemistry of the blown metal is 0.3–0.9% C, 0.05–0.1% Mn, 0.001–0.003% Si, 0.01–0.03% S and 0.005-0.03% P.
6. The BOS vessel is tilted towards the deslagging side and the steel is poured through a tap hole into a steel ladle with basic refractory lining. This process is called *tapping* the steel. The steel is further refined in the ladle furnace, by adding alloying

materials to give the it special properties required by the customer. Sometimes argon or nitrogen is bubbled into the ladle to make mix the alloys correctly.

7. After the steel is poured off from the BOS vessel, the slag is poured into the slag pots through the BOS vessel mouth and dumped.

Variants

Earlier converters had false bottom, which can be detached and fixed again and they are still in use. Modern converters have fixed bottom with plugs for argon purging. Energy Optimization Furnace is a BOF variant associated with a scrap preheater where the sensible heat in the off-gas is used for preheating scrap, located above the furnace roof.

Lance used for blowing has undergone changes, slagless lances have been employed to avoid jamming of lance during blowing which has a long tapering copper tip. Post combustion lance tips burn the CO generated during blowing into CO2 which is an asphyxiant. For slag free tapping, darts, refractory balls and slag detectors are employed. Modern converters are fully automated with auto blowing patterns and sophisticated control systems.

Secondary Steelmaking

Secondary steelmaking is most commonly performed in ladles and often referred to as ladle (metallurgy). Some of the operations performed in ladles include de-oxidation (or "killing"), vacuum degassing, alloy addition, inclusion removal, inclusion chemistry modification, de-sulphurisation and homogenisation. It is now common to perform ladle metallurgical operations in gas stirred ladles with electric arc heating in the lid of the furnace. Tight control of ladle metallurgy is associated with producing high grades of steel in which the tolerances in chemistry and consistency are narrow.

HIsarna Steelmaking Process

The HIsarna steelmaking process is a process for primary steelmaking in which iron ore is processed almost directly into steel. The process is based around a new type of blast furnace called a *Cyclone Converter Furnace*, which makes it possible to skip the process of manufacturing pig iron pellets that is necessary for the basic oxygen steelmaking process. Without the necessity for this preparatory step the HIsarna process is more energy-efficient and has a lower carbon footprint than traditional steelmaking processes.

The HIsarna process was developed in stages and with pauses at Koninklijke Hoogovens / Corus IJmuiden, starting in 1986. The final stages were made possible through the Ultra-Low Carbon Dioxide Steelmaking (ULCOS) consortium and cooperation between Corus and the Rio Tinto Group. The latter contributed their HIsmelt technology to the final design of the installation, prompting the name HIsarna for the process ("HI" from HIsmelt and "sarna" from *Isarna*, a Celtic word for iron).

History

The very first attempts at applying cyclone oven technology in the reduction of iron ore took place at Koninklijke Hoogovens in the 1960s. Cyclone technology had already been used successfully in different industrial, chemical processes and designers at Hoogovens thought it might be a strategy for improvement for their process. However, at the time they couldn't get it to work properly and the experiment was quickly abandoned.

The first serious revival came in 1986, when Hoogovens sought a method of producing steel without having to produce pig iron pellets. At that time the desire was mostly a cost-cutting measure in order to make the process cheaper in trying economic times. The trying times did not last however and the project was put on the back burner until the early 1990s.

By the early 1990s the availability of cokes was becoming limited due to many of the major coking facilities in the West that produced cokes from coal were reaching the end of their economic life. Heavy environmental restrictions made it unattractive to build new facilities, so steel producers sought ways to reduce the need for cokes; Hoogovens started putting more effort into the cyclone technology as a solution to this problem and a test facility for the cyclone part proved capable of producing twenty tons of pig iron per hour. The rest of the process didn't work very well though, so when steel producers massively moved from coke bricks to powdered cokes injection and China started mass-producing cokes, the project lost momentum again. The steep drop in prices of commodities around 1999 caused the project to be halted.

In 2004 however, the European Union brought pressure on the steel industry to reduce its carbon footprint; the ULCOS consortium was founded as a result and in the period 2005-2007 the cyclone technology was selected as one of four high-potential technologies. A theoretical answer was found to the earlier problems of the post-cyclone part of the cyclone furnace and the Rio Tinto Group had industrial-scale experience with the required process, called HIsmelt. An agreement between them and ULCOS added the HIsmelt technology to the cyclone furnace and the result was the HIsarna process.

As of September 2010 the first pilot plant for HIsarna is under construction at Corus IJmuiden. If successful the pilot furnace will be capable of producing 60,000 tons of pig iron per year.

Process

The HIsarna process is a variation of the basic oxygen steelmaking process in which the production of pig iron takes place in *Cyclone Converter Furnace* (CCF).

A CCF is shaped like a wine bottle: a "bottle" at the bottom and a thin "neck" at the top. The geometry of this furnace causes a hot air cyclone to form in the neck when the furnace is heated. Crushed iron ore is injected into this cyclone together with oxygen (so oxygen is injected at the top rather than at the bottom). The heat of the cyclone causes the initial (partial) reduction reaction to take place that reduces iron ore to iron. Also, the centrifugal force of the cyclone separates the reduced iron from remaining impurities and flings the reduced iron droplets against the wall of the furnace.

The molten iron droplets then drip down the furnace wall to the place where the "neck" widens into the "bottle". Here the droplets fall from the wall into the molten iron bath in the bottom of the furnace. On the way down the droplets pass through a second set of heated oxygen injectors and then a set of coal powder cannons. The reduction reaction now continues "as normal" in the bottom of the furnace, with the partially reduced iron ore further reducing to regular pig iron and the whole separating into two molten layers (a top layer of slag and a bottom layer of molten pig iron). Both layers can be tapped individually and the pig iron can be used immediately in the remainder of the basic oxygen steelmaking process.

Advantages

In a technical sense, the advantage of the HIsarna process is that it removes the step of creating pig iron pellets to create a porous material for the blast furnace. In the traditional process one cannot use powdered coal throughout the heating process, since the oily components of coal would turn into sludge and ruin the porous structure of the pig iron. By comparison, in a CCF, the powder form of the coal and ore are an advantage because the increased surface area improves the speed and quality of the reduction reaction in the cyclone. This in turn means that the pig iron that is finally drained from the molten iron bath is also of somewhat better quality.

The main advantages of the process are derived from those mentioned above, however: the fact that the separate step of creating pig iron pellets disappears from the process makes the process more energy efficient and reduces its carbon footprint. This makes the process attractive to steelmakers, who are being pressured to make their processes more environmentally friendly — particularly in Europe, where government regulations are increasingly attaching a financial penalty to high carbon dioxide emissions.

Bessemer Process

The Bessemer process was the first inexpensive industrial process for the mass-production of steel from molten pig iron prior to the open hearth furnace. The process is named after its inventor, Henry Bessemer, who took out a patent on the process in 1855. The process was independently discovered in 1851 by William Kelly. The process had also been used outside of Europe for hundreds of years, but not on an industrial scale. The key principle is removal of impurities from the iron by oxidation with air being blown through the molten iron. The oxidation also raises the temperature of the iron mass and keeps it molten.

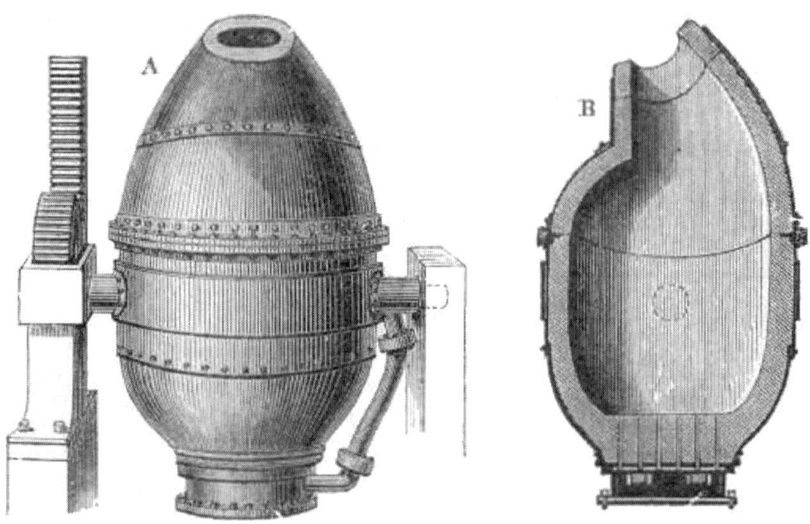

Figure: *Bessemer converter, schematic diagram*

The process using a basic refractory lining is known as the *basic Bessemer process* or *Gilchrist-Thomas process* after the discoverer Sidney Gilchrist Thomas.

Details

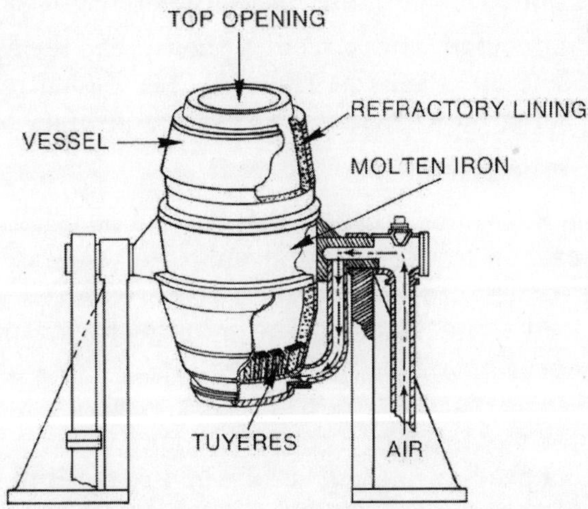

Figure: *Bessemer converter components.*

Oxidation

The oxidation process removes and skims off impurities such as silicon, manganese, and carbon in the form of oxides. These oxides either escape as gas or form a solid slag. The refractory lining of the converter also plays a role in the conversion—the clay lining is used in the *acid Bessemer*, in which there is low phosphorus in the raw material. Dolomite is used when the phosphorus content is high in the alkaline Bessemer (limestone or magnesite linings are also sometimes used instead of dolomite)—this is also known as a *Gilchrist-Thomas converter*, named after its inventor, Sidney Gilchrist Thomas. In order to give the steel the desired properties, other substances could be added to the molten steel when conversion was complete, such as spiegeleisen (a ferromanganese alloy).

Managing the Process

When the required steel has been formed, it is poured out into ladles and then transferred into moulds while the lighter slag is left behind. The conversion process, called the "blow", is completed in around twenty minutes. During this period the progress of the oxidation of the impurities is judged by the appearance of the flame issuing from the mouth of the converter: the modern use of photoelectric methods of recording the characteristics of the flame has greatly aided the blower in controlling the final quality of the product. After the blow, the liquid metal is recarburized to the desired point and other alloying materials are added, depending on the desired product.

A Bessemer converter can treat a "heat," the term for a batch of hot metal, of 5 to 30 tonnes at time [1]. They usually are operated in pairs; one being blown while another being filled or tapped.

Predecessor Processes

Before the Bessemer process, Britain had no practical method of reducing the carbon content of pig iron. Steel was manufactured by the reverse process of adding carbon to carbon-free wrought iron, usually imported from Sweden. The manufacturing process, called cementation process, consisted of heating bars of wrought iron together with charcoal for periods of up to a week in a long stone box. This produced blister steel. Up to 3 tons of expensive coke was burnt for each ton of steel produced. Such steel when rolled into bars was sold at £50 to £60 (approximately £3,390 to £4,070 in 2008) a long ton. The most difficult and work-intensive part of the process, however, was the production of wrought iron done in finery forges in Sweden.

This process was refined in the 18th century with the introduction of Benjamin Huntsman's crucible steel-making techniques, which added an additional three hours firing time and required additional large quantities of coke. In making crucible steel the blister steel bars were broken into pieces and melted in small crucibles each containing 20 kg or so. This produced higher quality crucible steel but increased the cost. The Bessemer process reduced the time needed to make steel of this quality to about half an hour while requiring only the coke needed to melt the pig iron initially. The earliest Bessemer converters produced steel for £7 a long ton, although it initially sold for around £40 a ton.

History

In 1740 Benjamin Huntsman developed the crucible technique for steel manufacture, at his workshop in the district of Handsworth in Sheffield. This process had an enormous impact on the quantity and quality of steel production.

Sir Henry Bessemer described the origin of his invention in his autobiography. According to him, at the time of the outbreak of the

Crimean War many English industrialists and inventors became interested in military technology. At the time steel was used to make only small items like cutlery and tools, but was too expensive for cannons. Starting in January 1855 he began working on a way to produce steel in the massive quantities required for artillery and by October he filed his first patent related to the Bessemer process.

According to his autobiography Bessemer was working with an ordinary reverbatory furnace but during a test, some pieces of pig iron were jostled off the side of the ladle, and were left above the ladle in the furnace's heat. When Bessemer went to push them into the ladle, he found that they were steel shells: the hot air alone had converted the outsides of the iron pieces to steel. This crucial discovery led him to completely redesign his furnace so that it would force high-pressure air through the molten iron using special air pumps. Intuitively this would seem to be folly because it would cool the iron. Instead, the oxygen in the forced air ignited silicon and carbon impurities in the iron, starting a positive feedback loop. As the iron became hotter, more impurities burned off, making the iron even hotter and burning off more impurities, producing a batch of hotter, purer, molten iron, which converts to steel more easily.

Bessemer licensed the patent for his process to four ironmasters, for a total of £27,000, but the licensees failed to produce the quality of steel he had promised—it was "rotten hot and rotten cold", according to his friend, William Clay—and he later bought them back for £32,500. His plan had been to offer the licenses to one company in each of several geographic areas, at a royalty price per ton that included a lower rate on a proportion of their output in order to encourage production, but not so large a proportion that they might decide to reduce their selling prices. By this method he hoped to cause the new process to gain in standing and market share.

He realised that the technical problem was due to impurities in the iron and concluded that the solution lay in knowing when to turn off the flow of air in his process so that the impurities were burned off but just the right amount of carbon remained. However, despite spending tens of thousands of pounds on experiments, he could not find the answer. Certain grades of steel are sensitive to the 78% nitrogen which was part of the air blast passing through the steel.

Bessemer was sued by the patent purchasers who couldn't get it to work. In the end Bessemer set up his own steel company because he knew how to do it, even though he could not convey it to his patent users. Bessemer's company became one of the largest in the world and changed the face of steel making. The solution was first discovered by English metallurgist Robert Forester Mushet, who had carried out thousands of experiments in the Forest of Dean. His method was to first burn off, as far as possible, *all* the impurities and carbon, then reintroduce carbon and manganese by adding an exact amount of spiegeleisen. This had the effect of improving the quality of the finished product, increasing its malleability—its ability to withstand rolling and forging at high temperatures and making it more suitable for a vast array of uses.

The first company to license the process was the Manchester firm of W & J Galloway, and they did so before Bessemer announced it at Cheltenham in 1856. They are not included in his list of the four to whom he refunded the license fees. However, they subsequently rescinded their license in 1858 in return for the opportunity to invest in a partnership with Bessemer and others. This partnership began to manufacture steel in Sheffield from 1858, initially using imported charcoal pig iron from Sweden. This was the first commercial production.

Sidney Gilchrist Thomas, a Londoner with a Welsh father, was an industrial chemist who decided to tackle the problem of phosphorus in iron, which resulted in the production of low grade steel. Believing that he had discovered a solution, he contacted his cousin, Percy Gilchrist, who was a chemist at the Blaenavon ironworks. The manager at the time, Edward Martin, offered Sidney equipment for large-scale testing and helped him draw up a patent that was taken out in May, 1878. Sidney Gilchrist Thomas's invention consisted of using dolomite or sometimes limestone linings for the Bessemer converter rather than clay, and it became known as the 'basic' Bessemer rather than the 'acid' Bessemer process. An additional advantage was that the processes formed more slag in the converter, and this could be recovered and used very profitably as a phosphate fertilizer.

Patent Battles

Patents of such value did not escape criticism, and invalidity was urged against them on various grounds. But Bessemer was able to maintain them intact without litigation, though he found it advisable to buy up the rights of one patentee.

In the case of Robert Forester Mushet, he was assisted by the patent being allowed to lapse in 1859 through non-payment of fees. Mushet's procedure was not essential and Bessemer proved this in 1865 by exhibiting a series of steel samples made using his process alone, but the value of the procedure was shown by its near universal adoption in conjunction with the Bessemer process. Whether or not Mushet's patents could have been sustained is not known, but in 1866 Robert Mushet's 16-year-old daughter travelled to London to confront Henry Bessemer at his offices, arguing that Bessemer's success was based on the results of her father's work. Bessemer decided to pay Mushet an annual pension of £300, a very considerable sum, which he paid for 25 years.

In 1866, Bessemer also provided finance for Zerah Colburn, the American locomotive engineer and journalist, to start a new weekly engineering newspaper called *Engineering* based in Bedford Street, London. It was not until many years later that the name of Colburn's benefactor was revealed. Prior to the launch of *Engineering*, Colburn, through the pages of *The Engineer*, had given support to Bessemer's work on steel and steelmaking.

Importance

The Bessemer process revolutionized steel manufacture by decreasing its cost, from £40 per long ton to £6–7 per long ton, along with greatly increasing the scale and speed of production of this vital raw material. The process also decreased the labour requirements for steel-making. Prior to its introduction, steel was far too expensive to make bridges or the framework for buildings and thus wrought iron had been used throughout the Industrial Revolution. After the introduction of the Bessemer process, steel and wrought iron became similarly priced, and most manufacturers turned to steel. Some claim the availability of cheap steel allowed large bridges to be built and enabled the construction of railroads, skyscrapers, and large ships; however, the largest ship of the 19th century was the iron SS *Great Eastern* launched in 1858, high pressure steam-powered locomotive railways appeared in 1825 with the opening of the Stockton and Darlington railway (the first passenger railway opened in 1830 between Liverpool and Manchester), and the first modern large suspension bridge was the Menai Suspension Bridge completed in 1826: the latter two decades before mass-produced steel. Other important steel products—also made using the open hearth process—were steel cable, steel rods and sheet steel which enabled large, high-pressure boilers and high-tensile strength steel for machinery which enabled much more powerful engines, gears and axles than were possible previously. With large amounts of steel it became possible to build much more powerful guns and carriages, tanks, armored fighting vehicles and naval ships. Industrial steel also made possible the building of giant turbines and generators thus making the harnessing of water and steam power possible. The introduction of the large scale steel production process paved the way to mass industrialisation as observed in the 19th–20th centuries.

As early as 1895 in the UK it was being noted that the heyday of the Bessemer process was over and that the open hearth method predominated. The *Iron and Coal Trades Review* said that it was "in a semi-moribund condition. Year after year, it has not only ceased to make progress, but it has absolutely declined." It has been suggested, both at that time and more recently, that the cause of this was the lack of trained personnel and investment in technology rather than anything intrinsic to the process itself. For example, one of the major causes of the decline of the giant ironmaking company Bolckow Vaughan of Middlesbrough was its failure to upgrade its technology. The basic process, Thomas-Gilchrist process, remained longer in use, especially in Continental Europe, where iron ores were of high phosphorus content and open hearth process was not able to remove all phosphorus; almost all inexpensive construction steel in Germany was produced with this method in the 1950s and 1960s [2]. It was eventually superseded by basic oxygen steelmaking.

The Bessemer Process in the United States

While visiting Europe to obtain information on shipbuilding, armor, and armaments from 1862-1863, Alexander Lyman Holley visited Bessemer's Sheffield works, and expressed interest in licensing the process for use in the US. Upon returning to the US, Holley met with the famous inventor John Ericsson, who referred Holley to a pair of businessmen who had helped him build the Civil War ironclad USS Monitor, John F. Winslow and John Augustus Griswold. With Winslow and Griswold's support, Holley returned to England in 1863, and paid Bessemer £10,000 to license the technology. The trio began setting up a mill in Troy, New York in 1865. The factory contained a number of Holley's innovations that greatly improved productivity over Bessemer's factory in Sheffield, and the owners gave a successful public exhibition in 1867. The Troy factory attracted the attention of the Pennsylvania Railroad, who wanted to use the new process to manufacture steel rail, and ended up funding Holley's second mill as part of its Pennsylvania Steel subsidiary. Between 1866 and 1877, the partners were able to license a total of 11 Bessemer steel mills. One of the investors they attracted was Andrew Carnegie, who saw great promise in the new steel technology after a visit to Bessemer in 1872, and saw it as a useful adjunct to his existing businesses, the Keystone

Bridge Company and the Union Iron Works. Holley built the new steel mill for Carnegie, and continued to improve and refine the process. The new mill, known as the Edgar Thomson Steel Works, opened in 1875, and sparked the growth of the United States as a major world steel producer.

Obsolescence

In the U.S., commercial steel production using this method stopped in 1968. It was replaced by processes such as the basic oxygen (Linz-Donawitz) process, which offered better control of final chemistry. The Bessemer process was so fast (10–20 minutes for a heat) that it allowed little time for chemical analysis or adjustment of the alloying elements in the steel. Bessemer converters did not remove phosphorus efficiently from the molten steel; as low-phosphorus ores became more expensive, conversion costs increased. The process permitted only limited amount of scrap steel to be charged, further increasing costs, especially when scrap was inexpensive. Use of electric arc furnace technology competed favourably with the Bessemer process resulting in its obsolescence.

Basic oxygen steelmaking is essentially an improved version of the Bessemer process (decarburization by blowing oxygen as gas into the heat rather than burning the excess carbon away by adding oxygen carrying substances into the heat). The advantages of pure oxygen blast over air blast was known to Henry Bessemer, but the 19th century technology was not advanced enough to allow for the production of the large quantities of pure oxygen to make it economically feasible for use.

Open Hearth Furnace

Open hearth furnaces are one of a number of kinds of furnace where excess carbon and other impurities are burnt out of pig iron to produce steel. Since steel is difficult to manufacture due its high melting point, normal fuels and furnaces were insufficient and the open hearth furnace was developed to overcome this difficulty. Compared to Bessemer steel, which it displaced, its main advantages were that it didn't expose the steel to excessive nitrogen (which would cause the steel to become brittle), was easier to control, and it permitted the melting and refining of large amounts of scrap iron and steel.

In 1865, the French engineer Pierre-Émile Martin took out a license from Siemens and first applied his regenerative furnace for making steel. Their process was known as the Siemens-Martin process, and the furnace as an "open-hearth" furnace. Most open hearth furnaces were closed by the early 1990s, not least because of their slow operation, being replaced by the basic oxygen furnace or electric arc furnace.

While arguably the first primitive open hearth furnace was the Catalan forge, invented in Spain in the 8th century, it is usual to confine the term to certain 19th-century and later steelmaking processes, thus excluding bloomeries (including the Catalan forge), finery forges, and puddling furnaces from its application.

Open Hearth Process

The open hearth process is batch process and a batch is called a "heat". The furnace is first inspected for possible damage. Once it is ready or repaired, it is charged with light scrap, such as sheet metal, shredded vehicles or waste metal. Once it has melted, heavy scrap, such as building, construction or steel milling scrap is added, together with pig iron from blast furnaces. Once all steel has melted, slag forming agents, such as limestone, are added. The oxygen in iron oxide and other impurities decarburize the pig iron by burning the carbon away, forming steel. To increase the oxygen contents of the heat, iron ore can be added to the heat.

The process is far slower than that of Bessemer converter and thus easier to control and take samples for quality control. Preparing a heat usually takes 8 h to 8 h 30 min to complete into steel. As the process is slow, it is not necessary to burn all the carbon away as in Bessemer process, but the process can be terminated at given point when desired carbon contents has been achieved.

The furnace is tapped the same way a blast furnace is tapped; a hole is drilled on the side of the hearth and the raw steel is let to flow out. Once all the steel has been tapped, the slag is skimmed away. The raw steel may be cast into ingots; this process is called *teeming*, or it may be used on continuous casting for the rolling mill.

The regenerators are the distinctive feature of the furnace and consist of fire-brick flues filled with bricks set on edge and arranged in such a way as to have a great number of small passages between them. The bricks absorb most of the heat from the outgoing waste gases and return it later to the incoming cold gases for combustion.

History

Sir Carl Wilhelm Siemens developed the Siemens regenerative furnace in the 1850s, and claimed in 1857 to be recovering enough heat to save 70–80% of the fuel. This furnace operates at a high temperature by using regenerative preheating of fuel and air for combustion. In regenerative preheating, the exhaust gases from the furnace are pumped into a chamber containing bricks, where heat is transferred from the gases to the bricks. The flow of the furnace is then reversed so that fuel and air pass through the chamber and are heated by the bricks. Through this method, an open-hearth furnace can reach temperatures high enough to melt steel, but Siemens did not initially use it for that.

In 1865, the French engineer Pierre-Émile Martin took out a license from Siemens and first applied his regenerative furnace for making steel. The most appealing characteristic of the Siemens regenerative furnace is the rapid production of large quantities of basic steel, used for example to construct high-rise buildings. The usual size of furnaces is 50 to 100 tons, but for some special processes they may have a capacity of 250 or even 500 tons.

The Siemens-Martin process complemented rather than replaced the Bessemer process. It is slower and thus easier to control. It also permits the melting and refining of large amounts of scrap steel, further lowering steel production costs and recycling an otherwise troublesome waste material. Its worst drawback is the fact that melting and refining a charge takes several hours. This was an advantage in the early 20th century, as it gave plant chemists time to analyze the steel and decide how much longer to refine it. But by about 1975, electronic instruments such as atomic absorption spectrophotometres had made analysis of the steel much easier and faster. The work environment around an open hearth furnace is said to be extremely dangerous, although that may be even more true of the environment around a basic oxygen or electric arc furnace.

Basic oxygen steelmaking eventually replaced the open hearth furnace. It rapidly superseded both the Bessemer process and Siemens-Martin process in Western Europe by the 1950s and in Eastern Europe by the 1980s. The open hearth steelmaking had superseded Bessemer process in UK by 1900, but elsewhere in Europe, especially in Germany, the Bessemer and Thomas processes were used until the late 1960s when they were superseded by basic oxygen steelmaking. The last European open hearth furnace in the former East Germany was stopped in 1993. In the US, steel production using the Bessemer process ended in 1968 and the open hearth furnaces had stopped by 1992. In Hunedoara steel works, Romania the last 420-tonne capacity open hearth furnace was shut down on 12 June 1999, demolished and scrapped between 2001 and 2003, but the eight smokestacks of the furnaces remained until February 2011. The last open hearth shop in China was shut down in 2001. The nation with the highest share of steel produced with open hearth furnaces (almost 50%) is Ukraine. The process is still in use in both India and Russia.

Electric Arc Furnace

An electric arc furnace (EAF) is a furnace that heats charged material by means of an electric arc.

Arc furnaces range in size from small units of approximately one ton capacity (used in foundries for producing cast iron products) up to about 400 ton units used for secondary steelmaking. Arc furnaces used in research laboratories and by dentists may have a capacity of only a few dozen grams. Industrial electric arc furnace temperatures can be up to 1,800 °C (3,272 °F), while laboratory units can exceed 3,000 °C (5,432 °F). Arc furnaces differ from induction furnaces in that the charge material is directly exposed to an electric arc, and the current in the furnace terminals passes through the charged material.

History

In the 19th century, a number of men had employed an electric arc to melt iron. Sir Humphry Davy conducted an experimental demonstration in 1810; welding was investigated by Pepys in 1815; Pinchon attempted to create an electrothermic furnace in 1853; and, in 1878–79, Sir William Siemens took out patents for electric furnaces of the arc type.

The first electric arc furnaces were developed by Paul Héroult, of France, with a commercial plant established in the United States in 1907. The Sanderson brothers formed The Sanderson Brothers steel Co. in Syracuse, New York, installing the first electric arc furnace in the U.S. This furnace is now on display at Station Square, Pittsburgh, Pennsylvania.

Initially "electric steel" was a speciality product for such uses as machine tools and spring steel. Arc furnaces were also used to prepare calcium carbide for use in carbide lamps. The *Stassano electric furnace* is an arc type furnace that usually rotates to mix the bath. The *Girod furnace* is similar to the *Héroult furnace.*

While EAFs were widely used in World War II for production of alloy steels, it was only later that electric steelmaking began to expand. The low capital cost for a mini-mill—around US$140–200 per ton of annual installed capacity, compared with US$1,000 per ton of annual installed capacity for an integrated steel mill—allowed mills to be quickly established in war-ravaged Europe, and also allowed them to successfully compete with the big United States steelmakers, such as Bethlehem Steel and U.S. Steel, for low-cost, carbon steel "long products" (structural steel, rod and bar, wire, and fasteners) in the U.S. market.

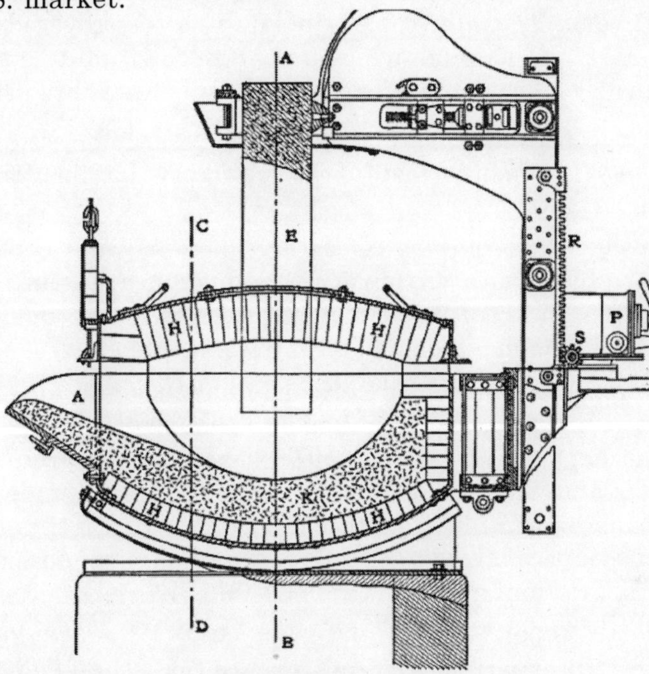

Transverse Section E F

Figure: *A schmatic cross section through a Heroult arc furnace. E is an electrode (only one shown), raised and lowered by the rack and pinion drive R and S. The interior is lined with refractory brick H, and K denotes the bottom lining. A door at A allows access to the interior. The furnace shell rests on rockers to allow it to be tilted for tapping.*

When Nucor—now one of the largest steel producers in the U.S.—decided to enter the long products market in 1969, they chose to start up a mini-mill, with an EAF as its steelmaking furnace, soon followed by other manufacturers. Whilst Nucor expanded rapidly in the Eastern US, the companies that followed them into mini-mill operations concentrated on local markets for long products, where the use of an EAF allowed the plants to vary production according to local demand. This pattern was also followed globally, with EAF steel production primarily used for long products, while integrated mills, using blast furnaces and basic oxygen furnaces, cornered the markets for "flat products"—sheet steel and heavier steel plate. In 1987, Nucor made the decision to expand into the flat products market, still using the EAF production method.

Construction

An electric arc furnace used for steelmaking consists of a refractory-lined vessel, usually water-cooled in larger sizes, covered with a retractable roof, and through which one or more graphite electrodes enter the furnace. The furnace is primarily split into three sections:

- the *shell*, which consists of the sidewalls and lower steel "bowl";
- the *hearth*, which consists of the refractory that lines the lower bowl;
- the *roof*, which may be refractory-lined or water-cooled, and can be shaped as a section of a sphere, or as a frustum (conical section). The roof also supports the refractory delta in its centre, through which one or more graphite electrodes enter.

The hearth may be hemispherical in shape, or in an eccentric bottom tapping furnace, the hearth has the shape of a halved egg. In modern meltshops, the furnace is often raised off the ground floor, so that ladles and slag pots can easily be maneuvered under either end of the furnace. Separate from the furnace structure is the electrode support and electrical system, and the tilting platform on which the furnace rests. Two configurations are possible: the electrode supports and the roof tilt with the furnace, or are fixed to the raised platform.

A typical alternating current furnace has three electrodes. Electrodes are round in section, and typically in segments with threaded couplings, so that as the electrodes wear, new segments can be added. The arc forms between the charged material and the electrode, the charge is heated both by current passing through the charge and by the radiant energy evolved by the arc. The electrodes are automatically raised and lowered by a positioning system, which may use either electric winch hoists or hydraulic cylinders. The regulating system maintains approximately constant current and power input during the melting of the charge, even though scrap may move under the electrodes as it melts. The mast arms holding the electrodes can either carry heavy busbars (which may be hollow water-cooled copper pipes carrying current to the electrode clamps) or be "hot arms", where the whole arm carries the current, increasing efficiency. Hot arms can be made from copper-clad steel or aluminium. Since the electrodes move up and down automatically for regulation of the arc, and are raised to allow removal of the furnace roof, large water-cooled cables connect the bus tubes/arms with the transformer located adjacent to the furnace. To protect the transformer from heat, it is installed in a vault and is itself cooled via pumped oil exchanging heat with the plant's water-cooling systems, as the electrical conditions for arc-furnace steelmaking are extremely stressful on the transformer.

The furnace is built on a tilting platform so that the liquid steel can be poured into another vessel for transport. The operation of tilting the furnace to pour molten steel is called "tapping". Originally, all steelmaking furnaces had a tapping spout closed with refractory that washed out when the furnace was tilted, but often modern furnaces have an eccentric bottom tap-hole (EBT) to reduce inclusion of nitrogen and slag in the liquid steel. These furnaces have a taphole that passes vertically through the hearth and shell, and is set off-centre in the narrow "nose" of the egg-shaped hearth. It is filled with refractory sand, such as olivine, when it is closed off. Modern plants may have two shells with a single set of electrodes that can be transferred between the two; one shell preheats scrap while the other shell is utilised for meltdown. Other DC-based furnaces have a similar arrangement, but have electrodes for each shell and one set of electronics.

AC furnaces usually exhibit a pattern of hot and cold-spots around the hearth perimeter, with the cold-spots located between the electrodes. Modern furnaces mount oxygen-fuel burners in the sidewall and use them to provide chemical energy to the cold-spots, making the heating of the steel more uniform. Additional chemical energy is provided by injecting oxygen and carbon into the furnace; historically this was done through lances in the slag door, now this is mainly done through multiple wall-mounted injection units that combine the oxygen-fuel burners and the oxygen or carbon injection systems into one unit.

A mid-sized modern steelmaking furnace would have a transformer rated about 60,000,000 volt-amperes (60 MVA), with a secondary voltage between 400 and 900 volts and a secondary current in excess of 44,000 amperes. In a modern shop such a furnace would be expected to produce a quantity of 80 metric tonnes of liquid steel in approximately 50 minutes from charging with cold scrap to tapping the furnace. In comparison, basic oxygen furnaces can have a capacity of 150–300 tonnes per batch, or "heat", and can produce a heat in 30–40 minutes. Enormous variations exist in furnace design details and operation, depending on the end product and local conditions, as well as ongoing research to improve furnace efficiency. The largest scrap-only furnace (in terms of tapping weight and transformer rating) is a DC furnace operated by Tokyo Steel in Japan, with a tap weight of 420 metric tonnes and fed by eight 32MVA transformers for 256MVA total power. To produce a ton of steel in an electric arc furnace requires approximately 400 kilowatt-hours per short ton or about 440 kWh per metric tonne; the theoretical minimum amount of energy required to melt a tonne of scrap steel is 300 kWh (melting point 1520°C/2768°F). Therefore, a 300-tonne, 300 MVA EAF will require approximately 132 MWh of energy to melt the steel, and a "power-on time" (the time that steel is being melted with an arc) of approximately 37 minutes. Electric arc steelmaking is only economical where there is plentiful electricity, with a well-developed electrical grid. In many locations, mills operate during off-peak hours when utilities have surplus power generating capacity and the price of electricity is less.

Operation

Scrap metal is delivered to a scrap bay, located next to the melt shop. Scrap generally comes in two main grades: shred (whitegoods, cars and other objects made of similar light-gauge steel) and heavy melt (large slabs and beams), along with some direct reduced iron (DRI) or pig iron for chemical balance. Some furnaces melt almost 100% DRI.

The scrap is loaded into large buckets called baskets, with "clamshell" doors for a base. Care is taken to layer the scrap in the basket to ensure good furnace operation; heavy melt is placed on top of a light layer of protective shred, on top of which is placed more shred. These layers should be present in the furnace after charging. After loading, the basket may pass to a scrap pre-heater, which uses hot furnace off-gases to heat the scrap and recover energy, increasing plant efficiency.

The scrap basket is then taken to the melt shop, the roof is swung off the furnace, and the furnace is charged with scrap from the basket. Charging is one of the more dangerous operations for the EAF operators. A lot of potential energy is released by multiple tonnes of falling metal; any liquid metal in the furnace is often displaced upwards and outwards by the solid scrap, and the grease and dust on the scrap is ignited if the furnace is hot, resulting in a fireball erupting.

In some twin-shell furnaces, the scrap is charged into the second shell while the first is being melted down, and pre-heated with off-gas from the active shell. Other operations are continuous charging—pre-heating scrap on a conveyor belt, which then discharges the scrap into the furnace proper, or charging the scrap from a shaft set above the furnace, with off-gases directed through the shaft. Other furnaces can be charged with hot (molten) metal from other operations. After charging, the roof is swung back over the furnace and meltdown commences.

The electrodes are lowered onto the scrap, an arc is struck and the electrodes are then set to bore into the layer of shred at the top of the furnace. Lower voltages are selected for this first part of the operation to protect the roof and walls from excessive heat and damage from the arcs. Once the electrodes have reached the heavy melt at the base of the furnace and the arcs are shielded by the scrap, the voltage can be increased and the electrodes raised slightly, lengthening the arcs and increasing power to the melt. This enables a molten pool to form more rapidly, reducing tap-to-tap times. Oxygen is blown into the scrap, combusting or cutting the steel, and extra chemical heat is provided by wall-mounted oxygen-fuel burners. Both processes accelerate scrap meltdown. Supersonic nozzles enable oxygen jets to penetrate foaming slag and reach the liquid bath.

An important part of steelmaking is the formation of slag, which floats on the surface of the molten steel. Slag usually consists of metal oxides, and acts as a destination for oxidised impurities, as a thermal blanket (stopping excessive heat loss) and helping to reduce erosion of the refractory lining. For a furnace with basic refractories, which includes most carbon steel-producing furnaces, the usual slag formers are calcium oxide (CaO, in the form of burnt lime) and magnesium oxide (MgO, in the form of dolomite and magnesite). These slag formers are either charged with the scrap, or blown into the furnace during meltdown. Another major component of EAF slag is iron oxide from steel combusting with the injected oxygen. Later in the heat, carbon (in the form of coke or coal) is injected into this slag layer, reacting with the iron oxide to form metallic iron and carbon monoxide gas, which then causes the slag to foam, allowing greater thermal efficiency, and better arc stability and electrical efficiency. The slag blanket also covers the arcs, preventing damage to the furnace roof and sidewalls from radiant heat.

Once the scrap has completely melted down and a flat bath is reached, another bucket of scrap can be charged into the furnace and melted down, although EAF development is moving towards single-charge designs. After the second charge is completely melted, refining operations take place to check and correct the steel chemistry and superheat the melt above its freezing temperature in preparation for tapping. More slag formers are introduced and more oxygen is blown into the bath, burning out impurities such as silicon, sulfur, phosphorus, aluminium, manganese, and calcium, and removing their oxides to the slag. Removal of carbon takes place after these elements have burnt out first, as they have a greater affinity for oxygen. Metals that have a poorer affinity for oxygen than iron, such as nickel and copper, cannot be removed through oxidation and must be controlled through scrap chemistry alone, such as introducing the direct reduced iron and pig iron mentioned earlier. A foaming slag is maintained throughout, and often overflows the furnace to pour out of the slag door into the slag pit. Temperature sampling and chemical sampling take place via automatic lances. Oxygen and carbon can be automatically measured via special probes that dip into the steel, but for all other elements, a "chill" sample—a small, solidified sample of the steel—is analysed on an arc-emission spectrometre.

Once the temperature and chemistry are correct, the steel is tapped out into a preheated ladle through tilting the furnace. For plain-carbon steel furnaces, as soon as slag is detected during tapping the furnace is rapidly tilted back towards the deslagging side, minimising slag carryover into the ladle. For some special steel grades, including stainless steel, the slag is poured into the ladle as well, to be treated at the ladle furnace to recover valuable alloying elements. During tapping some alloy additions are introduced into the metal stream, and more lime is added on top of the ladle to begin building a new slag layer. Often, a few tonnes of liquid steel and slag is left in the furnace in order to form a "hot heel", which helps preheat the next charge of scrap and accelerate its meltdown. During and after tapping, the furnace is "turned around": the slag door is cleaned of solidified slag, repairs may take place, and electrodes are inspected for damage or lengthened through the addition of new segments; the taphole is filled with sand at the completion of tapping. For a 90-tonne, medium-power furnace, the whole process will usually take about 60–70 minutes from the tapping of one heat to the tapping of the next (the tap-to-tap time).

Advantages of Electric Arc Furnace for Steelmaking

The use of EAFs allows steel to be made from a 100% scrap metal feedstock. This greatly reduces the energy required to make steel when compared with primary steelmaking from ores. Another benefit is flexibility: while blast furnaces cannot vary their production by much and can remain in operation for years at a time, EAFs can be rapidly started and stopped, allowing the steel mill to vary production according to demand. During the peak of global financial meltdown in 2009, an estimated quantity of only 1 million tonne was produced in USA employing EAF technique. Although steelmaking arc furnaces generally use scrap steel as their primary feedstock, if hot metal from a blast furnace or direct-reduced iron is available economically, these can also be used as furnace feed. As EAFs require large quantities of electrical power, many companies schedule their operations to take advantage of off peak electricity pricing.

A typical steelmaking arc furnace is the source of steel for a mini-mill, which may make bars or strip product. Mini-mills can be sited relatively near to the markets for steel products, and the transport requirements are less than for an integrated mill, which would commonly be sited near a harbour for access to shipping.

Environmental Issues

Although the modern electric arc furnace is a highly efficient recycler of steel scrap, operation of an arc furnace shop can have adverse environmental effects. Much of the capital cost of a new installation will be devoted to systems intended to reduce these effects, which include:

- Enclosures to reduce high sound levels
- Dust collector for furnace off-gas
- Slag production
- Cooling water demand
- Heavy truck traffic for scrap, materials handling, and product
- Environmental effects of electricity generation

Because of the very dynamic quality of the arc furnace load, power systems may require technical measures to maintain the quality of power for other customers; flicker and harmonic distortion are common side-effects of arc furnace operation on a power system.

Other Electric Arc Furnaces

For steelmaking, direct current (DC) arc furnaces are used, with a single electrode in the roof and the current return through a conductive bottom lining or conductive pins in the base. The advantage of DC is lower electrode consumption per ton of steel produced, since only one electrode is used, as well as less electrical harmonics and other similar problems.

The size of DC arc furnaces is limited by the current carrying capacity of available electrodes, and the maximum allowable voltage. Maintenance of the conductive furnace hearth is a bottleneck in extended operation of a DC arc furnace.

In a steel plant, a ladle furnace (LF) is used to maintain the temperature of liquid steel during processing after tapping from EAF or to change the alloy composition. The ladle is used for the first purpose when there is a delay later in the steelmaking process. The ladle furnace consists of a refractory roof, a heating system, and, when applicable, a provision for injecting argon gas into the bottom of the melt for stirring. Unlike a scrap melting furnace, a ladle furnace does not have a tilting or scrap charging mechanism.

Electric arc furnaces are also used for production of calcium carbide, ferroalloys and other non-ferrous alloys, and for production of phosphorus. Furnaces for these services are physically different from steel-making furnaces and may operate on a continuous, rather than batch, basis. Continuous process furnaces may also use paste-type, Søderberg electrodes to prevent interruptions due to electrode changes. Such a furnace is known as a submerged arc furnace because the electrode tips are buried in the slag/charge, and arcing occurs through the slag, between the matte and the electrode. A steelmaking arc furnace, by comparison, arcs in the open. The key is the electrical resistance, which is what generates the heat required: the resistance in a steelmaking furnace is the atmosphere, while in a submerged-arc furnace the slag or charge forms the resistance. The liquid metal formed in either furnace is too conductive to form an effective heat-generating resistance.

Amateurs have constructed a variety of arc furnaces, often based on electric arc welding kits contained by silical blocks or flower pots. Though crude, these simple furnaces can melt a wide range of materials, create calcium carbide, etc.

Plasma Arc Furnace

A plasma arc furnace (PAF) uses plasma torches instead of graphite electrodes. Each of these torches consists of a casing provided with a nozzle and an axial tubing for feeding a plasma-forming gas (either nitrogen or argon), and a burnable cylindrical graphite electrode located within the tubing. Such furnaces can be referred to as "PAM" (Plasma Arc Melt) furnaces. They are used extensively in the titanium melt industry and similar speciality metals industries.

Vacuum Arc Remelting

Vacuum arc remelting (VAR) is a secondary remelting process for vacuum refining and manufacturing of ingots with improved chemical and mechanical homogeneity.

In critical military and commercial aerospace applications, material engineers commonly specify VIM-VAR steels. VIM means Vacuum Induction Melted and VAR means Vacuum Arc Remelted. VIM-VAR steels become bearings for jet engines, rotor shafts for military helicopters, flap actuators for fighter jets, gears in jet or helicopter transmissions, mounts or fasteners for jet engines, jet tail hooks and other demanding applications.

Most grades of steel are melted once and are then cast or teemed into a solid form prior to extensive forging or rolling to a metallurgically sound form. In contrast, VIM-VAR steels go through two more highly purifying melts under vacuum. After melting in an electric arc furnace and alloying in an argon oxygen decarburization vessel, steels destined for vacuum remelting are cast into ingot molds. The solidified ingots then head for a vacuum induction melting furnace. This vacuum remelting process rids the steel of inclusions and unwanted gases while optimizing the chemical composition. The VIM operation returns these solid ingots to the molten state in the contaminant-free void of a vacuum. This tightly controlled melt often requires up to 24 hours. Still enveloped by the vacuum, the hot metal flows from the VIM furnace crucible into giant electrode molds. A typical electrode stands about 15 feet (5 m) tall and will be in various diameters. The electrodes solidify under vacuum.

For VIM-VAR steels, the surface of the cooled electrodes must be ground to remove surface irregularities and impurities before the next vacuum remelt. Then the ground electrode is placed in a VAR furnace. In a VAR furnace the steel gradually melts drop-by-drop in the vacuum-sealed chamber. Vacuum arc remelting further removes lingering inclusions to provide superior steel cleanliness and further remove gases such as oxygen, nitrogen and hydrogen. Controlling the rate at which these droplets form and solidify ensures a consistency of chemistry and microstructure throughout the entire VIM-VAR ingot. This in turn makes the steel more resistant to fracture or fatigue. This refinement process is essential to meet the performance characteristics of parts like a helicopter rotor shaft, a flap actuator on a military jet or a bearing in a jet engine.

For some commercial or military applications, steel alloys may go through only one vacuum remelt, namely the VAR. For example, steels for solid rocket cases, landing gears or torsion bars for fighting vehicles typically involve the one vacuum remelt.

Vacuum arc remelting is also used in production of titanium and other metals which are reactive or in which high purity is required.

Electro-Slag Remelting

Electroslag remelting (ESR), also known as electro-flux remelting, is a process of remelting and refining steel and other alloys for mission-critical applications in aircraft, thermal power stations, nuclear power plants, military technology, et al.

The electroslag remelting (ESR) process is used to remelt and refine steels and various super-alloys, resulting in high-quality ingots. This process can be started up through vacuum induction melting. The ESR process uses the as-cast alloy as a consumable electrode. Electric current (generally AC) is passed between the electrode and the new ingot, which is formed in the bottom of a water-cooled copper mold.

The new ingot is covered in an engineered slag that is superheated by the electric current. The electrode tip is slowly melted from contact with the slag. These metal droplets travel through the slag to the bottom of the water-cooled mold and slowly freeze as the ingot is directionally solidified upwards from the bottom of the mold. The slag pool floats above the refined alloy, continuously floating upwards as the alloy solidifies. The molten metal is cleaned of impurities that chemically react with the slag or otherwise float to the top of the molten pool as the molten droplets pass through the slag.

Electroslag remelting uses highly reactive slags (calcium fluoride, lime, alumina, or other oxides are usually the main components) to reduce the amount of type-A sulfide present in biometal alloys. It is a common practice in European industries. ESR reduces other types of inclusions as well, and is seen as an alternative to the vacuum arc remelting method that is prevalent in US industries.

Vacuum Arc Remelting

Vacuum arc remelting (VAR) is a secondary melting process for production of metal ingots with elevated chemical and mechanical homogeneity for highly demanding applications. The VAR process has revolutionized the speciality traditional metallurgical techniques industry, and has made possible incredibly controlled materials used in the biomedical, aviation, and aerospace fields.

Overview

VAR is used most frequently in high value applications. Essentially it is an additional processing step to improve the quality of metal. Because it is both time consuming and expensive, a majority of commercial alloys do not employ the process. Nickel, titanium, and speciality steels are materials most often processed with this method. The conventional path for production of titanium alloys includes single, double or even triple VAR processing. Use of this technique over traditional methods presents several advantages:

- The solidification rate of molten material can be tightly controlled. This allows a high degree of control over the microstructure as well as the ability to minimize segregation
- The gases dissolved in liquid metal during melting metals in open furnaces, such as nitrogen, oxygen and hydrogen are considered to be detrimental to the majority of steels and alloys. Under vacuum conditions these gases escape from liquid metal to the vacuum chamber.
- Elements with high vapor pressure such as carbon, sulfur, and magnesium (frequently contaminants) are lowered in concentration.
- Centreline porosity and segregation are eliminated.
- Certain metals and alloys, such as Ti, cannot be melted in open air furnaces

Process Description

The alloy to undergo VAR is formed into a cylinder typically by vacuum induction melting (VIM) or ladle refining (airmelt). This cylinder, referred to as an electrode is then put into a large cylindrical enclosed crucible and brought to a metallurgical vacuum (0.001–0.1 mmHg [0.13–13.33 Pa]). At the bottom of the crucible is a small amount of the alloy to be remelted, which the top electrode is brought close to prior to starting the melt. Several kiloamperes of DC current are used to start an arc between the two pieces, and from there, a continuous melt is derived.

The crucible (typically made of copper) is surrounded by a water jacket used to cool the melt and control the solidification rate. To prevent arcing between the electrode and the crucible side walls, the diameter of the crucible is larger than that of the electrode. As a result, electrode must be lowered as the melt consumes it. Control of the current, cooling water, and electrode gap is essential to effective control of the process, and production of defect free material.

Ideally, the melt rate stays constant throughout the process cycle, but control of the process is not simple. This is because there is very complex heat transfer going on involving conduction, radiation, convection (within the liquid metal), and advection (caused by the Lorentz Force). Ensuring the consistency of the melt process in terms of pool geometry, and melt rate is pivotal in ensuring the best possible properties from the alloy.

Materials and Applications

The VAR process is used on many different materials, however certain applications almost always use a material that has been VAR treated. A list of materials that may be VAR treated include:

- Steel
  - 9310
  - 4340 & 4330+V
  - 300M
  - 316
  - AF1410
  - Aermet 100
  - M50
  - BG42
  - Nitralloy
  - 16NCD13
  - 35NCD16
  - HY-100
  - HY-180
  - HY-TUF
  - D6AC
  - Maraging steels
  - UT-18
  - HP 9-4-30
- Titanium
  - Ti-6Al-4V
  - Ti-10V-2Al-3Fe
  - Ti-5Al-5V-5Mo-3Cr
- Invar
- Nitinol
- Nickel superalloys
  - Inconel alloys
  - Rene alloys
  - RR1000

- Zirconium
- Niobium
- Platinum
- Tantalum
- Rhodium

Note that pure titanium and most titanium alloys are double or triple VAR processed. Nickel-based super alloys for aerospace applications are usually VAR processed. Zirconium and niobium alloys used in the nuclear industry are routinely VAR processed. Pure platinum, tantalum, and rhodium may be VAR processed.

Argon Oxygen Decarburization

Argon oxygen decarburization (AOD) is a process primarily used in stainless steel making and other high grade alloys with oxidizable elements such as chromium and aluminum. After initial melting the metal is then transferred to an AOD vessel where it will be subjected to three steps of refining; decarburization, reduction, and desulphurization. AOD was invented in 1954 by the Lindé Division of The Union Carbide Corporation (which became known as Praxair in 1992).

Process

Decarburization:

Prior to the decarburization step, one more step should be taken into consideration: *de-siliconization*, which is a very important factor for refractory lining and further refinement.

The decarburization step is controlled by ratios of oxygen to argon or nitrogen to remove the carbon from the metal bath. The ratios can be done in any number of phases to facilitate the reaction.

The gases are usually blown through a top lance (oxygen only) and tuyeres in the sides/bottom (oxygen with an inert gas shroud). The stages of blowing remove carbon by the combination of oxygen and carbon forming CO gas.

$$4\ Cr_{(bath)} + 3\ O_2 \rightarrow 2\ Cr_2O_{3(slag)}$$

$$Cr_2O_{3(slag)} + 3\ C_{(bath)} \rightarrow 3\ CO_{(gas)} + 2\ Cr_{(bath)}$$

To drive the reaction to the forming of CO, the partial pressure of CO is lowered using argon or nitrogen. Since the AOD vessel isn't externally heated, the blowing stages are also used for temperature control. The burning of carbon increases the bath temperature.

By the end of this process around 97% of Cr is retained in the steel.

Reduction

After a desired carbon and temperature level have been reached the process moves to reduction. Reduction recovers the oxidized elements such as chromium from the slag. To achieve this, alloy additions are made with elements that have a higher affinity for oxygen than chromium, using either a silicon alloy or aluminum. The reduction mix also includes lime (CaO) and fluorspar (CaF_2). The addition of lime and fluorspar help with driving the reduction of Cr_2O_3 and managing the slag, keeping the slag fluid and volume small.

Desulphurization

Desulphurization is achieved by having a high lime concentration in the slag and a low oxygen activity in the metal bath.

$$S_{(bath)} + CaO_{(slag)} \rightarrow CaS_{(slag)} + O_{(bath)}$$

So, additions of lime are added to dilute sulfur in the metal bath. Also, aluminum or silicon may be added to remove oxygen. Other trimming alloy additions might be added at the end of the step. After sulfur levels have been achieved the slag is removed from the AOD vessel and the metal bath is ready for tapping. The tapped bath is then either sent to a stir station for further chemistry trimming or to a caster for casting.

Pattern Welding

Pattern welding is the practice in sword and knife making of forming a blade of several metal pieces of differing composition that are forge-welded together and twisted and manipulated to form a pattern. Often called (Modern) Damascus steel, blades forged in this manner often display bands of slightly different patterning along their entire length. These bands can be highlighted for cosmetic purposes by proper polishing or acid etching. Pattern welding was an outgrowth of laminated or piled steel, a similar technique used to combine steels of different carbon contents, providing a desired mix of hardness and toughness. Although modern steelmaking processes negate the need to blend different steels, pattern welded steel is still used by custom knifemakers for the cosmetic effects it produces.

History

Pattern welding developed out of the necessarily complex process of making blades that were both hard and tough from the erratic and unsuitable output from early iron smelting in bloomeries. The bloomery does not generate temperatures high enough to melt iron and steel, but instead reduces the iron oxide ore into particles of pure iron, which then weld into a mass of sponge iron, consisting of lumps of impurities in a matrix of relatively pure iron, which is too soft to make a good blade. Carburizing thin iron bars or plates forms a layer of harder, high carbon steel on the surface, and early bladesmiths would forge these bars or plates together to form relatively homogeneous bars of steel. This laminating different types of steels together produces patterns that can be seen in the surface of the finished blade, and this forms the basis for pattern welding.

Pattern Welding in Europe

By the 2nd and 3rd century AD, the Celts were commonly using pattern welding for decoration in addition to structural reasons. Alternating layers of steel would be forged into rods, which would then be twisted to form complex patterns when forged into a blade. By the 6th and 7th centuries, pattern welding had reached a level where thin layers of patterned steel were being overlaid onto a soft iron core, making the swords far better as the iron gave them a flexible and springy core that would take any shock from sword blows to stop the blade bending or snapping. By the end of the Viking era, pattern welding fell out of use in Europe.

During the Middle ages, Damascus steel was being produced in India and brought back to Europe. The similarities in the markings led many to believe it was the same process being used, and pattern welding was revived by European smiths who were attempting to duplicate the Damascene steel. While the methods used by Damascene smiths to produce their blades was lost, recent efforts by metallurgists and bladesmiths (such as Verhoeven and Pendray) to reproduce steel with identical characteristics have yielded a process that does not involve pattern welding. However even these attempts have not been a huge success.

A similar technique was also employed by Scandinavian Medieval swordsmiths. The Mora knife is today manufactured with a similar technique. Today the traditional crucible steel is seldom used, but the high carbon steel is usually tool steel or stainless steel.

Modern Decorative Use

The ancient swordmakers exploited the aesthetic qualities of pattern welded steel. The Vikings in particular were fond of twisting bars of steel around each other, welding the bars together by hammering and then repeating the process with the resulting bars, to create complex patterns in the final steel bar. Two bars twisted in opposite directions created the common chevron pattern. Often, the centre of the blade was a core of soft steel, and the edges were solid high carbon steel, similar to the laminates of the Japanese.

The American Bladesmith Society's Master Smith test, for example, requires a 300 layer blade to be forged. Large numbers of layers are generally produced by folding, where a small number of layers are welded together, then the blank is cut in half, stacked, and welded again, with each operation doubling the number of layers. Starting with just two layers, eight folding operations will yield 512 layers in the blank. A blade ground from such a blank will show a grain much like an object cut from a block of wood, with similar random variations in pattern. Some manufactured objects can be re-purposed into pattern welded blanks. "Cable Damascus", forged from high carbon multi-strand cable, is a popular item for bladesmiths to produce, producing a finely grained, twisted pattern, while chainsaw chains produce a pattern of randomly positioned blobs of colour.

Some modern bladesmiths have taken pattern welding to new heights, with elaborate applications of traditional pattern welding techniques, as well as with new technology. A layered billet of steel rods with the blade blank cut perpendicular to the layers can also produce some spectacular patterns, including mosaics or even writing. Powder metallurgy allows alloys that would not normally be compatible to be combined into solid bars. Different treatments of the steel after it is ground and polished, such as bluing, etching, or various other chemical surface treatments that react differently to the different metals used can create bright, high-contrast finishes on the steel. Some master smiths go as far as to use techniques such as electrical discharge machining to cut interlocking patterns out of different steels, fit them together, then weld the resulting assembly into a solid block of steel.

Crucible Steel

Crucible steel describes a number of different techniques for making steel in a crucible. Its manufacture is essentially a refining process in which another form of steel that had been consolidated by hammering, such as shear steel, is melted to give a more uniform material. Crucible steel has aroused considerable interest for well over a thousand years and there is a sizable body of work concerning its nature and production.

The *crucible steel* of the Industrial Revolution is referred to in this article as English crucible steel, which was developed for making watch springs. It was also used in other applications such as scissors, axes and swords.

The puddling process, developed in the 1840s, produced a satisfactory grade of wrought iron for many applications at a cost far below crucible steel; however, while puddled iron was suitable for structural applications, it did not replace steel in cutting tools. By the end of the 19th century new steel making processes such as Bessemer and the open hearth furnace were able to produce quality steel at a cost even below puddling, making the crucible process obsolete. The crucible process continued to be used for speciality steels, but is today obsolete. Basic oxygen furnaces and electric arc furnaces have largely replaced all previous ones.

Methods of Crucible Steel Production

Various methods were used to produce crucible steel. According to Islamic texts such as al-Tarsusi and Abu Rayhan Biruni, three methods are described for indirect production of steel. The medieval Islamic historian Abu Rayhan Biruni (c. 973–1050) provides us with the earliest reference of the production of Damascus steel. He describes only three methods for producing steel. The first two methods have a long history in Central Asia and in the Indian subcontinent while the third is exclusive to South Asia. These three methods are generally considered to have originated from the Indian Subcontinent. The first method and the most common traditional method is solid state carburization of wrought iron. This is a diffusion process in which wrought iron is packed in crucibles or a hearth with charcoal, then heated to promote diffusion of carbon into the iron to produce steel. Carburization is the basis for the wootz process of steel. The second method is the decarburization of cast iron by removing carbon from the cast iron. Another indirect method uses wrought iron and cast iron. In this process, wrought iron and cast iron may be heated together in a crucible to produce steel by fusion. In regard to this method Abu Rayhan Biruni states: "this was the method used in Hearth". It is proposed that the Indian method refers to Wootz carburization method i.e. the Mysore or Tamil processes.

Variations of co-fusion process have been found preliminary in Persia and Central Asia but have also been found in Hyderabad, India called Deccani or Hyderabad process. For the carbon, a variety of organic materials are specified by the contemporary Islamic authorities, including pomegranate rinds, acorns, fruit skins like orange peel, leaves as well as the white of egg and shells. Slivers of wood are mentioned in some of the Indian sources, but significantly none of the sources mention charcoal.

Ancient History of South Asian Production

Crucible steel is generally attributed to production centres in India and Sri Lanka where it was produced using the so-called "wootz" process and it is assumed that its appearance in other locations was due to long distance trade. Only recently it has become apparent that places in Central Asia like Merv in Turkmenistan and Akhsiket in Uzbekistan were important centres of production of crucible steel. The Central Asian finds are all from excavations and date from the 8th to 12th centuries AD, while the Indian/Sri Lankan material is as early as 300 BC. In addition India's superior iron ore has trace vanadium and other rare earths which unintentionally leads to the formation of carbon nano tubes in Indian crucible steel which was famous throughout the middle east for its ability to retain its edge even after prolonged usage.

South India/Sri Lanka

There are many ethnographic accounts of Indian crucible steel production, however, scientific investigations of crucible steel remains have only been published from four regions: three in India and one in Sri Lanka. Indian/Sri Lankan crucible steel is commonly referred to as wootz. It is generally agreed that wootz is an English corruption of the word ukko or hookoo. European accounts from the 17th century onwards have referred to the repute and manufacture of 'wootz', a traditional crucible steel made specially in parts of southern India in the former provinces of Golconda, Mysore and Salem. As yet the scale of excavations and surface surveys is too limited to link the literary accounts to archaeometallurgical evidence.

South India

The known sites of crucible steel production in south India, i.e. at Konasamudram and Gatihosahalli, date from at least the late medieval period, 16th century. One of the earliest known sites, which shows some promising preliminary evidence that may be linked to ferrous crucible processes in Kodumanal, near Coimbatore in Tamil Nadu. The site is dated between the third century BC and the third century AD. By the seventeenth century the main centre of crucible steel production seems to have been in Hyderabad. The process was apparently quite different from that recorded elsewhere. Wootz from Hyderabad or the Decanni process for making watered blades involved a co-fusion of two different kinds of iron- one was low in carbon and the other was a high-carbon steel or cast iron. Wootz steel was widely exported and traded throughout ancient Europe, China, the Arab world, and became particularly famous in the Middle East, where it became known as Damascus steel.

Sri Lanka

Recent archaeological investigations have suggested that Sri Lanka also supported innovative technologies for iron and steel production in antiquity. The Sri Lankan system of crucible steel making was partially independent of the various Indian and Middle Eastern systems. Their method was something similar to the method of carburization of wrought iron. The earliest confirmed crucible steel site is located in the knuckles range in the northern area of the Central Highlands of Sri Lanka dated to 6th –10th centuries AD. In twelfth century the land of Serendib (Sri Lanka) seems to have been the main supplier of crucible steel, but over the centuries slipped back, and by the nineteenth century just a small industry survived in the Balangoda district of the central southern highlands.

West-Facing Process

A series of excavations at Samanalawewa indicated the unexpected and previously unknown technology of west-facing smelting sites, which are different types of steel production. These furnaces were used for direct smelting to steel. Because of their location on the western sides of hilltops for use of wind in the smelting process they are named west-facing. Sri Lankan furnace steels were known and traded between the 9th and 11th centuries and earlier, but apparently not later. These sites were dated to the 7th–11th centuries. The coincidence of this dating with the 9th century Islamic reference to Sarandib is of great importance. The crucible process existed in India at the same time that the west- facing technology was operating in Sri Lanka.

Early History of Central Asian Production

Central Asia

Central Asia has a rich history of crucible steel production, beginning during the late 1st millennium AD. From the sites in modern Uzbekistan and Merv in Turkmenistan, there exists good archaeological evidence for the large scale production of crucible steel. They all belong in broad terms to the same early medieval period between the late 8th or early 9th and the late 12th century AD Contemporary with the early crusades.

Uzbekistan

The two most prominent crucible steel sites in eastern Uzbekistan carrying the Ferghana Process are Akhsiket and Pap in the Ferghana Valley, whose position within the Great Silk Road has been historically and archaeologically proved. The material evidence of the sites consists of large number of archaeological finds relating to steel making from 9th–12th centuries AD in the form of hundreds of thousands of fragments of crucibles often with massive slag cakes. Archaeological work at Akhsiket, has identified that the crucible steel process was of the carburization of iron ore.

This process appears to be typical of and restricted to the Ferghana Valley in eastern Uzbekistan, and it is therefore called the Ferghana Process. This process lasts in that region for roughly four centuries..

Turkmenistan

Evidences of the production of crucible steel have been found in Merv, Turkmenistan, a major city on the 'Silk Road'. The Islamic scholar, al-Kindi (AD 801–866) mentions that during the ninth century the region of Khorasan, the area to which the cities Nishapur, Merv, Herat and Balkh belong, is a steel manufacturing centre. Evidence from a metallurgical workshop at Merv, dated to the ninth- early tenth century A.D., provides an illustration of the co-fusion method of steel production in crucibles, about 1000 years earlier than the distinctly different wootz process. The crucible steel process at Merv might be seen as technologically related to what Bronson (1986, 43) calls Heyderabad process, a variation of the wootz process, after the location of the process documented by Voysey in the 1820s.

Historical Accounts and Investigations

In the first centuries of the Islamic period, there appear some scientific studies on swords and steel. The best known of these are by Jabir ibn Hayyan 8th century, al-Kindi 9th century, Abu Rayhan Biruni in the early 11th century, Murda al Tarsusi in the late 12th century, and Fakhr-i- Mudabbir 13th century. Any of these contains far more information about Indian and damascene steels than appears in the entire literature of classical Greece and Rome.

The first European references to crucible steel seem to be no earlier than the Post Medieval period. European experiments with "Damascus" steels go back to at least the sixteenth century, but it was not until the 1790s that laboratory researchers began to work with steels that were specifically known to be Indian/wootz. At this time,

Europeans knew of India's ability to make crucible steel from reports brought back by travellers who had observed the process at several places in southern India.

From the mid- 17th century onwards, there are numerous vivid eyewitness accounts of the production of steel by European travellers to the Indian subcontinent. These include accounts by Jean Baptist Tavernier in 1679, Francis Buchanan in 1807, and H.W. Voysey in 1832. The 18th, 19th and early 20th century saw a heady period of European interest in trying to understand the nature and properties of wootz steel. Indian wootz engaged the attention of some of the best-known scientists. One was Michael Faraday who was fascinated by wootz steel. 'It was probably the investigations of George Pearson in 1795 reported at the Royal Society, which had the most far- reaching impact in terms of kindling interest amongst European scientist in wootz. He was the first of these scientists to publish his results and, incidentally, the first to use the word "wootz" in print.

Another investigator, David Mushet, was able to infer that wootz was made by fusion. David Mushet patented his process in 1800. He made his report in 1805. But the first successful European process had been developed by Benjamin Huntsman some 50 years previously in the 1740s.

Early Modern Steel in Europe

Blister steel:

- Watch video: Blister steel:1
- Watch video: Shear steel:2

It was possible to produce quality steel in Europe, by importing the highly valued Swedish iron. Although it was not understood at the time, the Swedish ore contained very low levels of common impurities, leading to higher quality irons and steels from otherwise identical techniques applied to other ores. Swedish bar iron was packed into stone boxes in layers with charcoal in between them and heated in a furnace for an entire week.

The result was a bar of metal known as *blister steel* – the surface of the bars became uneven from a multitude of blisters (or blebs) – which varied in quality from one bar to the next and within each bar. A number of blister rods were then wrapped into a larger bundle and re-heated and hammer-forged to mix together and even out the carbon content, resulting in the final product, *shear steel*. Germany was particularly well invested in this process, largely due to being physically close to Sweden, and became a major steel exporter in the 18th century. The technique was the cementation process.

English Crucible Steel

A new technique was developed in England by Benjamin Huntsman, a clockmaker in search of a better steel for clock springs. It was only in 1740 after he moved to Handsworth near Sheffield, and after years of experimenting in secret he perfected his process. Huntsman's system used a coke-fired furnace capable of reaching 1,600 °C, into which ten or twelve clay crucibles, each holding about 15 kg of iron, were placed. When the pots are at a white heat they are charged with blister steel broken into lumps of about ½ kg, and a flux to help remove impurities. The pots are removed after about 3 hours in the furnace, impurities skimmed off, and the molten steel poured into ingots. Sheffield's Abbeydale Industrial Hamlet has preserved a water-wheel powered, scythe-making works dating from Huntsman's times, which is still operated for the public, several times per year using crucible steel made on the Abbeydale site.

Before the introduction of Huntsman's technique, Sheffield produced about 200 tonnes of steel per year based on Swedish wrought iron. The introduction of Huntsman's technique changed this radically; one hundred years later the amount had risen to over 80,000 tonnes per year – almost half of Europe's total production. This discovery enabled Sheffield to develop from a small township into one of Europe's leading industrial cities.

The steel was cast in a specialised workshop called a 'crucible furnace', which consisted of a workshop at ground level and a subterranean cellar. The furnace buildings varied in size and architectural style, gradually becoming larger towards the latter part of the 19th century as technological developments enabled multiple pots to be fired in one melt and gas was gradually introduced as a means of fuel to heat the crucibles. Each workshop had a series of standard features, such as rows of melting holes, teaming pits, roof vents, rows of shelving for the crucible pots and annealing furnaces to prepare each pot prior to firing. Additional ancillary rooms for the weighing each charge and for the manufacture of the clay crucibles were either attached to the workshop, or located within the cellar complex.

Crucible Steel Elsewhere

Another form of crucible steel was developed in 1837 by the Russian engineer, Pavel Anosov. His technique relied less on the heating and cooling, and more on the quenching process of rapidly cooling the molten steel when the right crystal structure had formed within. He called his steel bulat; its secret died with him. In the United States crucible steel was pioneered by William Metcalf.

Conclusion

Crucible steels remained the world's best, although very expensive, for some time. The introduction of the Bessemer process and open hearth steelmaking processes gradually replaced it, being able to produce steel of similar (or better) quality on a much larger scale more quickly and cheaply. The Bessemer process took a couple of decades of development until it was able to produce quality steel; however, nitrogen introduced with the blown air causes it to become brittle with age. The other part of the problem was lack of control of the final carbon content, which was finally solved by completely de-carburizing the iron and then adding back an alloy ingot containing the desired amount of carbon plus some manganese to tie up any sulfur. It was also found that iron ores other than hematite could be used by lining the converter with limestone, which reacted with the phosphorus in the iron.

The Bessemer process was eventually displaced by the open hearth process, which although requiring much more time to make a batch of steel was much easier to control for quality. Today's basic oxygen furnaces produce steel at a much faster rate than old style open hearth furnaces.

Tatara (Furnace)

The tatara is the traditional Japanese furnace used for smelting iron and steel. The word later also came to mean the entire building housing the furnace.

The steel, or tamahagane , used in the forging of Japanese swords (*nihontô* (åe,gR?), commonly known as *katana* (R?)) by contemporary Japanese forge masters like Kihara Akira and Gassan Sadatoshi is still smelted in a *tatara*. One of the few remaining *tatara* is the Nittoho *tatara* in Shimane Prefecture, Japan.

Etymology

It is generally agreed that the word *tatara* is foreign to Japan, originating in India or Central Asia. There is a possibility that the word came from a place in ancient Korea, where the two dynasties would often meet. According to the *Kojiki* (one of the oldest Japanese historical texts), this meeting place was called Tatara-ba or Tataratsu, and the word was possibly imported with iron-working technology. Japanese scholar Tokutaro Yasuda suggests that the word may be from the Sanskrit word *taatara,* meaning "heat," noting that the Sanskrit word for steel is *sekeraa,* which is very similar to the word used in Japan for the steel bloom which the tatara produces. The two Chinese characters used when the word has the original meaning are Ž′— and, besides as *tatara,* they can be also read as *fumifuigo,* or foot bellows.

Process

Figure: *The structure of a* tatara. *On the two sides of the furnace are visible two bellows*

The smelting process used is different from that of the modern mass production of steel. A clay vessel about 1.1 metres (4 feet) tall, 3 metres (12 feet) long, and 1.1 metres (4 feet wide) is constructed.

This is the *tatara.* After the clay tub has dried, it is fired until dry. A charcoal fire is started from soft pine charcoal, then the smelter will wait for the fire to reach the correct temperature. At that point, he will direct the addition of ironsand, known as *satetsu.* This will be layered in with more charcoal and more ironsand over the next 72 hours. 4 or 5 people need to constantly work on this process.

It takes about a week to build the *tatara* and complete the iron conversion to steel. When the process is done, the clay tub is broken and the steel bloom, known as a *kera,* is removed. At the end of the process, the *tatara* will have consumed about 10 tons of satetsu and 12 tons of charcoal, leaving about 2.5 tons of *tamahagane.*

The *Nittoho* Tatara

In 1977, the Japanese Society for Preservation of Japanese Art Swords (Nittoho), together with the Japanese government's Agency for Cultural Affairs and Hitachi Works, built in Shimane Prefecture the so-called Nittoho *Tatara* to provide the steel necessary for the production of Japanese swords. The Nittoho *Tatara* is managed jointly with Yasugi Works, a subsidiary company of Hitachi Metals, and is operational only during the winter.

Cementation Process

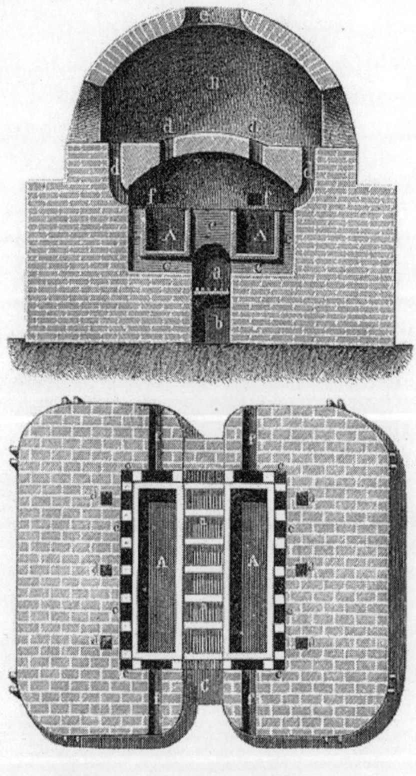

Figure: *Cementation furnace*

The cementation process is an obsolete technique for making steel by carburization of iron. Unlike modern steelmaking, it increased the amount of carbon in the iron. It was apparently developed before the 17th century. Derwentcote Steel Furnace, built in 1720, is the earliest surviving example of a cementation furnace. Another example in the UK is the Cementation furnace in Doncaster Street, Sheffield.

Origins

The process was described in a treatise published in Prague in 1574. It was again invented by Johann Nussbaum of Magdeburg, who began operations at Nuremberg (with partners) in 1601. The process was patented in England by William Ellyot and Mathias Meysey in 1614. At that date, the "invention" could consist merely of the introduction of a new industry or product, or even a mere monopoly. They evidently soon transferred the patent to Sir Basil Brooke, but he was forced to surrender it in 1619.

A clause in the patent prohibiting the import of steel was found to be undesirable because he could not supply as much good steel as was needed. Brooke's furnaces were probably in his manor of Madeley at Coalbrookdale (which certainly existed before the English Civil War) where two cementation furnaces have been excavated. He probably used bar iron from the Forest of Dean, where he was a partner in farming the King's ironworks in two periods.

By 1631, it was recognised that Swedish iron was the best raw material and then or later particularly certain marks (brands) such as *double bullet* (so called from the mark OO) from Österby and *hoop L* from Leufsta (now Lövsta), whose mark consisted of an L in a circle, both belonging to Louis De Geer and his descendants.

These were among the first ironworks in Sweden to use the Walloon process of fining iron, producing what was known in England as oregrounds iron. It was so called from the Swedish port of Öregrund, north of Stockholm, in whose hinterland most of the ironworks lay. The ore used came ultimately from the Dannemora mine.

Process

The process begins with wrought iron and charcoal. It uses one or more long stone *pots* inside a furnace. Typically, in Sheffield, each was 14 feet by 4 feet and 3.5 feet deep. Iron bars and charcoal are packed in alternating layers, with a top layer of charcoal and then refractory matter to make the pot or "coffin" airtight.

Some manufacturers used a mix of powdered charcoal, soot and mineral salts, called *cement powder*. In larger works up to 16 tons of iron was treated in each cycle.

Depending on the thickness of the iron bars, the pots were then heated from below for a week or more. Bars were regularly examined and when the correct condition was reached the heat was withdrawn and the pots were left until cool—usually around fourteen days. The iron had "gained" a little over 1% in mass from the carbon in the charcoal, and had become heterogeneous bars of *blister steel.*

The bars were then shortened, bound, heated and hammered, pressed or rolled to become *shear steel.* Alternatively they could be broken up and melted in a crucible using a crucible furnace with a flux to become *crucible steel* or *cast steel*, a process devised by Benjamin Huntsman in the 1740s.

Cementation Process for Brass

In the early modern period, brass, an alloy of copper and zinc, was usually produced by a cementation process in which metallic copper was heated with calamine, a zinc ore.

Alloy Steel

Alloy steel is steel that is alloyed with a variety of elements in total amounts between 1.0% and 50% by weight to improve its mechanical properties. Alloy steels are broken down into two groups: low-alloy steels and high-alloy steels.

The difference between the two is somewhat arbitrary: Smith and Hashemi define the difference at 4.0%, while Degarmo, et al., define it at 8.0%. Most commonly, the phrase "alloy steel" refers to low-alloy steels.

Every steel is truly an alloy, but not all steels are called "alloy steels". Even the simplest steels are iron (Fe) (about 99%) alloyed with carbon (C) (about 0.1% to 1%, depending on type).

However, the term "alloy steel" is the standard term referring to steels with *other* alloying elements *in addition to* the carbon. Common alloyants include manganese (the most common one), nickel, chromium, molybdenum, vanadium, silicon, and boron. Less common alloyants include aluminum, cobalt, copper, cerium, niobium, titanium, tungsten, tin, zinc, lead, and zirconium.

The following is a range of improved properties in alloy steels (as compared to carbon steels): strength, hardness, toughness, wear resistance, corrosion resistance, hardenability, and hot hardness. To achieve some of these improved properties the metal may require heat treating.

Some of these find uses in exotic and highly-demanding applications, such as in the turbine blades of jet engines, in spacecraft, and in nuclear reactors.

Because of the ferromagnetic properties of iron, some steel alloys find important applications where their responses to magnetism are very important, including in electric motors and in transformers.

Low-Alloy Steels

Low-alloy steels are usually used to achieve better hardenability, which in turn improves its other mechanical properties. They are also used to increase corrosion resistance in certain environmental conditions.

With medium to high carbon levels, low-alloy steel is difficult to weld. Lowering the carbon content to the range of 0.10% to 0.30%, along with some reduction in alloying elements, increases the weldability and formability of the steel while maintaining its strength. Such a metal is classed as a high-strength low-alloy steel.

Some common low alloy steels are:

- D6AC
- 300M
- 256A

Principal Low-Alloy Steels

SAE Designation	***Composition***
13xx	Mn 1.75%
40xx	Mo 0.20% or 0.25% or 0.25% Mo & 0.042% S
41xx	Cr 0.50% or 0.80% or 0.95%, Mo 0.12% or 0.20% or 0.25% or 0.30%
43xx	Ni 1.82%, Cr 0.50% to 0.80%, Mo 0.25%
44xx	Mo 0.40% or 0.52%
46xx	Ni 0.85% or 1.82%, Mo 0.20% or 0.25%
47xx	Ni 1.05%, Cr 0.45%, Mo 0.20% or 0.35%
48xx	Ni 3.50%, Mo 0.25%
50xx	Cr 0.27% or 0.40% or 0.50% or 0.65%
50xxx	Cr 0.50%, C 1.00% min
50Bxx	Cr 0.28% or 0.50%, and added boron
51xx	Cr 0.80% or 0.87% or 0.92% or 1.00% or 1.05%
51xxx	Cr 1.02%, C 1.00% min
51Bxx	Cr 0.80%, and added boron
52xxx	Cr 1.45%, C 1.00% min
61xx	Cr 0.60% or 0.80% or 0.95%, V 0.10% or 0.15% min
86xx	Ni 0.55%, Cr 0.50%, Mo 0.20%
87xx	Ni 0.55%, Cr 0.50%, Mo 0.25%
88xx	Ni 0.55%, Cr 0.50%, Mo 0.35%
92xx	Si 1.40% or 2.00%, Mn 0.65% or 0.82% or 0.85%, Cr 0.00% or 0.65%
94Bxx	Ni 0.45%, Cr 0.40%, Mo 0.12%, and added boron
ES-1	Ni 5%, Cr 2%, Si 1.25%, W 1%, Mn 0.85%, Mo 0.55%, Cu 0.5%, Cr 0.40%, C 0.2%, V 0.1%

Material Science

Alloying elements are added to achieve certain properties in the material. As a guideline, alloying elements are added in lower percentages (less than 5%) to increase strength or hardenability, or in larger percentages (over 5%) to achieve special properties, such as corrosion resistance or extreme temperature stability.

Manganese, silicon, or aluminum are added during the steelmaking process to remove dissolved oxygen, sulfur and phosphorus from the melt. Manganese, silicon, nickel, and copper are added to increase strength by forming solid solutions in ferrite. Chromium, vanadium, molybdenum, and tungsten increase strength by forming second-phase carbides.

Nickel and copper improve corrosion resistance in small quantities. Molybdenum helps to resist embrittlement. Zirconium, cerium, and calcium increase toughness by controlling the shape of inclusions. Manganese sulfide, lead, bismuth, selenium, and tellurium increase machinability.

The alloying elements tend to form either compounds or carbides. Nickel is very soluble in ferrite; therefore, it forms compounds, usually Ni_3Al.

Aluminium dissolves in the ferrite and forms the compounds Al_2O_3 and AlN. Silicon is also very soluble and usually forms the compound $SiO_2 \cdot M_xO_y$.

Manganese mostly dissolves in ferrite forming the compounds MnS, $MnO \cdot SiO_2$, but will also form carbides in the form of $(Fe,Mn)_3C$. Chromium forms partitions between the ferrite and carbide phases in steel, forming $(Fe,Cr_3)C$, Cr_7C_3, and $Cr_{23}C_6$.

The type of carbide that chromium forms depends on the amount of carbon and other types of alloying elements present. Tungsten and molybdenum form carbides if there is enough carbon and an absence of stronger carbide forming elements (i.e., titanium & niobium), they form the carbides W_2C and Mo_2C, respectively. Vanadium, titanium, and niobium are strong carbide forming elements, forming vanadium carbide, titanium carbide, and niobium carbide, respectively.

Alloying elements also have an effect on the eutectoid temperature of the steel. Manganese and nickel lower the eutectoid temperature and are known as *austenite stabilizing elements*. With enough of these elements the austenitic structure may be obtained at room temperature. Carbide-forming elements raise the eutectoid temperature; these elements are known as *ferrite stabilizing elements*.

Principal effects of major alloying elements for steel		
Element	***Percentage***	***Primary function***
Aluminium	0.95–1.30	Alloying element in nitriding steels
Bismuth	-	Improves machinability
Boron	0.001–0.003	A powerful hardenability agent
Chromium	0.5–2	Increases hardenability
	4–18	Increases corrosion resistance
Copper	0.1–0.4	Corrosion resistance
Lead	-	Improved machinability
Manganese	0.25–0.40	Combines with sulfur and with phosphorus to reduce the brittleness. Also helps to remove excess oxygen from molten steel.
	>1	Increases hardenability by lowering transformation points and causing transformations to be sluggish
Molybdenum	0.2–5	Stable carbides; inhibits grain growth. Increases the toughness of steel, thus making molybdenum a very valuable alloy metal for making the cutting parts of machine tools and also the turbine blades of turbojet engines. Also used in rocket motors.
Nickel	2–5	Toughener
	12–20	Increases corrosion resistance
Silicon	0.2–0.7	Increases strength
	2.0	Spring steels
	Higher percentages	Improves magnetic properties
Sulfur	0.08–0.15	Free-machining properties
Titanium	-	Fixes carbon in inert particles; reduces martensitic hardness in chromium steels
Tungsten	-	Also increases the melting point.
Vanadium	0.15	Stable carbides; increases strength while retaining ductility; promotes fine grain structure. Increases the toughness at high temperatures

Maraging Steel

Maraging steels (a portmanteau of "martensitic" and "aging") are steels (iron alloys) which are known for possessing superior strength and toughness without losing malleability, although they cannot hold a good cutting edge. *Aging* refers to the extended heat-treatment process. These steels are a special class of low-carbon ultra-high-strength steels which derive their strength not from carbon, but from precipitation of inter-metallic compounds. The principal alloying element is 15 to 25 wt.% nickel. Secondary alloying elements are added to produce intermetallic precipitates, which include cobalt, molybdenum, and titanium. Original development was carried out on 20 and 25 wt.% Ni steels to which small additions of Al, Ti, and Nb were made.

The common, non-stainless grades contain 17–19 wt.% nickel, 8–12 wt.% cobalt, 3–5 wt.% molybdenum, and 0.2–1.6 wt.% titanium. Addition of chromium produces stainless grades resistant to corrosion. This also indirectly increases hardenability as they require less nickel: high-chromium, high-nickel steels are generally austenitic and unable to transform to martensite when heat treated, while lower-nickel steels can transform to martensite. Alternative variants of Ni-reduced maraging steels are based on alloys of Fe and Mn plus minor additions of Al, Ni, and Ti where compositions between Fe-9wt.% Mn to Fe-15wt.% Mn have been used.

The Mn has a similar effect as Ni, i.e. it stabilizes the austenite phase. Hence, depending on their Mn content, Fe-Mn maraging steels can be fully martensitic after quenching them from the high temperature austenite phase or they can contain retained austenite. The latter effect enables the design of maraging-TRIP steels where TRIP stands for Transformation-Induced-Plasticity.

Properties

Due to the low carbon content maraging steels have good machinability. Prior to aging, they may also be cold rolled to as much as 90% without cracking. Maraging steels offer good weldability, but must be aged afterward to restore the original properties to the heat affected zone.

When heat-treated the alloy has very little dimensional change, so it is often machined to its final dimensions. Due to the high alloy content maraging steels have a high hardenability. Since ductile FeNi martensites are formed upon cooling, cracks are non-existent or negligible. The steels can be nitrided to increase case hardness, and polished to a fine surface finish.

Non-stainless varieties of maraging steel are moderately corrosion-resistant, and resist stress corrosion and hydrogen embrittlement. Corrosion-resistance can be increased by cadmium plating or phosphating.

Heat Treatment Cycle

The steel is first annealed at approximately 820 °C (1,510 °F) for 15–30 minutes for thin sections and for 1 hour per 25 mm thickness for heavy sections, to ensure formation of a fully austenitized structure.

This is followed by air cooling to room temperature to form a soft, heavily-dislocated iron-nickel lath (untwinned) martensite. Subsequent aging (precipitation hardening) of the more common alloys for approximately 3 hours at a temperature of 480 to 500 °C produces a fine dispersion of $Ni_3(X,Y)$ intermetallic phases along dislocations left by martensitic transformation, where X and Y are solute elements added for such precipitation. Overaging leads to a reduction in stability of the primary, metastable, coherent precipitates, leading to their dissolution and replacement with semi-coherent Laves phases such as Fe_2Ni/Fe_2Mo. Further excessive heat-treatment brings about the decomposition of the martensite and reversion to austenite. Newer compositions of maraging steels have revealed other intermetallic stoichiometries and crystallographic relationships with the parent martensite, including rhombohedral and massive complex $Ni_{50}(X,Y,Z)_{50}$ ($Ni_{50}M_{50}$ in simplified notation).

Uses

Maraging steel's strength and malleability in the pre-aged stage allows it to be formed into thinner rocket and missile skins than other steels, reducing weight for a given strength. Maraging steels have very stable properties, and, even after overaging due to excessive temperature, only soften slightly. These alloys retain their properties at mildly elevated operating temperatures and have maximum service temperatures of over 400 °C (752 °F).

They are suitable for engine components, such as crankshafts and gears, and the firing pins of automatic weapons that cycle from hot to cool repeatedly while under substantial load. Their uniform expansion and easy machinability before aging make maraging steel useful in high-wear components of assembly lines and dies. Other ultra-high-strength steels, such as AerMet alloys, are not as machinable because of their carbide content.

In the sport of fencing, blades used in competitions run under the auspices of the Fédération Internationale d'Escrime are usually made with maraging steel. Maraging blades are superior for foil and épée because crack propagation in maraging steel is 10 times slower than in carbon steel, resulting in less blade breakage and fewer injuries. Stainless maraging steel is used in bicycle frames and golf club heads. It is also used in surgical components and hypodermic syringes, but is not suitable for scalpel blades because the lack of carbon prevents it from holding a good cutting edge.

Maraging steel production, import, and export by certain states, such as the United States, is closely monitored by international authorities because it is particularly suited for use in gas centrifuges for uranium enrichment; lack of maraging steel significantly hampers this process. Older centrifuges used aluminum tubes; modern ones, carbon fiber composite.

Physical Properties

- Density: 8.1 g/cm^3 (0.29 lb/in^3)
- Specific heat, mean for 0–100 °C (32–212 °F): 813 J/kg·K (0.108 Btu/lb·°F)
- Melting point: 2,575 °F, 1,413 °C
- Thermal conductivity: 25.5 W/m·K
- Mean coefficient of thermal expansion: 11.3×10
- Yield tensile strength: typically 1,030–2,420 MPa (149,000–351,000 psi)
- Ultimate tensile strength: typically 1.6–2.5 GPa (230,000–360,000 psi). Grades exist up to 3.5 GPa (510,000 psi)
- Elongation at break: up to 15%
- K_{IC} fracture toughness: up to 175 MPa·m$^{-D1}{}_2$
- Young's modulus: 210 GPa (30,000,000 psi)
- Shear modulus: 77 GPa (11,200,000 psi)
- Bulk modulus: 140 GPa (20,000,000 psi)
- Hardness (aged): 50 HRC (grade 250); 54 HRC (grade 300); 58 HRC (grade 350)

Weathering Steel

Weathering steel, best-known under the trademark COR-TEN steel and sometimes written without the hyphen as "Corten steel", is a group of steel alloys which were developed to eliminate the need for painting, and form a stable rust-like appearance if exposed to the weather for several years.

United States Steel Corporation (USS) holds the registered trademark on the name COR-TEN. Although USS sold its discrete plate business to International Steel Group (now Arcelor-Mittal) in 2003, it still sells COR-TEN branded material in strip-mill plate and sheet forms.

The original COR-TEN received the standard designation A 242 ("COR-TEN A") from the ASTM International standards group. Newer ASTM grades are A 588 ("COR-TEN B") and A 606 fol thin sheet. All alloys are in common production and use.

Properties

"Weathering" means that due to their chemical compositions, these steels exhibit increased resistance to atmospheric corrosion compared to other steels. This is because the steel forms a protective layer on its surface under the influence of the weather.

The corrosion-retarding effect of the protective layer is produced by the particular distribution and concentration of alloying elements in it. The layer protecting the surface develops and regenerates continuously when subjected to the influence of the weather. In other words, the steel is allowed to rust in order to form the 'protective' coating.

Chemical composition of Cor-ten grades									
Grade	*C*	*Si*	*Mn*	*P*	*S*	*Cr*	*Cu*	*V*	*Ni*
Cor-ten A	0.12	0.25-0.75	0.20-0.50	0.01-0.20	0.030	0.50-1.25	0.25-0.55		0.65
Cor-ten B	0.16	0.30-0.50	0.80-1.25	0.030	0.030	0.40-0.65	0.25-0.40	0.02-0.10	0.40

The mechanical properties of weathering steels depend on which alloy and how thick the material is.

ASTM A 242

The original A 242 alloy has a yield strength of 50,000 pounds per square inch (340,000 kPa) and ultimate tensile strength of 70,000 psi (480,000 kPa) for light-medium rolled shapes and plates up to 0.75 inches (19 mm) thick. It has yield strength of 46,000 psi (320,000 kPa) and ultimate strength of 67,000 psi (460,000 kPa) for medium weight rolled shapes and plates from 0.75–1 inch (19–25 mm) thick. The thickest rolled sections and plates – from 1.5–4 in (38–102 mm) thick have yield strength of 42,000 psi (290,000 kPa) and ultimate strength of 63,000 psi (430,000 kPa).

ASTM A 588

A 588 has a yield strength of at least 50,000 psi (340,000 kPa), and ultimate tensile strength of 70,000 psi (480,000 kPa) for all rolled shapes and plate thicknesses up to 4 in (100 mm) thick. Plates from 4–5 in (102–127 mm) have yield strength at least 46,000 psi (320,000 kPa) and ultimate tensile strength at least 67,000 psi (460,000 kPa), and plates from 5–8 in (127–203 mm) thick have yield strength at least 42,000 psi (290,000 kPa) and ultimate tensile strength at least 63,000 psi (430,000 kPa).

Use

Weathering steel is popularly used in outdoor sculptures, such as in the large Chicago Picasso sculpture, which stands in the plaza of the Daley Centre Courthouse in Chicago, which is also constructed of the same COR-TEN steel and as exterior facades, for its rustic antique appearance. Examples include Barclays Centre, Brooklyn, New York, The Angel of the North, Gateshead, UK and the Humanities and Arts complex at Leeds Metropolitan University - Broadcasting Place - Leeds, UK.

It is also used in bridge and other large structural applications such as the New River Gorge Bridge, the newer span of the Newburgh–Beacon Bridge, and the creation of the Australian Centre for Contemporary Art (ACCA) and Mona. It is very widely used in marine transportation, in the construction of Intermodal containers as well as visible sheet piling along recently widened sections of London's M25 motorway.

The first use of COR-TEN for architectural applications was the John Deere World Headquarters in Moline, Illinois. The building was designed by architect Eero Saarinen, and completed in 1964. The main buildings of Odense University, designed by Knud Holscher and Jørgen Vesterholt and built 1971–1976, are clad in COR-TEN steel, earning them the nickname *Rustenborg*. In 1977, Robert Indiana created a Hebrew version of the *Love* sculpture using the four-letter word ahava (àáää, "love" in Hebrew) made from COR-TEN for the Israel Museum Art Garden in Jerusalem, Israel. In Denmark, all masts for supporting the catenary on electrified railways are made of COR-TEN for aesthetic reasons.

COR-TEN was used in 1971 for an order of electric railcars built by the St. Louis Car Company for Illinois Central Railroad. The use of COR-TEN was seen as a cost-cutting move in comparison with the contemporary railcar standard of stainless steel. A subsequent order in 1979 was built to similar specs, including COR-TEN bodies, by Bombardier. The cars were painted, a standard practice for COR-TEN railcars. However, the durability of COR-TEN did not live up to expectations, with rust holes appearing in the railcars. Ironically, painting may have contributed to the problem, as painted weathering steel is no more corrosion-resistant than conventional steel, because the protective patina will not form in time to prevent corrosion over a localized area of attack such as a small paint failure. Most of these railcars still operate out of Chicago.

COR-TEN was used to build the exterior of Barclays Centre, made up of 12,000 pre-weathered steel panels engineered by Dissimilar Metal Design. The *New York Times* says of the material, "While it can look suspiciously unfinished to the casual observer, it has many fans in the world of art and architecture".

Disadvantages

Using weathering steel in construction presents several challenges. Ensuring that weld-points weather at the same rate as the other materials may require special welding techniques or material. Weathering steel is not rustproof in itself. If water is allowed to accumulate in pockets, those areas will experience higher corrosion rates, so provision for drainage must be made. Weathering steel is sensitive to humid subtropical climates. In such environments, it is possible that the protective patina may not stabilize but instead continue to corrode.

For example, the former Omni Coliseum, built in 1972 in Atlanta, never stopped rusting, and eventually large holes appeared in the structure. This was a major factor in the decision to demolish it just 25 years after construction. The same thing can happen in environments laden with sea salt. Hawaii's Aloha Stadium, built in 1975, is one example of this. Weathering steel's normal surface weathering can also lead to rust stains on nearby surfaces.

The U.S. Steel Tower in Pittsburgh, Pennsylvania was constructed by U.S. Steel in part to showcase COR-TEN steel. The initial weathering of the material resulted in a discolouration of the surrounding city sidewalks, as well as other nearby buildings. A cleanup effort was orchestrated by the corporation once weathering was complete to clean the markings. A few of the nearby sidewalks were left uncleaned, and remain a rust colour.

This problem has been reduced in newer formulations of weathering steel. Staining can be prevented if the structure can be designed so that water does not drain from the steel onto concrete where stains would be visible.

Tool Steel

Tool steel refers to a variety of carbon and alloy steels that are particularly well-suited to be made into tools. Their suitability comes from their distinctive hardness, resistance to abrasion, their ability to hold a cutting edge, and/or their resistance to deformation at elevated temperatures (red-hardness). Tool steel is generally used in a heat-treated state. Many high carbon tool steels are also more resistant to corrosion due to their higher ratios of elements such as vanadium and niobium.

With a carbon content between 0.7% and 1.5%, tool steels are manufactured under carefully controlled conditions to produce the required quality. The manganese content is often kept low to minimize the possibility of cracking during water quenching. However, proper heat treating of these steels is important for adequate performance, and there are many suppliers who provide tooling blanks intended for oil quenching.

Tool steels are made to a number of grades for different applications. Choice of grade depends on, among other things, whether a keen cutting edge is necessary, as in stamping dies, or whether the tool has to withstand impact loading and service conditions encountered with such hand tools as axes, pickaxes, and quarrying implements. In general, the edge temperature under expected use is an important determinant of both composition and required heat treatment. The higher carbon grades are typically used for such applications as stamping dies, metal cutting tools, etc.

Tool steels are also used for special applications like injection molding because the resistance to abrasion is an important criterion for a mold that will be used to produce hundreds of thousands of parts.

The AISI-SAE grades of tool steel is the most common scale used to identify various grades of tool steel. Individual alloys within a grade are given a number; for example: A2, O1, etc.

Water-Hardening Types

W-type tool steel gets its name from its defining property of having to be water quenched. W-grade steel is essentially high carbon plain-carbon steel. This type of tool steel is the most commonly used tool steel because of its low cost compared to other tool steels. They work well for small parts and applications where high temperatures are not encountered; above 150 °C (302 °F) it begins to soften to a noticeable degree. Hardenability is low so W-grade tool steels must be quenched in water. These steels can attain high hardness (above HRC 66) and are rather brittle compared to other tool steels. W steels are still sold, especially for springs, but are much less widely used than they were in the 19th and early 20th centuries. This is partly because W steels warp and crack much more in quench than oil-quenched or air hardeneing steels.

The toughness of W-type tool steels are increased by alloying with manganese, silicon and molybdenum. Up to 0.20% of vanadium is used to retain fine grain sizes during heat treating.

Typical applications for various carbon compositions are:

- 0.60–0.75% carbon: machine parts, chisels, setscrews; properties include medium hardness with good toughness and shock resistance.

- 0.76–0.90% carbon: forging dies, hammers, and sledges.
- 0.91–1.10% carbon: general purpose tooling applications that require a good balance of wear resistance and toughness, such as rasps, drills, cutters, and shear blades.
- 1.11–1.30% carbon: files, small drills, lathe tools, razor blades, and other light-duty applications where more wear resistance is required without great toughness. Steel of about 0.8 % C gets as hard as steel with more carbon, but the free iron carbide particles in 1% or 1.25% carbon steel make it hold an edge better. However, the fine edge probably rusts off faster than it wears off, if it is used to cut acidic or salty materials.

Cold-Working Types

These tool steels are used on larger parts or parts that require minimal distortion during hardening. The use of oil quenching and air hardening helps reduce distortion as opposed to higher stress caused by quicker water quenching. More alloying elements are used in these steels, as compared to water-hardening grades.

These alloys increase the steels' hardenability, and thus require a less severe quenching process. These steels are also less likely to crack and are often used to make knife blades.

Oil-Hardening Types

Grade	*Composition*	*Notes*
O1	0.90% C, 1.0–1.4% Mn, 0.50% Cr, 0.50% Ni, 0.50% W	It is a very good cold work steel and also makes very good knives. It can be hardened to about 57-61 HRC.

Air-Hardening Types

The first air hardening grade tool steel was mushet steel, which was known as *air-hardening steel* at the time.

Modern air-hardening steels are characterized by low distortion during heat treatment because of their high-chromium content. They also harden in air because they have less alloyants than oil-hardening grades.

Their machinability is good for tool steels and they have a balance of wear resistance and toughness (i.e. between the D- and shock-resistant grades).

Grade	*Composition*	*Notes*
A2	1.0% C, 1.0% Mn, 5.0% Cr, 0.3% Ni, 1.0% Mo, 0.15–0.50% V	A common general purpose tool steel; it is the most commonly used variety of air-hardening steel. It is commonly used for blanking and forming punches, trimming dies, thread rolling dies, and injection molding dies.
A3	1.25% C, 0.5% Mn, 5.0% Cr, 0.3% Ni, 0.9–1.4% Mo, 0.8–1.4% V	
A4	1.0% C, 2.0% Mn, 1.0% Cr, 0.3% Ni, 0.9–1.4% Mo	
A6	0.7% C, 1.8–2.5% Mn, 0.9–1.2% Cr, 0.3% Ni, 0.9–1.4% Mo	This type of tool steel air-hardens at a relatively low temperature (approximately the same temperature as oil-hardening types) and is dimensionally stable. Therefore it is commonly used for dies, forming tools, and gauges that do not require extreme wear resistance but do need high stability.
A7	2.00–2.85% C, 0.8% Mn, 5.00–5.75% Cr, 0.3% Ni, 0.9–1.4% Mo, 3.9–5.15% V, 0.5–1.5 W	
A8	0.5–0.6% C, 0.5% Mn, 4.75–5.50% Cr, 0.3% Ni, 1.15–1.65% Mo, 1.0–1.5 W	
A9	0.5% C, 0.5% Mn, 0.95–1.15% Si, 4.75–5.00% Cr, 1.25–1.75% Ni, 1.3–1.8% Mo, 0.8–1.4% V	
A10	1.25–1.50% C, 1.6–2.1% Mn, 1.0–1.5% Si, 1.55–2.05% Ni, 1.25–1.75% Mo	This grade contains a uniform distribution of graphite particles to increase machinability and provide self-lubricating properties. It is commonly used for gauges, arbors, shears, and punches.

D-Types

D-type tool steels contain between 10% and 18% chromium. These steels retain their hardness up to a temperature of 425 °C (797 °F). Common applications for these grade of tool steel is forging dies, die-casting die blocks, and drawing dies. Due to high chromium content, certain D-grade tool steel grades are often considered stainless or semi-stainless tool steels. However their corrosion resistance was very limited due to the precipitation of major amount of chromium and carbon as carbides.

Grade	*Composition*	*Notes*
D2	1.5% C, 11.0–13.0% Cr; additionally 0.45% Mn, 0.030% P, 0.030% S, 1.0% V, 0.9% Mo, 0.30% Si	D2 is very wear resistant but not as tough as lower alloyed steels. Mechanical properties of D2 is very sensitive to heat treatment. It is widely used for shear blades, planer blades and industrial cutting tools, sometimes used for knives.

1.2767

ISO 1.2767, also known as DIN X 45 NiCrMo 4, AISI 6F7, and BS EN 20 B, is an air hardening tool steel with a primary alloying element of nickel. It possesses good toughness, stable grains, and is highly polishable. It is primarily used for dies in plastic injection molding application that involve high stresses. Other applications include blanking dies, forging dies, and industrial blades.

Shock Resisting Types

S-type tool steel are designed to resist shock at both low and high temperatures. A low carbon content is required for the necessary toughness (approximately 0.5% carbon). Carbide-forming alloys provide the necessary abrasion resistance, hardenability, and hot-working characteristics. This family of steels displays very high impact toughness and relatively low abrasion resistance, it can attain relatively high hardness (HRC 58/60). This type of steel is used in applications such as jackhammer bits. In the US, toughness usually comes from 1 to 2% silicon and 0.5-1% molybdenum. In Europe, shock steels often contained .5-.6 % carbon and around 3% nickel. 1.75% to 2.75% nickel is still used in some shock resisting and high strength low alloy steels (HSLA), such as L6, 4340, and Swedish saw steel, but it is expensive.

High Speed Types

T-type and M-type tool steels are used for cutting tools where strength and hardness must be retained at temperatures up to or exceeding 760 °C (1,400 °F). M-type tool steels were developed to reduce the amount of tungsten and chromium required.

T1 (also known as 18-4-1) is a common T-type alloy. Its composition is 0.7% carbon, 18% tungsten, 4% chromium, and 1% vanadium. M2 is a common M-type alloy.

Hot-Working Types

H-type tool steels were developed for strength and hardness during prolonged exposure to elevated temperatures. All of these tool steels use a substantial amount of carbide forming alloys. H1 to H19 are based on a chromium content of 5%; H20 to H39 are based on a tungsten content of 9-18% and a chromium content of 3–4%; H40 to H59 are molybdenum based.

Special Purpose Types

- P-type tool steel is short for plastic mold steels. They are designed to meet the requirements of zinc die casting and plastic injection molding dies.

- L-type tool steel is short for low alloy special purpose tool steel. L6 is extremely tough.
- F-type tool steel is water hardened and substantially more wear resistant than W-type tool steel.

Comparison

AISI-SAE Tool Steel Grades

Defining Property	*AISI-SAE Grade*	*Significant Characteristics*
Water-hardening	W	
Cold-working	O	Oil-hardening
	A	Air-hardening; medium alloy
	D	High carbon; high chromium
Shock resisting	S	
High speed	T	Tungsten base
	M	Molybdenum base
Hot-working	H	H1–H19: chromium baseH20–H39: tungsten base H40–H59: molybdenum base
Plastic mold	P	
Special purpose	L	Low alloy
	F	Carbon tungsten

Steel Design

Steel design, or more specifically, structural steel design, is an area of knowledge of structural engineering used to design steel structures. The structures can range from schools to homes to bridges.

In structural engineering, a structure is a body or combination of pieces of rigid bodies in space to form a fitness system for supporting loads. Structures such as buildings, bridges, aircraft and ships are all examples under steel structure.

The effects of loads on structures are determined through structural analysis. Steel structure is steel construction material, a profile, formed with a specific shape or cross section and certain standards of chemical composition and mechanical properties.

There are currently two common methods of steel design: The first (and older) method is the Allowable Strength Design (ASD) method. The second (newer) is the Load and Resistance Factor Design (LRFD) method.

Design for Strength

ASD: In this method, the engineer uses the ASD load combinations (below) to determine the required strength of a member and arranges for the allowable strength to satisfy this equation:

$$R_a \leq \frac{R_n}{\Omega}$$

where:

- R_a = required strength,
- R_n = nominal strength, specified in Chapters B through K of the AISC SCM,
- Ω = safety factor, specified in Chapters B through K of the AISC SCM,
- R_n/Ω = allowable strength.

LRFD

In this method, the engineer uses the Load and Resistance Factor Design (LRFD) load combinations (below) to determine the required strength of a member and arranges for the allowable strength to satisfy this equation:

$$R_u \leq \phi * R_n$$

where:

- R_u = required strength,
- R_n = nominal strength, specified in Chapters B through K of the AISC SCM,
- Φ = resistance factor, specified in Chapters B through K of the AISC SCM,
- $\Phi \cdot R_n$ = allowable strength.

ASD versus LRFD

As per the AISC SCM, 14th ed., either design method is allowed by the AISC SCM 14th edition. A common misconception about the two methods is that ASD gives a more conservative value. In reality, ASD is more conservative in designs with a live to dead load ratio of 3 or lower. With a higher ratio, LRFD is more conservative. The two design methods are related through the Ω factor of ASD and the φ factor of LRFD. While these factors have different uses, they are always related by the following expression:

$$\Omega = \frac{1.5}{\phi}$$

The value of these factors vary according to the country codes.

Load Combination Equations

Allowable Strength Design: For ASD, the required strength, R_a, is determined from the following load combinations (according to the AISC SCM, 13 ed.) and:

$D + F$

$D + H + F + L + T$

$D + H + F + (L_r \text{ or } S \text{ or } R)$

$D + H + F + 0.75(L + T) + 0.75(L_r \text{ or } S \text{ or } R)$

$D + H + F \pm (W \text{ or } 0.7E)$

$D + H + F + (0.75W \text{ or } 0.7E) + 0.75L + 0.75(L_r \text{ or } S \text{ or } R)$

$0.6D + W + H$ $0.6D \pm (W \text{ or } 0.7E)$

where:

- D = dead load,
- D_i = weight of Ice,
- E = earthquake load,
- F = load due to fluids with well-defined pressures and maximum heights,
- F_a = flood load,
- H = load due to lateral earth pressure, ground water pressure, or pressure of bulk materials,
- L = live load due to occupancy,
- L_r = roof live load,
- S = snow load,
- R = nominal load due to initial rainwater or ice, exclusive of the ponding contribution,
- T = self straining load,
- W = wind load,
- W_i = wind on ice.

Special Provisions exist for accounting flood loads and atmospheric loads i.e. D_i and W_i

Load and Resistance Factor Design

For LRFD, the required strength, R_u, is determined from the following factored load combinations:

$1.4(D + F)$

$1.2(D + F + T) + 1.6(L + H) + 0.5(L_r \text{ or } S \text{ or } R)$

$1.2D + 1.6(L_r \text{ or } S \text{ or } R) + (L \text{ or } 0.8W)$

$1.2D + 1.6W + L + 0.5(L_r \text{ or } S \text{ or } R)$

$1.2D \pm 1.0E + L + 0.2S + 0.9D + 1.6W + 1.6H$

$0.9D + 1.6\ H \pm (1.6W \text{ or } 1.0E)$

where the letters for the loads are the same as for ASD.

For the wind consideration, the ASCE allows a "position correction factor" which turns the coefficient of wind action to 1,36:

1,2D + 1,36W + ... the same above or 0,9D - 1,36W

AISC Steel Construction Manual

American Institute of Steel Construction (AISC), Inc. publishes the *AISC Manual of Steel Construction* (Steel construction manual, or SCM), which is currently in its 14th edition. Structural engineers use this manual in analyzing, and designing various steel structures. Some of the chapters of the book are as follows.

- Dimensions and properties of various types of steel sections available on the market (W, S, C, WT, HSS, etc.)
- General Design Considerations
- Design of Flexural Members
- Design of Compression Members
- Design of Tension members
- Design of Members Subject to Combined Loading
- Design Consideration for Bolts
- Design Considerations for Welds
- Design of Connecting Elements
- Design of Simple Shear Connections
- Design of Flexure Moment Connections
- Design of Fully Restrained (FR) Moment Connections
- Design of Bracing Connections and Truss Connections
- Design of Beam Bearing Plates, Column Base Plates, Anchor Rods, and Column Splices
- Design of Hanger Connections, Bracket Plates, and Crane-Rail Connections
- Specifications and Codes
- Miscellaneous Data and Mathematical Information
- General Nomenclature

CISC Handbook of Steel Construction

Canadian Institute of Steel Construction publishes the "CISC Handbook of steel Construction". CISC is a national industry organisation representing the structural steel, open-web steel joist and steel plate fabrication industries in Canada. It serves the same purpose as the AISC manual, but conforms with Canadian standards.

Design Actions

Steel Material Properties

The properties of structural steel result from both its chemical composition and its method of manufacture , including processing during fabrication. Product standards define the limits for composition, quality and performance and these limits are used or presumed by structural designers. This article reviews the principal properties that are of interest to the designer and indicates the relevant standards for particular products

Material Properties Required for Design

The properties that need to be considered by designers when specifying steel construction products are:

- Strength
- Toughness
- Ductility
- Weldability
- Durability.

For design, the mechanical properties are derived from minimum values specified in the relevant product standard. Weldability is determined by the chemical content of the alloy, which is governed by limits in the product standard. Durability depends on the particular alloy type - 'ordinary' carbon steel, 'weathering steel' or stainless steel.

Factors that Influence Mechanical Properties

Steel derives its mechanical properties from a combination of chemical composition, heat treatment and manufacturing processes. While the major constituent of steel is iron the addition of very small quantities of other elements can have a marked effect upon the properties of the steel. The strength of steel can be increased by the addition of alloys such as manganese, niobium and vanadium. However, these alloy additions can also adversely affect other properties, such as ductility, toughness and weldability.

Minimizing the sulphur level can enhance ductility , and toughness can be improved by the addition of nickel. The chemical composition for each steel specification is therefore carefully balanced and tested during its production to ensure that the appropriate properties are achieved.

The alloying elements also produce a different response when the material is subjected to heat treatments involving cooling at a prescribed rate from a particular peak temperature. The manufacturing process may involve combinations of heat treatment and mechanical working that are of critical importance to the performance of the steel.

Mechanical working takes place as the steel is being rolled or formed. The more steel is rolled, the stronger it becomes. This effect is apparent in the material standards, which tend to specify reducing levels of yield strength with increasing material thickness.

The effect of heat treatment is best explained by reference to the various production process routes that can be used in steel manufacturing, the principal ones being:

- As-rolled steel
- Normalized steel
- Normalized-rolled steel
- Thermomechanically rolled (TMR) steel
- Quenched and tempered (Q&T) steel.

Steel cools as it is rolled, with a typical rolling finish temperature of around 750°C. Steel that is then allowed to cool naturally is termed 'as-rolled' material. Normalizing takes place when as-rolled material is heated back up to approximately 900°C, and held at that temperature for a specific time, before being allowed to cool naturally.

This process refines the grain size and improves the mechanical properties, specifically toughness. Normalized-rolled is a process where the temperature is above 900°C after rolling is completed. This has a similar effect on the properties as normalizing, but it eliminates the extra process of reheating the material. Normalized and normalized-rolled steels have an 'N' designation.

The use of high tensile steel can reduce the volume of steel needed but the steel needs to be tough at operating temperatures, and it should also exhibit sufficient ductility to withstand any ductile crack propagation. Therefore, higher strength steels require improved toughness and ductility, which can be achieved only with low carbon clean steels and by maximizing grain refinement.

The implementation of the thermomechanical rolling process (TMR) is an efficient way to achieve this. Thermomechanically rolled steel utilises a particular chemistry of the steel to permit a lower rolling finish temperature of around 700°C. Greater force is required to roll the steel at these lower temperatures, and the properties are retained unless reheated above 650°C. Thermomechanically rolled steel has an 'M' designation.

The process for Quenched and Tempered steel starts with a normalized material at 900°C. It is rapidly cooled or 'quenched' to produce steel with high strength and hardness, but low toughness. The toughness is restored by reheating it to 600°C, maintaining the temperature for a specific time, and then allowing it to cool naturally (Tempering). Quenched and tempered steels have a 'Q' designation.

Quenching involves cooling a product rapidly by immersion directly into water or oil. It is frequently used in conjunction with tempering which is a second stage heat treatment to temperatures below the austenitizing range. The effect of tempering is to soften previously hardened structures and make them tougher and more ductile.

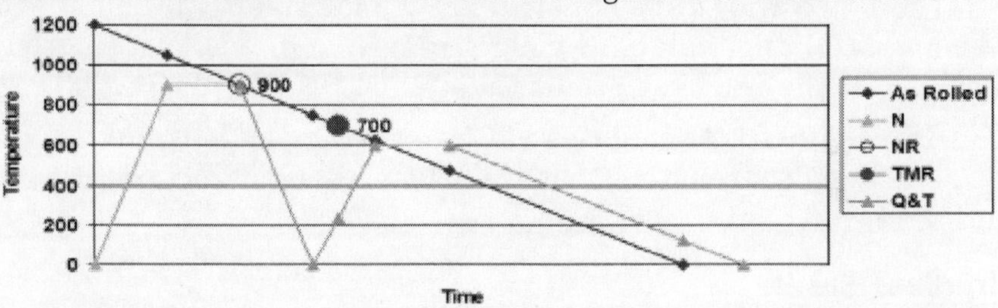

Figure: *Schematic temperature / time graph of rolling processes*

Strength

Yield Strength: Yield strength is the most common property that the designer will need as it is the basis used for most of the rules given in design codes . In European Standards for structural carbon steels (including weathering steel), the primary designation relates to the yield strength, e.g. S275 steel is a structural steel with a specified minimum yield strength of 275 N/mm^2.

The product standards also specify the permitted range of values for the ultimate tensile strength (UTS). The minimum UTS is relevant to some aspects of design.

Hot Rolled Steels: For hot rolled carbon steels, the number quoted in the designation is the value of yield strength for material up to 16 mm thick. Designers should note that yield strength reduces with increasing plate or section thickness (thinner plate is worked more than thick plate and working increases the strength). For the two most common grades of steel used in UK, the specified minimum yield strengths and the minimum tensile strength are shown in table below for steels to BS EN 10025-2.

Minimum yield and tensile strength for common steel grades						
Grade	*Yield strength (N/mm²) for nominal thickness t (mm)*				*Tensile strength (N/mm²) for nominal thickness t (mm)*	
	$t \leq 16$	$16 < t \leq 40$	$40 < t \leq 63$	$63 < t \leq 80$	$3 < t \leq 100$	$100 < t \leq 150$
S275	275	265	255	245	410	400
S355	355	345	335	325	470	450

The UK National Annex to BS EN 1993-1-1 allows the minimum yield value for the particular thickness to be used as the nominal (characteristic) yield strength f_y and the minimum tensile strength f_u to be used as the nominal (characteristic) ultimate strength.

Similar values are given for other grades in other parts of BS EN 10025 and for hollow sections to BS EN 10210-1 .

Cold Formed Steels

There is a wide range of steel grades for steels suitable for cold forming. Minimum values of yield strength and tensile strength are specified in the relevant product standard BS EN 10346:2009.

BS EN 1993-1-3 tabulates values of basic yield strength f_{yb} and ultimate tensile strength f_u that are to be used as characteristic values in design.

Stainless Steels

Grades of stainless steel are designated by a numerical 'steel number' (such as 1.4401 for a typical austenitic steel) rather than the 'S' designation system for carbon steels. The stress-strain relationship does not have the clear distinction of a yield point and stainless steel 'yield' strengths for stainless steel are generally quoted in terms of a proof strength defined for a particular offset permanent strain (conventionally the 0.2% strain).

The strengths of commonly used structural stainless steels range from 170 to 450 N/mm². Austenitic steels have a lower yield strength than commonly used carbon steels; duplex steels have a higher yield strength than common carbon steels. For both austenitic and duplex stainless steels, the ratio of ultimate strength to yield strength is greater than for carbon steels. BS EN 1993-1-4 tabulates nominal (characteristic) values of yield strength f_y and ultimate minimum tensile strength f_u for steels to BS EN 10088-1 for use in design.

Toughness

It is in the nature of all materials to contain some imperfections. In steel these imperfections take the form of very small cracks. If the steel is insufficiently tough, the 'crack' can propagate rapidly, without plastic deformation and result in a 'brittle fracture'. The risk of brittle fracture increases with thickness, tensile stress, stress raisers and at colder temperatures.

The toughness of steel and its ability to resist brittle fracture are dependent on a number of factors that should be considered at the specification stage. A convenient measure of toughness is the Charpy V-notch impact test. This test measures the impact energy required to break a small notched specimen, at a specified temperature, by a single impact blow from a pendulum.

The various product standards specify minimum values of impact energy for different sub-grades of each strength grade. For non-alloy structural steels the designations of the subgrades are JR, J0, J2 and K2. For fine grain steels and quenched and tempered steels (which are generally tougher, with higher impact energy) different designations are used. A summary of the toughness designations is given in the table below.

Specified Minimum Impact Energy for Carbon Steel Sub-Grades

Standard	*Subgrade*	*Impact Strength*	*Test Temperature*
BS EN 10025-2 BS EN 10025-5 BS EN 10210-1	JR	27J	20°C
	J0	27J	0°C
	J2	27J	-20°C
	K2	40J	-20°C
BS EN 10025-3	N	40J	-20°c
	NL	27J	-50°c
BS EN 10025-4	M	40J	-20°c
	ML	27J	-50°c
BS EN 10025-6	Q	30J	-20°c
	QL	30J	-40°c
	QL1	30J	-60°c

For thin gauge steels for cold forming, no impact energy requirements are specified for material less than 6 mm thick.

The selection of an appropriate sub-grade, to provide adequate toughness in design situations is given in BS EN 1993 1 10. The rules relate the exposure temperature, stress level etc, to a 'limiting thickness' for each sub-grade of steel. Guidance on selection of an appropriate sub-grade is given in ED007 *Selection of steel subgrades in accordance with the Eurocodes.*

Stainless steels are generally much tougher than carbon steels; minimum values are specified in BS EN 10088-4. BS EN 1993-1-4 states that austenitic and duplex steels are adequately tough and not susceptible to brittle fracture for service temperatures down to -40°C.

Ductility

Ductility is a measure of the degree to which a material can strain or elongate between the onset of yield and eventual fracture under tensile loading as demonstrated in the figure below. The designer relies on ductility for a number of aspects of design, including redistribution of stress at the ultimate limit state, bolt group design, reduced risk of fatigue crack propagation and in the fabrication processes of welding, bending and straightening.

The various standards for the grades of steel in the above table insist on a minimum value for ductility so the design assumptions are valid and if these are specified correctly the designer can be assured of their adequate performance.

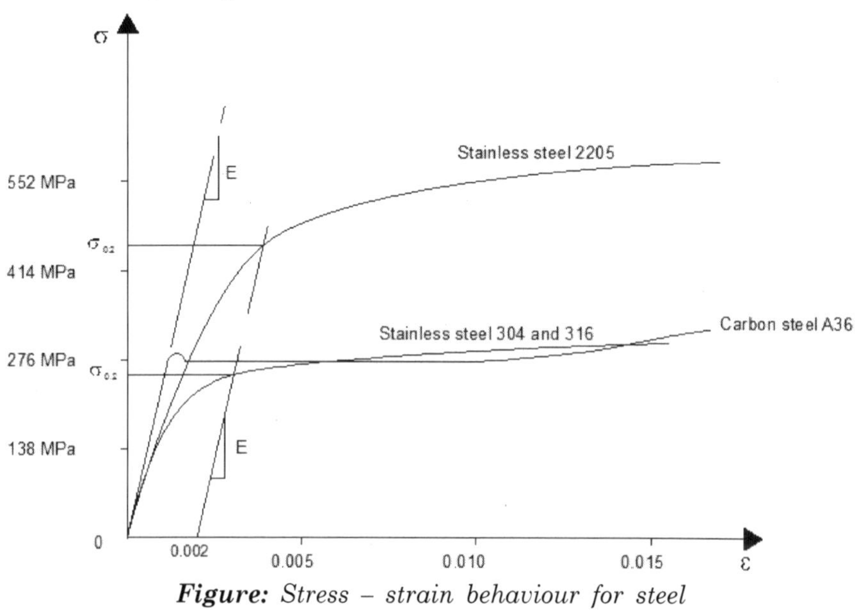

Figure: *Stress – strain behaviour for steel*

Weldability

All structural steels are essentially weldable. However, welding involves locally melting the steel, which subsequently cools. The cooling can be quite fast because the surrounding material, e.g. the beam, offers a large 'heat sink' and the weld (and the heat introduced) is usually relatively small.

This can lead to hardening of the 'heat affected zone' (HAZ) and to reduced toughness. The greater the thickness of material, the greater the reduction of toughness.

The susceptibility to embrittlement also depends on the alloying elements principally, but not exclusively, the carbon content. This susceptibility can be expressed as the 'Carbon Equivalent Value' (CEV), and the various product standards for carbon steels standard give expressions for determining this value.

BS EN 10025 sets mandatory limits for CEV for all structural steel products covered, and it is a simple task for those controlling welding to ensure that welding procedure specifications used are qualified for the appropriate steel grade, and CEV.

Other Mechanical Properties of Steel

Other mechanical properties of structural steel that are important to the designer include:

- Modulus of elasticity, $E = 210{,}000$ N/mm^2
- Shear modulus, $G = E/[2(1 + \nu)]$ N/mm^2, often taken as 81,000 N/mm^2
- Poisson's ratio, $\nu = 0.3$
- Coefficient of thermal expansion, $\alpha = 12 \times 10^{-6}$/°C (in the ambient temperature range).

Durability

A further important property is that of corrosion prevention. Although special corrosion resistant steels are available these are not normally used in building construction. The exception to this is weathering steel.

The most common means of providing corrosion protection to construction steel is by painting or galvanizing. The type and degree of coating protection required depends on the degree of exposure, location, design life, etc. In many cases, under internal dry situations no corrosion protection coatings are required other than appropriate fire protection. Detailed information on the corrosion protection of structural steel is available.

Weathering Steel

Weathering steel is a high strength low alloy steel that resists corrosion by forming an adherent protective rust 'patina', that inhibits further corrosion. No protective coating is needed. It is extensively used in the UK for bridges and has been used externally on some buildings. It is also used for architectural features and sculptural structures such as the Angel of the North.

Stainless Steel

Stainless steel is a highly corrosion-resistant material that can be used structurally, particularly where a high-quality surface finish is required. Suitable grades for exposure in typical environments are given below.

The stress-strain behaviour of stainless steels differs from that of carbon steels in a number of respects. The most important difference is in the shape of the stress-strain curve.

While carbon steel typically exhibits linear elastic behaviour up to the yield stress and a plateau before strain hardening is encountered, stainless steel has a more rounded response with no well-defined yield stress.

Therefore, stainless steel 'yield' strengths are generally defined for a particular offset permanent strain (conventionally the 0.2% strain), as indicated in the figure on the right which shows typical experimental stress-strain curves for common austenitic and duplex stainless steels.

The curves shown are representative of the range of material likely to be supplied and should not be used in design.

Specified mechanical properties of common stainless steels to EN 10088-4

Description	*Grade*	*Minimum 0.2% proof strength (N/mm²)*	*Ultimate tensile strength (N/mm²)*	*Elongation at fracture (%)*
Basic chromium-nickel austenitic steels	1.4301	210	520 – 720	45
	1.4307	200	500 – 700	45
Molybdenum-chromiumnickel austenitic steels	1.4401	220	520 – 670	45
	1.4404	220	520 – 670	45
Duplex steels	1.4162	450	650 – 850	30
	1.4462	460	640 – 840	25

The mechanical properties apply to hot rolled plate. For cold rolled and hot rolled strip, the specified strengths are 10-17% higher.

Guidelines for stainless steel selection		
ISO 9223 Atmospheric Corrosion Class	*Typical outdoor environment*	*Suitable stainless steel*
C1 (Very low)	Deserts and arctic areas (very low humidity)	1.4301/1.4307, 1.4162
C2 (Low)	Arid or low pollution (rural)	1.4301/1.4307, 1.4162
C3 (Medium)	Coastal areas with low deposits of salt Urban or industrialised areas with moderate pollution	1.4401/1.4404, 1.4162 (1.4301/1.4307)
C4 (High)	Polluted urban and industrialised atmosphere Coastal areas with moderate salt deposits Road environments with de-icing salts	1.4462, (1.4401/1.4404), other more highly alloyed duplexes or austenitics
C5 (Very high)	Severely polluted industrial atmospheres with high humidity Marine atmospheres with high degree of salt deposits and splashes	1.4462, other more highly alloyed duplexes or austenitics

Materials suitable for a higher class may be used for lower classes but might not be cost effective. Materials within brackets might be considered if some moderate corrosion is acceptable. Accumulation of corrosive pollutants and chlorides will be higher in sheltered locations; hence it might be necessary to choose a recommended grade from the next higher corrosion class.

Design

The design process encompasses the architectural design, the development of the structural concept, the analysis of the steel structure and the verification of members. Steel solutions are lighter than their concrete equivalents, with the opportunity to provide more column-free flexible floor space, less foundations and a fast, safe construction programme.

For the designer, a steel solution means reliable materials, known material and section properties, precise off-site manufacture and extensive support including software, design guides and easy to use resistance tables.

Design Process

The fundamental process of structural design commences with the preparation of a structural concept, which is itself based on an architectural design for the structure.

For simple, common forms of structure, it will be possible to prepare a concept design directly from the architectural design - typical solutions are well understood.

For more complex structures, or innovative designs, best practice is to develop the structural concept in conjunction with the architectural scheme, so that an efficient, appropriate solution can be developed.

Once the concept design has been established, the structural design can be completed, involving determination of loads, analysis and member verification.

Steel Design

Steel is ideally suited for design. Material properties are known and member properties are accurate, meaning that analysis is precise. Design rules are clear mature, without undue conservatism, having been developed over many decades. There is a wealth of support resources, including software, to facilitate efficient design.

Concept Design

The choice and design of the primary structure is a fundament part of the concept design of buildings, and ideally should be integrated with the development of the architectural design. Meeting client, planning and Building Regulation requirements are paramount, but there will be a range of structural forms that meet these requirements, each with its own advantages. The merits of different structural forms should be reviewed against the requirements for the structure. Key considerations include:

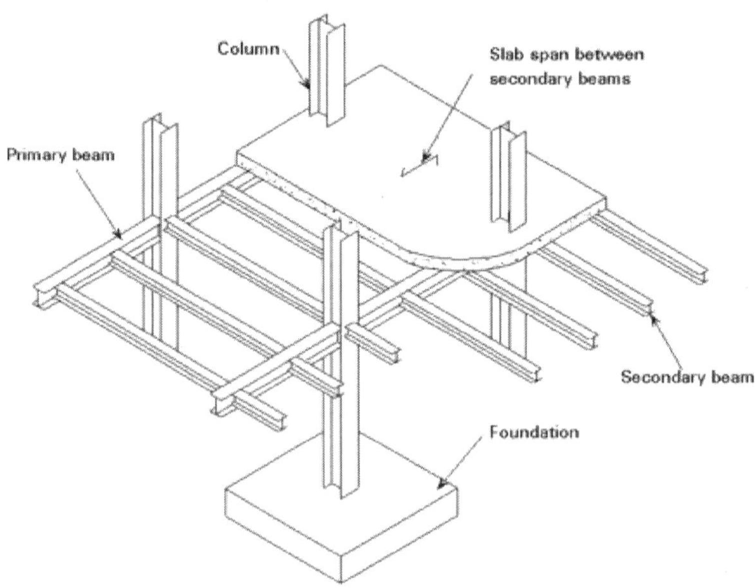

Figure: *Principal structural elements of a multi-storey building*

- Cost and speed of construction
- Building height and plot ratio
- Future flexibility and adaptability

- Site constraints including ground conditions
- The need for special structural arrangements in public spaces or circulation areas
- Floor grids and dimensional coordination with the planning grid
- Structural construction depth (ceiling to floor level)
- Servicing strategy and its coordination and integration with the structure (horizontally and vertically)
- Floor loadings
- Fire resistance
- Sustainability requirements.

The principal structural elements of a typical multi-storey building comprise floors, beams and columns. A wide variety of alternative forms and arrangements can be used in multi-storey steel framed structures to deliver the benefits of:

- Economy
- Shallow floor construction
- Integration of services
- Flexible, column-free floor space
- Reduced foundations
- Rapid on-site construction.

Factors Affecting Choice of Structural System

Main articles: Long-span beams, Floor systems, Service integration, Braced frames, Continuous frames, Simple connections, Moment resisting connections, Steel construction products, Composite construction, Facades and interfaces

Floor grids define the spacing of the columns in orthogonal directions, which are influenced by:

- The planning grid (typically based on multiples of 0.6 m, 1.2 m or 1.5 m)
- The column spacing on the facade, to suit the external envelope (typically 5.4 m to 7.5 m)
- The intended use of the internal space
- The requirements for building services distribution.

The British Council for Offices (BCO) Guide to Specification provides extensive guidance on the preparation of scheme designs.

For naturally ventilated offices, a building width of 12 m to 15 m is typically used, which can be achieved by two spans of 6 m to 7.5 m, with a column placed adjacent to a central column. Natural lighting also plays a role in choice of the width of floor plate. In larger buildings, a long-span solution provides a considerable enhancement in flexibility of layout. For air conditioned offices, a clear span of 15 m to 18 m is often used. Floor-to-floor height will be an important consideration at the concept design stage. The table below gives typical floor to floor heights for buildings of different use.

Typical Floor-to-Floor Heights	
Prestige office	4 - 4.2 m
Speculative office	3.6 - 4.0 m
Renovation project	3.5 - 3.9 m

If planning restrictions limit overall building height, shallow floor solutions may be adopted, or solutions that involve integrating the services in the structural depth. Typical structural depths for different types of construction are given in the table below.

Typical Structural Depths (Ceiling to Floor)	
Composite beam construction	800 mm - 1200 mm
Cellular beams (with service integration)	800 mm - 1100 mm
Precast concrete floors (7.5 m span)	1200 mm - 1200 mm
Precast concrete floors (14 m span)	1450 mm - 1450 mm
Slim floor or integrated beams	600 mm - 800 mm

Stability Systems

External Envelope: Cladding systems that are used in multi-storey buildings depend on the building height and the degree of fenestration. Fully glazed facades are widely used, although provision for solar shading generally has to be made. Cladding systems include:

Brickwork: Ground supported up to 3 storeys. Supported by stainless steel angles attached to edge beams for taller buildings

Glazing Systems: Generally triple glazing or double layer facades supported on aluminium posts or glass fins.

Curtain Walling: Aluminium or other lightweight facade that is attached to the perimeter steelwork

Insulated Render or Tiles: Cladding system supported on light steel infill walls, mainly used in public sector buildings and residential buildings.

The external skin of a multi-storey building is usually supported off the structural frame. In most high quality commercial buildings, the cost of external cladding systems greatly exceeds the cost of the primary structure. This influences the design and construction of the structural system in the following ways:

- Reducing the floor zone may be cost-effective as it reduces the area of cladding.
- The perimeter structure must provide a satisfactory platform to support the cladding system and be sufficiently stiff to meet any deflection criteria.

Structural Principles

Once the structural concept has been fixed, the structural design may be completed. The structural design process involves the following steps:

- Calculating the permanent actions and determining the applied actions
- Identifying the load paths that carry the applied actions (vertical and horizontal) to the foundations
- Selecting preliminary sizes for members
- Analysis of the structure, if necessary, to determine design effects on individual members and deflections
- Analysis of the structure to assess sensitivity to second-order effects, allowing for these if necessary
- Verifying the members by ensuring the design resistance exceeds the design effects
- Verification of the members, and the frame, by ensuring the deflections do not exceed the limits given in the design Standard, or those specified by the Client.

Actions

Characteristic actions include:

- Imposed floor loads
- Imposed roofs on floors
- Wind actions
- Snow loads
- Actions on structures exposed to fire
- Actions during execution of buildings (for example during concreting of composite slabs and beams)

- Accidental actions (used when considering the requirement to avoid disproportionate collapse)

Characteristic actions should be determined from the appropriate parts of BS EN 1991-1 and the relevant UK National Annexes.

Analysis

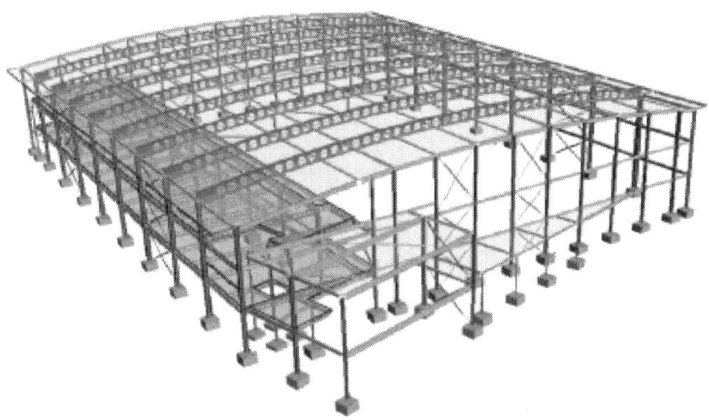

Figure: *Model of a building constructed with cellular beams (Fastrak model courtesy of CSC)*

For 'simple construction', frame analysis will not be necessary at the Ultimate Limit State (ULS), as the members can be designed in isolation. It may be necessary (or convenient) to use analysis software to determine the lateral deflections when at the Serviceability Limit State (SLS) and when assessing frame stability. Analysis will be required for continuous frames.

Generally, elastic analysis is used. Plastic analysis (and elastic-plastic analysis) is generally only used for the design of portal frames .

Although manual methods of analysis may be used, most designers find it convenient to use readily available software .

Modelling Members: Most software contains libraries of all the standard steel sections, with the associated member properties used in the analysis. Non-standard members may be modelled with equivalent section properties. Tapered or haunched members may be modelled with a number of short elements, each with different section properties. Curved members may be modelled with a series of straight members.

Modelling Joints: Common UK practice is to assume joints are either nominally pinned (and modelled as perfectly pinned) or nominally rigid (and modelled as perfectly rigid). It is then important to ensure that the physical details correspond to the design assumptions.

Sensitivity to Second-Order Effects

All frames experience second-order effects, typically because under lateral loads (or simply due to frame imperfections), the vertical loads are no longer concentric with the bases. The effect of this displacement is not accounted for in a first-order analysis. Some frames are sufficiently stiff such that second-order effects are small enough to be ignored. When second-order effects must be accounted for, this can be achieved by using second-order analysis, or by a simple amplifier of the lateral loads.

All frames must be assessed for sensitivity to second-order effects, and these effects allowed for if necessary.

Design Standards

The over-arching requirement for design in the UK is to satisfy the Building Regulations. Practically, design of steel structures will generally be in accordance with BS 5950 or the Eurocodes.

Building Regulations

Building Regulations require that a safe structure be constructed. Although a series of design Standards are cited in the Regulations, there is no absolute requirement that the listed Standards are used for design.

BS 5950

Although BS 5950 was withdrawn in March 2010, it is likely to be used for steel building design for a number of years. The advantages of design to BS 5950 include:

- Familiarity with the Standard enjoyed by experienced designers
- The extensive use of look-up tables within the Standard
- The specific sections of the Standard devoted to particular building types, such as portal frames.

Eurocodes

The Eurocodes are a set of unified structural design standards for use across Europe, developed by CEN (European Committee for Standardisation), to cover the design of all types of structures in steel, concrete, timber, masonry and aluminium.

In the UK, the Eurocodes are published by BSI under the designations BS EN 1990 to BS EN 1999; each of these ten Eurocodes is published in several Parts and each Part is accompanied by a UK National Annex (which must be used for construction in the UK) that adds certain UK specific provisions.

In addition to the Eurocodes and the National Annex, non-contradictory complementary information (NCCI) is provided, to provide further guidance on the application of the Eurocodes.

National Annexes

The National Annex (NA) is an essential document when using any Eurocode Part; the relevant NA covers the country where the construction will take place. Where the opportunity is given in the text of the Eurocode, the National Annex will:

- Specify the value of factors, modify limiting values or formulae
- Specify which design method may be used
- Specify which options may be used, if they are given in the Eurocode
- State whether an informative annex may be used - such annexes are explicitly identified as being open to National acceptance.
- Identify NCCI.

NCCI

The National Annex may give references to publications and other guidance containing non contradictory complementary information (NCCI) that will assist the designer when designing a structure to the Eurocodes.

The Eurocodes omit some design guidance where it is considered to be readily available in text books or other established sources. Publications that contain such design guidance may be referenced in the National Annex as NCCI.

Several 'Published Documents' (PDs) are available, published by BSI . These NCCI documents are informative, without the status of a Standard, but are generally helpful reference documents for designers.

Basis of Structural Design

BS EN 1990 can be considered as the 'core' document of the structural Eurocode system as it establishes the principles and requirements for the safety, serviceability and durability of structures. It also describes the basis for structural design and verification.

The main sections of BS EN 1990 include:

- Requirements
- Principles of limit state design
- Basic variables
- Structural analysis and design assisted by testing
- Verification by the partial safety factor method.

Designers will refer to BS EN 1990 (and its National Annex) for load factors and combination factors, used when determining design combinations of actions (ULS loads).

BS EN 1993-1 (Eurocode 3)

BS EN 1993-1 Eurocode 3: Design of steel structures comprises a set of general rules in twelve parts (BS EN 1993-1-1 to BS EN 1993-1-12) for all types of steel buildings. The commonly used Parts include:

- BS EN 1993-1-1. This Part provides most of the general rules used in the design of steel buildings, including material properties, guidance on analysis, the assessment of second-order effects and the calculation of member resistances.
- BS EN 1993-1-5. This Part provides rules for members fabricated from plate, but also provides the rules for calculating the resistance of webs - typically verified under concentrated loads.
- BS EN 1993-1-8. This Part provides the design rules for joints. It includes rules for the resistance of joint components, such as bolts and welds.

Additional rules are provided in separate Parts for other structures, e.g. BS EN 1993-2 provides design rules for bridges.

A comprehensive range of design guidance is available for use in the UK, all incorporating the influence of the UK National Annexes.

BS EN 1994 (Eurocode 4)

BS EN 1994 covers the design of composite structures and elements.

Guidance is available covering the design of composite floors in accordance with Eurocode 4 in SCI publication *Composite design of steel framed buildings* (P359).

Common Structural Systems

Building frames may be broadly classified by their stability system, as braced or continuous frames . A portal frame is a particular type of a continuous frame. For multi-storey frames, a braced frame is likely to be most economical, because the fabrication effort for the joints in braced frames is generally much less than for joints in continuous frames. Continuous frames must be used when bracing cannot be provided within the structure.

Within both types of frame, a wide range of floor systems is available. Several floor systems utilise the benefits of composite construction , which will be the de facto solution for many structures. The floor systems in common use are briefly described in the following sections

Composite Construction

Figure: *Trapezoidal decking on downstand steel beams (Image courtesy of Richard Lees Steel Decking Ltd)*

Composite construction is the dominant form of construction for the multi-storey building sector. Its success is due to the strength and stiffness that can be achieved, with minimum use of materials, utilising the compressive strength of concrete and the tensile strength of steel. Composite floors offer significant advantages related to speed of construction and reduced overall construction depth.

Composite floor slabs generally use either relatively shallow profiled steel decking , typically spanning up to 3.75 m, or deep deck systems, spanning up to 9 m (if propped during construction). Composite floor slabs may also be constructed using pre-cast planks as the permanent formwork. Floor slabs may be formed from pre-cast planks, but still allow the supporting beam to be designed as a composite member.

Composite beams involve the transfer of force between the steel section and the concrete which it supports, to prevent slip and thus ensure the two elements perform as a composite whole.

For beams located wholly under the slab (known as "downstand" beams) the transfer of force is commonly achieved using headed shear studs, which are attached to the upper flange of the steel beam. Studs are usually welded on site, through the decking, to the (unpainted) top flange of the beam. Alternatively, smaller shear connectors may be shot-fired to the steel beam. In some forms of construction, the shear bond between the steel member and the encasing concrete is sufficient to provide composite action without additional shear connectors.

Different types of composite beam are available:

- Downstand beams - the steel beam is wholly below the concrete slab
- Shallow floor solutions - the steel beam is at least partially integrated within the depth of the floor slab.

Types of composite beams

Figure: *Conventional "downstand" beam*

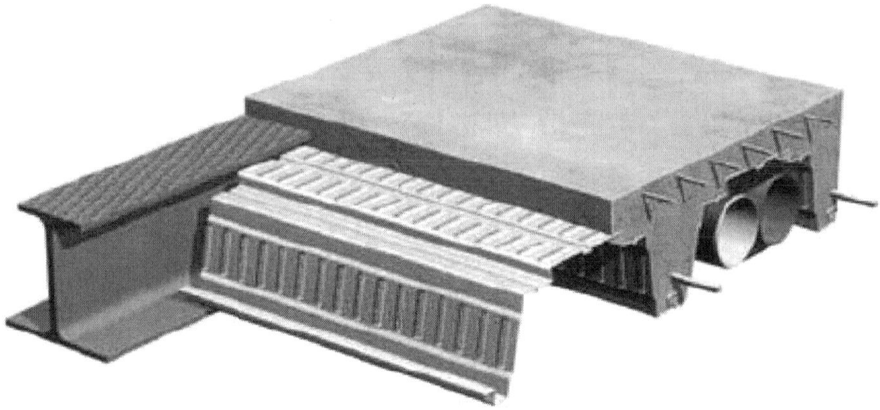

Figure: *Integrated beam with deep decking*

Precast Concrete Units

Precast concrete units may be used in conjunction with steel beams. The units may be solid or hollow-core, and with tapered or bluff ends. They are normally prestressed. The steel beams and precast units may be designed as a composite member, provided specific detailing rules are satisfied to ensure that composite behaviour.

Figure: *Precast concrete floor slabs on a steel frame (Image courtesy of Atlas Ward Structures Ltd.)*

Integrated Floor Solutions

Figure: *Slimdek 'shallow floor' system*

Integrated (or "shallow") floors offer a range of benefits, including a reduced construction depth (compared to an orthodox "downstand" solution) and, in some solutions, a virtually flat soffit, allowing easy location of services.

A number of integrated floor solutions are available, including a range of rolled and fabricated options providing shallow depth members with wide bottom flanges, so that precast planks or steel decking can be placed on the bottom flange.

Tata Steel produce a range of asymmetric beams (ASBUK), which are rolled beams with a wider bottom flange than the top. The ASBUK may be combined with deep decking in the Slimdek system.

Long-Span Beams

Generally long spans result in flexible, column free internal spaces, reduced substructure costs, and reduced erection times. This broad range of benefits means that they are commonly used in a wide range of building types.

Long-span beam options include:

- Composite beams with web openings
- Cellular composite beams
- Tapered girders.

The design of long-span steel and (steel-concrete) composite beams is generally carried out in accordance with BS 5950, BS EN 1993 orBS EN 1994.

For some types of beam this codified guidance is complemented by specific design guidance, such as that on the design of beams with large web openings, or manufacturers' software.

Trusses

A truss is essentially a triangulated system of straight interconnected structural elements. The most common use of trusses is in buildings, where support to roofs, the floors and internal loading such as services and suspended ceilings, is readily provided. Trusses are commonly used in a range of buildings including airport terminals, aircraft hangers, sports stadia roofs, auditoriums and other leisure buildings. Trusses are also used to provide large column-free spaces in commercial buildings.

The main reasons for using trusses are:

- To span large distances
- To provide a lightweight solution
- To control deflection
- To support heavy loads.

Trusses may be exposed within the structure and be fabricated from hollow sections for aesthetic appeal. Large loads may require the use of open sections (UKC, typically) or sections built up from plate.

In all cases, the additional fabrication costs make trusses more expensive than conventional beam and column structures.

Portal Frames

Portal frames are low-rise structures, comprising columns and rafters, connected by moment-resisting connections. The resistance to lateral and vertical actions is provided by the rigidity of the connections and the bending stiffness of the members, which is increased by a substantial haunch to reinforce the rafter sections.

This form of continuous frame is stable in its plane and provides a clear span that is unobstructed by bracing. Bracing is provided in the longitudinal direction, between frames.

Portal frames are structurally efficient and lightweight, accounting for over 90% of the single-storey market in the UK. Portal frames are used for industrial, storage, retail and commercial applications.

In addition to a primary steel structure, portal framed buildings also commonly include:

- Secondary, thin gauge cold-rolled steel members (purlins and side rails) which restrain the primary steel members and support the cladding
- Profiled steel wall and roof cladding.

Member Design

Steel member design is based on the requirements given in BS EN 1993-1-1.

The overall process in member design involves:

- Classification of the cross section
- Cross-sectional resistance
- Member buckling resistance
- Resistance to combined axial loading and bending, where applicable.

Member design is often completed using software, of by reference to the resistance tables in the 'Blue Book'.

Connections

The cost of connections is a significant proportion of the overall cost of the structure and the use of standardised, simple details is recommended wherever possible. Moment resisting connections will undoubtedly be more expensive than the nominally pinned connections used in braced construction.

Design of joints in steel structures in the UK is covered by BS EN 1993-1-8 and its UK National Annex

BS EN 1993-1-1 requires that connection behaviour be accounted for in frame analysis, if it is significant. Nominally pinned connections and continuous connections may be modelled as pinned and rigid (respectively) in the analysis, but semi-continuous joint behaviour should be taken into account.

BS EN 1993-1-8 provides numerical methods to classify a joint, but offers the option that joints may be classified on the basis of previous satisfactory performance.

The UK National Annex specifies that joints designed in accordance with the industry standard guides (on nominally pinned and moment-resisting connections) are simple and rigid respectively. Classification in accordance the guidance in the UK NA is recommended.

Simple Connections

Simple connections are nominally pinned connections that transmit end shear only and do not transfer significant moments. This assumption underpins the design of multi-storey braced frames in the UK, in which the beams are designed as simply-supported and the columns are designed for axial load and the small nominal bending moments induced by the end reactions from the beams.

Simple connections include:

- Beam-to-beam and beam-to-column connections using:
  - o Partial depth end plates
  - o Full depth end plates
  - o Fin plates
- Column splices (bolted cover plates and end plates)
- Column bases

Nominally Pinned Connections

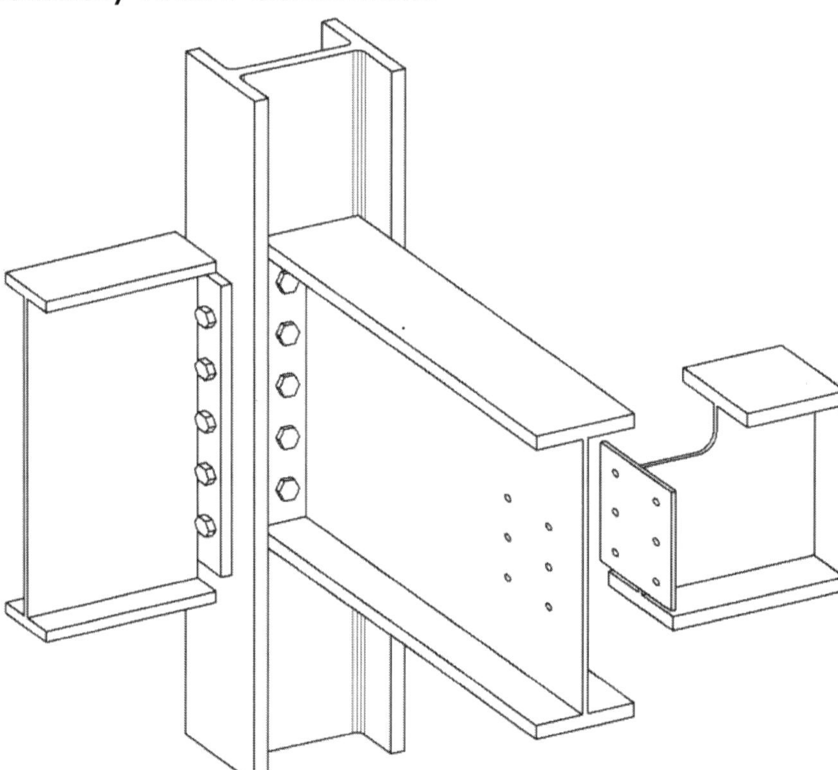

Figure: *End plate connections*

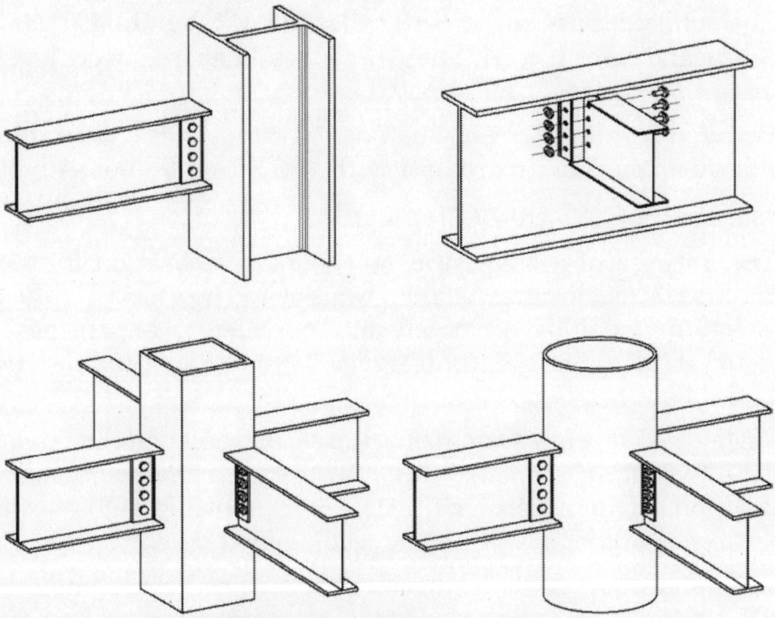

Figure: *Fin plate connection*

Moment-Resisting Connections

Moment resisting connections are used in continuous frames . Connections in multi-storey frames are most likely to be full depth end plate connections and extended end plate connections.

Connections in portal frames will be haunched at the eaves, often with stiffeners in the column member. The apex connection may have a small haunch or a simple extended end plate.

Moment-resisting connections may also be provided by providing a welded connection between members.

Structural Robustness

Building Regulations require that all buildings should be designed to avoid disproportionate collapse . Commonly, this is achieved by designing the joints in a steel frame (the beam-to-column connections and the column splices) for tying forces. Guidance on the design values of tying forces is given in BS EN 1991-1-7 Annex A, and its UK National Annex .

Two generic types of strategy for designing robust structures for accidental actions are provided in BS EN 1991-1-7 :

1. Strategies based on identified accidental actions.
2. Strategies based on unidentified accidental actions.

The robustness requirements relate to the building Class (the type, size and use of a building) and establish required levels of robustness for different building Classes.

Detailed guidance on robustness strategies for steel framed buildings designed in accordance with the Eurocode is available.

Specification of Structural Steelwork

The purpose of a structural steelwork specification is to state what materials and products should be used and how work (fabrication and erection) should be carried out, in order to ensure that the completed structure meets the designer's assumptions and the client's needs.

Under the Eurocode system, fabrication and erection should comply with the execution standard BS EN 1090 which covers material specification, component specification, workmanship and tolerances.

In the UK, the National Structural Steelwork Specification (NSSS) is recommended for use with steel structures. The NSSS complements BS EN 1090, by providing additional requirements on transfer of information, weld inspection etc, and other requirement which are not addressed in BS EN 1090.

3

Structural Analysis

Understanding Structural Behaviour

It is recognised that with the universal presence of computer analysis, an intuitive understanding becomes increasingly important, both in the creation of analysis models and critically, in the appraisal of the analysis results, such as the deflected shape, distribution of moments or distribution of reactions.

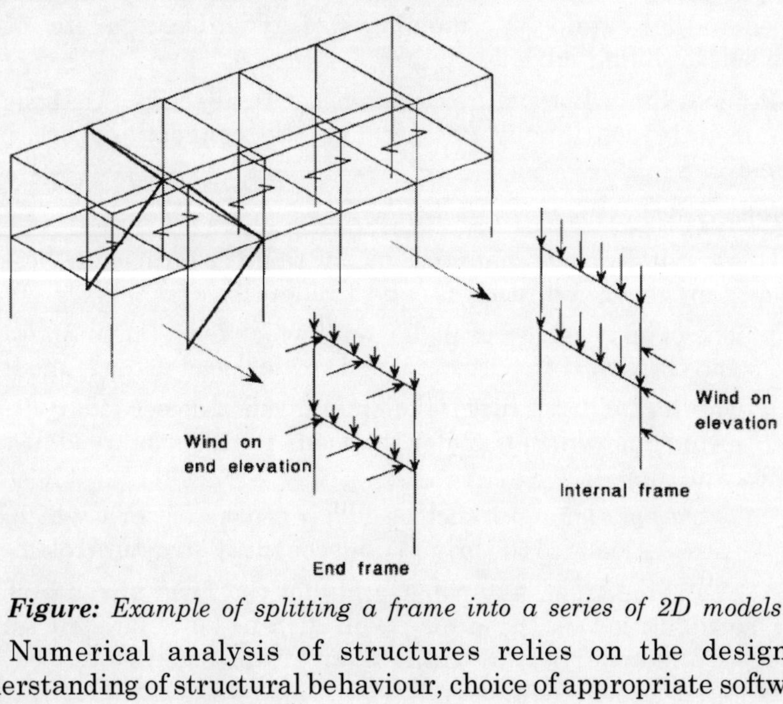

Figure: *Example of splitting a frame into a series of 2D models*

Numerical analysis of structures relies on the designer's understanding of structural behaviour, choice of appropriate software, method of analysis and above all the use of engineering judgement to know when the answers are reasonable.

An intuitive approach uses broader, more dynamic reasoning skills to evaluate the behaviour of any particular structure. The key principles involved in developing this kind of understanding of structural behaviour are:

- To consider the deformed shape of a structure
- To use statically determinate simple systems, so that good appreciation of the behaviour of the real structure with all its complexities can be gained.

It is helpful to use the graphic options of the software to review input data, such as loads, and output data, such as deflections and bending moments.

Most orthodox buildings are a series of repeating 2-D frames and often it is convenient to model in this way. Most steel sections are highly efficient in one primary direction and moment resisting connections to the minor axis can be difficult and expensive.

However, many multi-storey buildings are modelled in 3-D, as it is very effective to copy and repeat similar floors together with defined load patterns. 3-D modelling is also useful for analysing complex frames and for cataloguing member size, type, location, etc. within the whole building model.

Pinned, braced structures are most cost effective. Analysis can accommodate continuous design, but the connections are more expensive.

Modelling

It is generally convenient to consider first the form of the building frame in orthogonal directions, and to identify:

- The primary structural elements which form the main frames and transfer both horizontal and vertical load to the foundations
- The secondary structural elements, such as secondary beams or purlins, which transfer the loads to the primary structural elements
- The other elements, such as cladding or partitions, which only transfer loads to the primary or secondary structural elements.

At the same time, any constraints on the form of the building must be identified as these may well dictate how the structure is modelled, and in particular, which (if any) frames may be braced, and which must be modelled as rigid.

The objective for the designer is (within the constraints of the specification and any architectural requirements) to provide a safe, economical structure. The definition of an economical structure is not straightforward, and it may be necessary to investigate several forms of framing before undertaking the detailed analysis and design. However, it is possible to provide general guidance based mainly on the understanding that moment resisting connections are more expensive than nominally pinned connections. Thus in order of preference, the designer should consider:

- 'Simple' construction - i.e. braced frames with nominally pinned joints
- Rigid frames in one direction, with ties and bracing in the other
- Rigid frames in two directions.

It must be emphasised that, in most cases, there is more than one option for the form of the building frame. Further advice on structural form can be found in the Steel Designers' Manual.

Multi-Storey Buildings

Steel frame multi-storey structures in the UK are typically analysed and designed for two types of loading - gravity and lateral. For a structure where floor grids repeat, greater levels of repeatability within the structure, and thus a more economic design, can be achieved by analysing the structure in the following order:

1. Analyse and design the structure to resist gravity actions (self-weight, imposed actions, snow loads, etc). This structure comprises floors, often composite floor decks acting compositely or non-compositely with steel beams and columns. It is recommended to model:
   - (i) One typical floor first - ensuring common member sizes are used where possible to maximize standardization
   - (ii) Using this floor, replicate it within the building as many times as possible; design all floors and the columns for gravity combination of actions.
2. Analyse and design for lateral actions (arising from wind and initial imperfections etc.) and design the lateral load resisting system. This system can consist of one or more of the following:

- Braced frames - with bays containing diagonal braces or cross-bracing which resist the lateral loading in tension and/or compression

- Continuous frames with bays resisting lateral load due to frame action and moment-resisting connections between beams and columns
- Concrete shear walls which are typically planar elements or groups of planar elements (cores) which resist the lateral load in shear or shear and bending respectively.

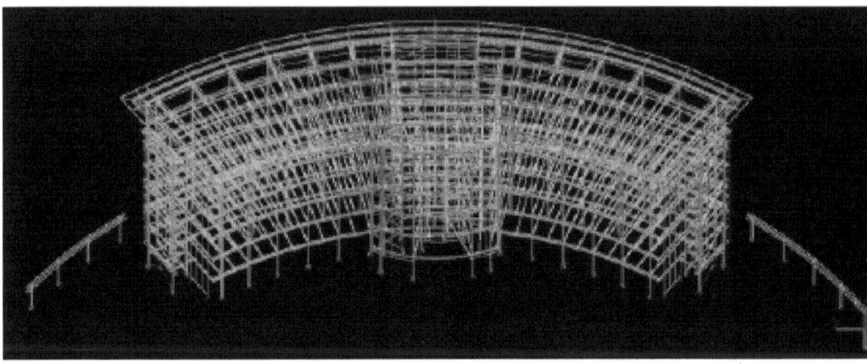

Figure: *Model of a multi-storeybuilding (Image courtesy of Graitec UK Ltd.)*

A number of simplifying assumptions are made when modelling the building for analysis:

- Analysis elements are aligned with the tops of steel beams in floors thereby ignoring the small offsets in centre line between beams of different depth
- The horizontal offset of edge beams is usually small enough to be ignored
- All columns are typically modelled as being co-linear along their centre line
- small offsets of columns from grids are typically ignored in design
- to ensure that all the lateral loading is carried by the braced or moment frames (continuous framing), it is typical to assume that all columns not in braced bays or moment frames are pinned at each floor level, so they do not attract lateral loads.

More information on modelling multi-storey buildings can be found in the Steel Designers' Manual.

Trusses and Lattice Girders

There are a variety of models which may be used for the analysis of a truss. These include:

- Pin jointed frames
- Continuous chords and pin jointed internal, i.e. web, members
- Rigid frames.

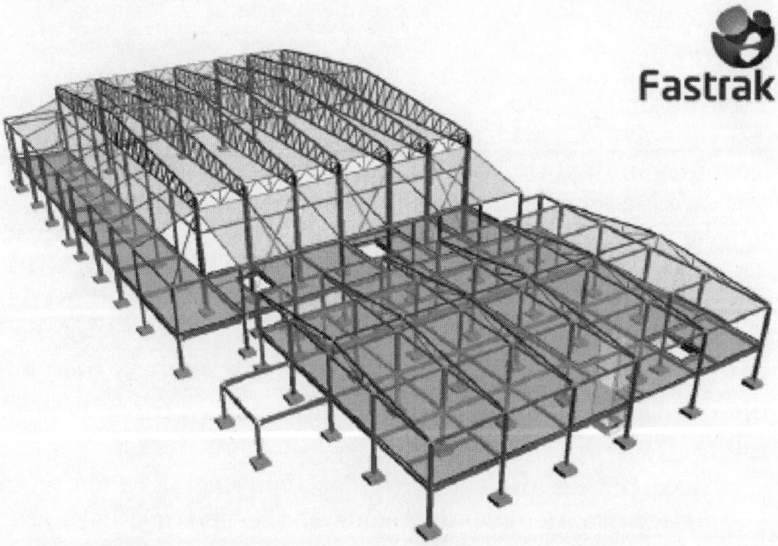

Figure: *Model of a building with trusses (Fastrak model courtesy of CSC)*

The first two options are preferred, since in most situations there will be no bending moments to be included in the joint capacity checks and connection design. Pin jointed frames are the traditional choice to model trusses, while the assumption of continuous chords and pin jointed internals usually better reflects the behaviour in practice.

In this case the chords resist some bending due to loading off 'panel points' but behave primarily as an axially loaded continuous beam. The internals are axially loaded only - moments due to self-weight are usually ignored. The analysis model should reflect this behaviour.

In trusses using hollow sections, despite the fact that the internals are fully welded to the chords, this type of connection is still assumed to be pinned. This is due to the relatively thin walls of hollow sections and the large deformations that such connections can sustain.

The third option is usually only particularly relevant to trusses that use Vierendeel action. In this case the behaviour of the truss, in the way it resists loads, is through bending action in both the chords and internals. This has the advantage that the diagonal internals are omitted but the efficacy of the hollow section is somewhat lost as the Virendeel truss acts as a continuous frame and significant moments develop at the ends of members. The connections between the hollow sections have to be stiffer since they have to be designed to resist the bending moments due to the Vierendeel action.

Portal Frames

Proprietary software dedicated to the analysis of portal frames generally involves an elastic analysis to check frame deflection at the serviceability limit state, and an elastic-plastic analysis to determine the forces and moments in the frame at the ultimate limit state. These methods have largely replaced the rigid-plastic method which is unable to account for the important second-order effects in portal frames. Nevertheless the rigid-plastic method continues to be allowed by BS EN 1993-1-1Cl 5.4.3 (1) providing certain criteria are met of the same standard - that ensure second-order effects can safely be disregarded.

BS EN 1993-1-1 provides three alternatives for 'plastic global analysis':

- Elastic-plastic analysis with plastified sections and/or joints as plastic hinges (referred to below as the 'elastic-plastic method')
- Non-linear plastic analysis considering the partial plastification of members in plastic zones. This method is not generally used in commercial portal frame design
- Rigid plastic analysis neglecting the elastic behaviour between hinges.

The approach to analysis in BS EN 1993-1-1 is to presuppose that second-order effects need to be allowed for in the design except under special circumstances when the second-order effects are small enough to be ignored.

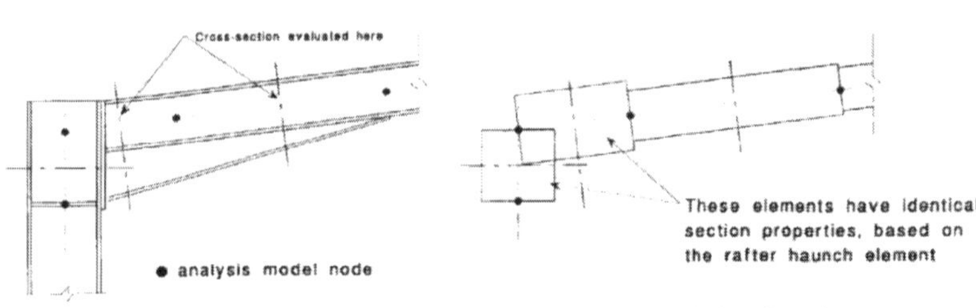

Figure: *Modelling the eaves haunch*

Haunches are frequently provided at the eaves and apex connections of a portal frame. Typically analysis software does not have the facility to use a 'tapered element'. In such cases it is acceptable to model tapered members as a series of uniform prismatic elements. The assumption that the neutral axis remains at the centre line of the rafter and does not descend towards the haunch is safe, since it tends to overestimate both the compression in the bottom flange, and the shear.

Special Members

Normal frame members are generally modelled as one (or more) straight elements, with associated section properties. Universal beams, universal columns, tees, angles, channels and hollow sections are modelled on this basis. Non-standard sections may require a different approach.

Curved Members

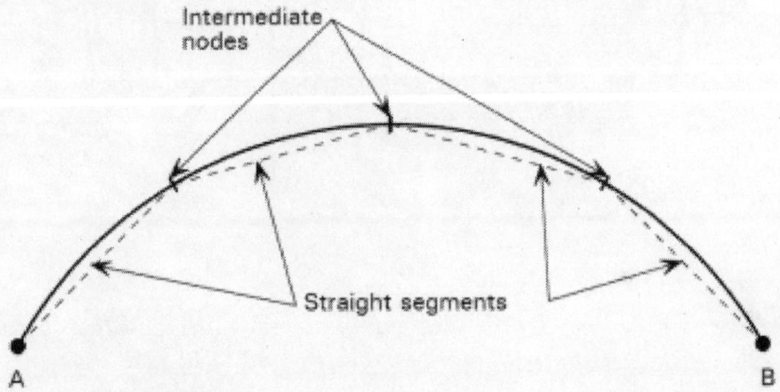

Figure: *Modelling curved members using straight segments*

Curved members are modelled as a series of short, straight elements. Modelling by using more, shorter elements, improves the accuracy of the results. As a general guide, a length of arc corresponding to 15° produces reasonable results. More information can be found in Design of Curved Steel (SCI P281).

Tapered Members

Tapered members can be simply modelled as a series of short elements, each with an inertia corresponding to the depth of the member at that position. Generally, three such sections give reasonable accuracy when modelling tapered members.

Castellated and Cellular Members

Many steelwork analysis programs provide libraries of standard section properties, and may also include the section properties for castellated beams. This will allow the structural designer to include castellated members in a frame model in the same way as standard sections. Whilst the frame bending moments produced by this approach will generally be satisfactory, the structural designer should note that the deflection of a castellated or cellular beam will be more than that predicted by Engineer's bending theory. This is due to the Vierendeel effect and to shear deflection.

Fastrak

Figure: *Model of a building constructed with cellular beams (Fastrak model courtesy of CSC)*

As a rule of thumb, the deflection of a cellular or castellated beam may be taken as 25% greater than the equivalent depth beam without openings. Additional deflection due to the Vierendeel effect becomes more significant with multiple, long openings. As a rule of thumb, the deflection of a beam with multiple, long openings may be taken as 35% greater than that of the equivalent depth beam without openings.

In some circumstances, the structural designer may conclude that the additional deflection may be ignored, or is not critical. Alternatively an allowance may be made for the additional deflection during design and checking of the members.

More information can be found in Design of composite and non-composite cellular beams (SCI P100).

Joints

Within a frame, joint behaviour affects the distribution of internal forces and moments and the overall deformation of the structure. In many cases, however, the effect of modelling a continuous joint as fully rigid, or a simple joint as perfectly pinned, compared to modelling the real behaviour, is sufficiently small to be neglected.

Figure: *Modelling joints and connections (Image courtesy of Graitec UK Ltd.)*

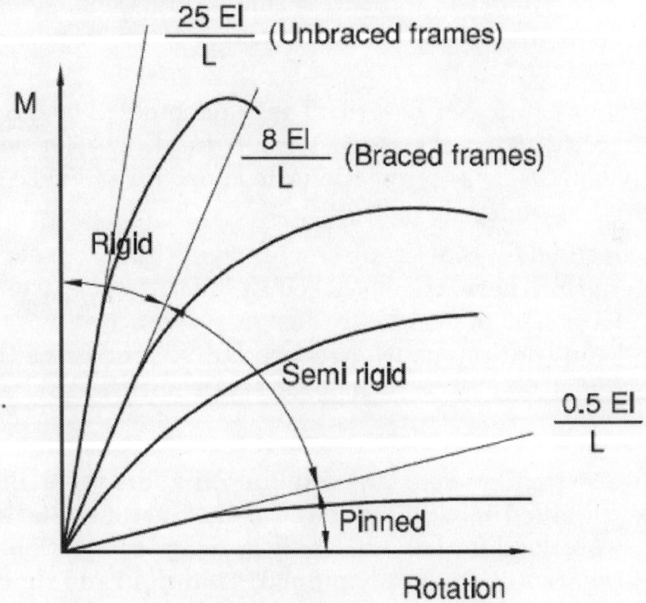

Figure: *Classification of joints*

Elastic analysis programs consider only the stiffness of the joint and it is convenient to define three types as follows:

- Simple - a joint which may be assumed not to transmit bending moments. Sometimes referred to as a pinned connection, it must also be sufficiently flexible to be regarded as a pin for analysis purposes.

- Continuous - a joint which is stiff enough for the effect of its flexibility on the frame bending moment diagram to be neglected. Sometimes referred to as 'rigid'; they are, by definition, moment-resisting.
- Semi-continuous - a joint which is too flexible to qualify as continuous, but is not a pin. The behaviour of this type of joint must be taken into account in the frame analysis.

BS EN 1993-1-8 requires that joints be classified according to their stiffness for analysis and should also have sufficient strength to transmit the forces and moments acting at the joint that result from the analysis.

When Classifying the Stiffness of a Joint: Simple joints - are described as 'nominally pinned' rather than pinned since it is accepted that some moment is transferred. In this definition these moments are insufficient to adversely affect the member design. The UK National Annex BS EN 1993-1-8 indicates that connections designed in accordance with the principles given in SCI P358 may be classified as nominally pinned.

Continuous joints - are described as 'rigid joints'. The UK National Annex refers the designer to SCI P398 in which, for many, but importantly not all, cases connections designed for strength alone can be considered as rigid.

Semi-continuous joints - are 'semi-rigid' and are usually also 'partial strength'. These are described as 'ductile connections' in UK practice and are used in plastically designed semi-continuous frames. For braced semi-continuous frames, the UK NA indicates that these may be designed using the principles given in SCI P183 .

Supports

The interaction between the foundation and supporting ground is complex. Detailed modelling of the soil-structure relationship is probably too involved for general analysis. Base connections fall into the same categories of nominally pinned, semi rigid and rigid as other joints, as BS EN 1993-1-8 has no specific recommendations covering their rotational stiffness.

More information can be found in SN045 .

Pinned and Rocker Bases

Where a true pin or rocker is used, the rotational stiffness is zero. The use of such bases is rarely justified in practice, as careful consideration needs to be given to the issues connected with shear transfer to the foundation and temporary stability of the column during erection.

Nominally Pinned Bases

If a column base is nominally pinned and the foundation design assumes the base moment is zero, it is recommended that:

- When using elastic global analysis to establish the design forces and moments at the Ultimate Limit State the base should be assumed to be pinned.
- When checking frame stability i.e. when checking whether the frame is susceptible to second-order effects the base may be assumed to have a stiffness equal to 10% of the column stiffness (which can be taken as 4EI/L)
- When calculating deflections at the Serviceability Limit State the base may be assumed to have stiffness equal to 20% of the column stiffness.

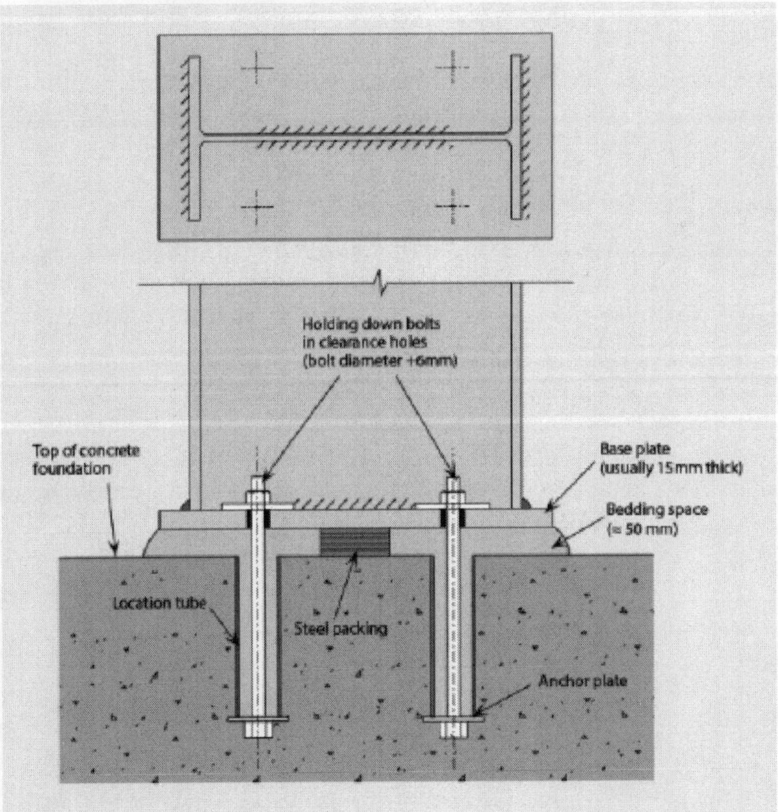

Figure: *Example of a nominally pinned base*

Nominally Rigid Bases

If a column is rigidly connected to a suitable foundation, the following recommendations should be applied:

- The stiffness of the base should be limited to the stiffness of the column when using elastic global analysis to establish the design forces and moments at the Ultimate Limit State
- The base may be assumed to be rigid when calculating deflections at the Serviceability Limit State
- For elastic-plastic global analysis, the assumed stiffness of the base must be consistent with the assumed moment capacity of the base but should not exceed the stiffness of the column. Any base moment capacity between zero and the plastic moment of resistance of the column may be assumed, provided that the foundation and base plate are designed to resist a moment equal to the assumed moment capacity, together with the forces obtained from the analysis.

Semi-Rigid Bases

A nominal base stiffness of up to 20% of the column stiffness may be assumed in elastic global analysis, provided that the foundation is designed for the moments and forces obtained from this analysis.

Modelling Base Stiffness

Bespoke software normally has the facility to select the recommended values of base stiffness. Unless such software is used, the base stiffness may be modelled by the use of a spring stiffness or dummy members at the column base.

When assessing the sensitivity of the frame to second-order effects (calculating α_{cr}) a nominally pinned base can be modelled with a spring stiffness equal to $0.4EI_{col}/L_{col}$. When calculating frame deflections at the serviceability limit state a nominally pinned base can be modelled with a spring stiffness equal to $0.8EI_{col}/L_{col}$.

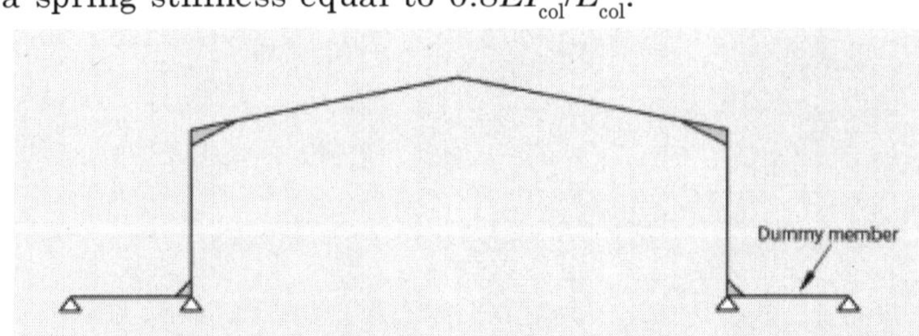

Figure: *Modelling base fixity by a dummy member*

If the computer program cannot accommodate a rotational spring, the base fixity may be modelled by a dummy member of equivalent stiffness as shown below. When modelling a nominally pinned base the second moment of area (I_y) of the dummy member should be taken as:

- $I_y = 0.1\, I_{y,col}$ when assessing frame stability
- $I_y = 0.2\, I_{y,col}$ when calculating deflections at SLS.

In both cases, the length of the dummy member is $L = 0.75\, L_{col}$, and modelled with a pinned support at the extreme end.

Results from analysis with the use of dummy members should not be used explicitly, as the provision of an additional support will affect the base reactions. The vertical base reaction should be taken as the axial force in the column.

Model Verification

The most important part in any analysis exercise is to review the output in order to confirm that an appropriate structural model has been used, and that the applied loads are correct. This is not to confirm that the execution of the analysis is correct! When using proven software the analysis will be correct - the exercise is to check the structural designer's input.

Software frequently contains default values for certain input data. Support fixity and restraint conditions are common examples of data which may have default values. Default values are intended to avoid the necessity for the structural engineer to enter data, and represent the 'usual' condition, which may be amended by the user. The structural designer must give due regard to the default values assumed by the program, and either satisfy himself that these are appropriate, or amend the value accordingly. All input data, whether default or user input, remains the responsibility of the designer. Default values are common in both analysis and design software.

Questions the designer should ask when reviewing the output from structural modelling software include:

Program Defaults	Are these correct? Are the default conditions appropriate for the physical details and the loading of the frame?
Loading	Viewed graphically, do the loads in each loadcase appear correct? Are any elements without load? Do loads on some elements appear to be orders of magnitude different from others? Are the loads applied in the correct orientation?

Deflected Form	Is the deflected form correct? Has the structure deflected as expected, and is the order of the overall deflection as expected?
Bending Moment Diagram	Is the form of the bending moment diagram as expected? Are moments shown where releases would have been appropriate? Does the overall envelope on a member equate to that calculated by a simplistic approach, typically $wL^2/8$ or $WL/4$?
Reactions	Checking by hand calculation, do the total reactions provided in the output (vertically and horizontally) equate to the applied loads? Do the reactions quoted for different unfactored loadcases differ by orders of magnitude? Is the distribution of load to the supports as expected?
Spring Stiffnesses	Are the spring stiffness values assumed for analysis appropriate to the members as designed?
Trusses	If the truss is simply supported and carrying a uniformly distributed load, do the maximum chord forces equate to $wL^2/8$, divided by the depth of the truss? Does the vertical component of the end internal member equate to the vertical end reaction? Are the displacements of the correct form and order? In the analysis model, is one end of the truss free to move longitudinally? Do redundant members at the ends of the truss (if included in the model) attract load, or have they been released?

Analysis of a Structure

For common building structures, analysis is concerned with determining the building displacements together with the internal forces and moments that result from the applied loading. The results are determined by mathematically combining the structural stiffness of the analysis model together with the actions applied.

Assumed Simplifications

The analytical model of the structure is usually created by defining the idealised geometry, material properties and the structural supports. Assuming perfectly straight members is common, however this is not always the case. Next the load model is created defining the location, magnitudes and directions of actions on the structure. These actions are typically grouped into design situations by type, e.g. permanent, variable, wind and snow, etc.

Finally, combinations of actions are created which add together the actions in the design situations multiplying them by relevant factors as defined in design standards.

When the material properties are non-linear, for tension only elements or compression only supports, a 'non linear' analysis is required.

Similarly, should the loads be other than static loads, for example time-dependant loading from a machine or an acceleration spectrum to model an earthquake, then a time history or a response spectrum analysis will be required.

As well as common modelling idealisations such as assuming perfectly elastic material, straight members, consistent section properties, actions applied at one point or uniformly distributed, there are also analysis idealisations and simplifications.

To determine the internal forces and moments in a building frame the following should be taken into account (if significant):

- Second-order P-Δ effects - effects of deformed geometry, that introduce additional forces caused by deformation of the frame
- Second-order P-δ effects - effects of deformed geometry, that introduce additional forces caused by deformation of the member
- Global imperfections in the structure - e.g. the 'lean' of out of plumb columns
- Local imperfections in members - e.g. initial bow of the member
- Residual stresses in members
- Flexural, shear and axial deformations
- Joint behaviour.

Analysis by Hand

The calculations required to obtain the shear forces and bending moments in simply-supported beams form the basis of many other calculations. Generally only the simplest analysis is undertaken by hand - software is pervasive.

There is no need to use software when designing simple columns or beams. Hand calculations are also useful for initial sizing of frames or continuous beams. Details regarding hand calculations can be found in many textbooks.

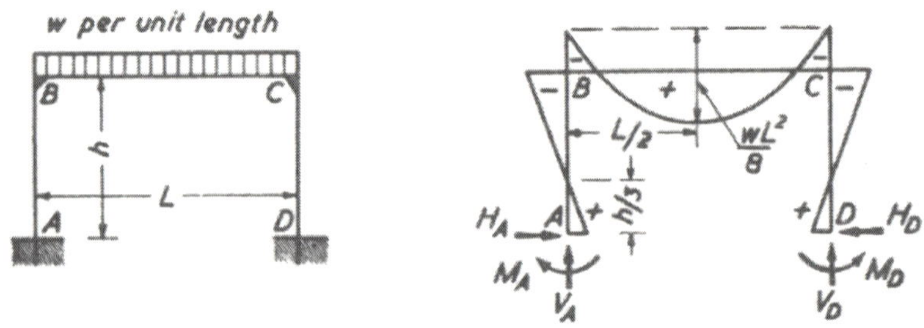

$$M_A = M_D = \frac{wL^2}{12N_1} \qquad M_B = M_C = -\frac{wL^2}{6N_1} = -2M_A$$

$$M_{max} = \frac{wL^2}{8} + M_B \qquad V_A = V_D = \frac{wL}{2} \qquad H_A = H_D = \frac{3M_A}{h}$$

Figure: *Example of design guidance for rigid frames*

Useful formulae for bending moments, shear and deflection are presented in The Steel Designer's Manual, Appendix: Design theory or in Steel Buildings .

Analysis by Software

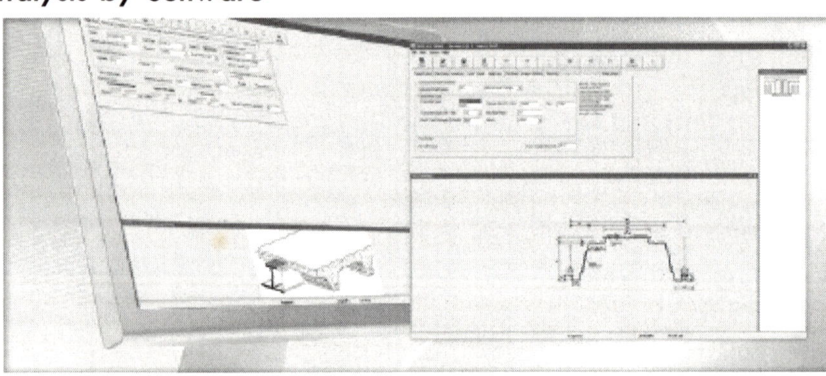

Figure: *Example of design software*

There are many analysis packages currently available on the market. They offer a wide range of features and analysis capabilities. The majority of analysis programs performs elastic analysis, but some also offer plastic or elastic-plastic analysis.

Analysis software is applicable to a wide range of structural forms: buildings, bridges, towers, masts, tented structures, etc. Some programs include the facility for different structure types, such as trusses , where the nodes are all pinned or portal frames, which allow the modelling of eaves haunches and restraint positions with ease. Usually, the engineer has to define structure geometry, member sizes, supports and actions before the analysis can be run.

The modern trend is moving towards the analysis being a subset of 'design software'. A user builds a 'physical model' in the design software defining the members and connections, the software selects initial sizes for members based on engineering rules of thumb and then creates an analysis model automatically. Many analysis programs also offer some design capacity, however for the design of particular elements such as concrete or composite floors designers often use specialist, bespoke software intended only for the design of these components.

Element Types

Analysis packages have a range of different element types. Some of the more common are:

- Inactive elements, which are not used in the analysis
- Linear elements like beam, truss, link element or rigid beam element applicable to all analysis types
- Non-linear elements such as tension/compression only elements, cables, non linear springs or gap elements used only in non-linear analysis
- 2D elements like membranes, plates or shells.

Software specific to steel design generally has libraries of default materials and member types, so as the member is selected its properties are automatically associated.

Joints

Joints are usually by default set up as fixed, rigid or spring. In the latter case the user is prompted to enter the rotational spring stiffness. Some programs may have pre-set standard defaults, for example the base stiffness values.

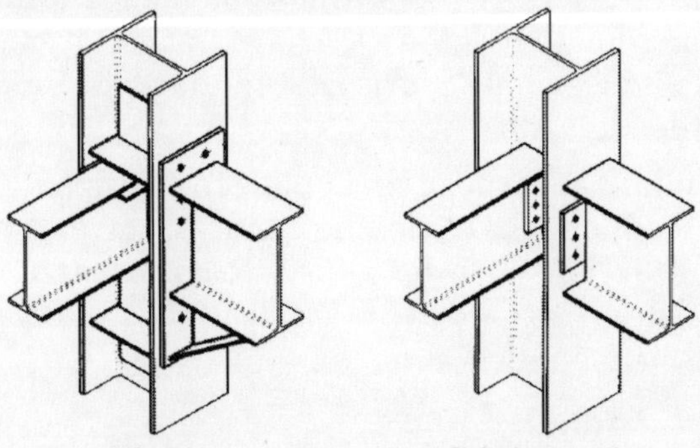

Figure: *Comparison of a rigid and pinned joint*

Determination of connection stiffness is described in BS EN 1993-1-8.

Analysis Types

Most commercially available software offers a multitude of analysis types. They can be broken down into a number of 'classes' which are described below:

- Static analysis - used to determine the nodal displacements, the element deflections together with the element forces, moments and stresses. This is the most common form for the analysis of building structures.
- Dynamic analysis - also called vibration analysis - used to determine the natural frequencies and corresponding mode shapes of vibration.
- Buckling analysis - used to determine the modes and associated load factors for buckling and assess whether or not the structure is prone to buckling at a higher or lower load than has been applied.
- Response spectrum analysis - used in earthquake situations to apply an acceleration spectrum to a structure and to determine from this the design shears and moments in the elements.
- Time history analysis - used to apply time-dependent loading to a structure.

In the design of the majority of building structures in the UK, the only analysis types that are likely to be used are first-order analysis (linear static) and second-order analysis (P-delta static) - the latter is for structures which are susceptible to second-order effects.

First-Order Analysis

In first-order analysis the stiffness of the structure is assumed to be constant and unaffected by changes in the geometry of the structure when it is loaded. This is the standard assumption of linear-elastic first-order analysis.

The principle of superposition applies to this approach. Where the analysis model remains the same, the results from analyses of different sets of applied actions can be added together and the results of individual design situations can be scaled. The analysis results are proportional to the applied actions.

Second-Order Analysis

In second-order analysis, the effective stiffness of the structure is changed by the action of the loads upon it. Examples of this are cable structures, where a cable becomes apparently stiffer as it straightens out. The principle of superposition does not apply, as effects of actions interact.

Second-order effects, often called P-delta effects are commonly illustrated by considering the additional displacements, forces and moments which arise from the application of actions on a deflecting structure. These are known as second-order effects. In some circumstances a first-order analysis may be used to approximate the results of a second-order analysis, by techniques such as the Amplified Sway Method, which is suitable for elastic frame analysis by computer.

Second-Order Effects

Second-order effects are non-linear effects that occur in every structure where elements are subject to axial load. P-delta is actually only one of many second-order effects. It is a genuine 'effect' that is associated with the magnitude of the applied axial compression (P) and a displacement (delta).

There are two kinds of second-order effects:

- P-Δ (P-'Big' delta) - a structure effect resulting from joint displacement
- P-δ (P-'little' delta) - a member effect resulting from deformation in member geometry.

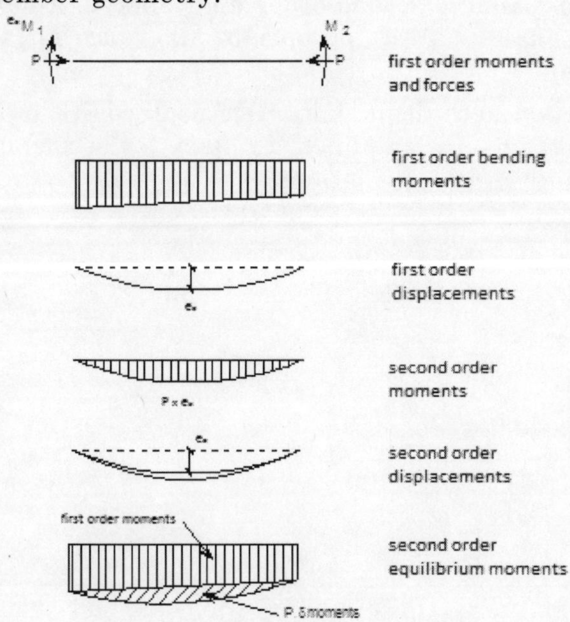

Figure: *P-δ (P-'little' delta) effects*

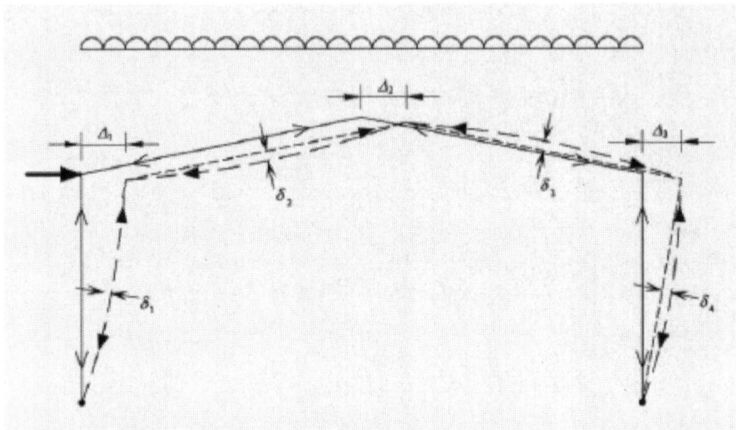

Figure: *P-Δ and P-δ effects in a portal frame*

Second-order effects increase the deflections, moments and forces beyond those calculated by first-order analysis. Sensitivity to second-order effects needs to be assessed for each designed structure.

When second-order effects are significant, two options are possible:

- a rigorous second-order analysis (i.e. in practice, using an appropriate second-order analysis software)
- an approximate second-order analysis (i.e. a modified first-order analysis with appropriate allowance for second-order effects).

In the second method, also known as 'modified first-order analysis', the applied actions are amplified, to allow for second-order effects while using first-order calculations.

Elastic Analysis

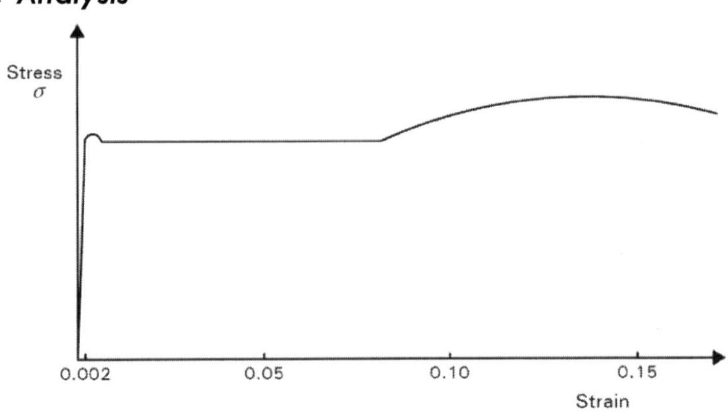

Figure: *Typical stress-strain curve for structural steel*

Elastic analysis programs are the most widely used for structural analysis and are based on the assumption that the material which is being modelled is linear-elastic., therefore the limiting stress is the value corresponding to the strain of 0.002, up to which steel behaves as a linear material. The appropriate value of the elastic modulus has to be provided in the analysis. BS EN 1993-1-1 allows the plastic cross-sectional resistance to be used with the results of elastic analysis, provided the section Class is Class 1 or Class 2.

Plastic Analysis

Plastic analysis relies on the formation and rotation of hinges, therefore it demands Class 1 sections. Plastic hinge rotations occur at sections where the bending moment reaches the plastic moment or resistance of the cross-section at load levels below the full ULS loading.

Plastic analysis is generally used only for portal frame design, where it results in a more economical frame than an elastic analysis. This is because plastic analysis allows relatively large redistribution of bending moments throughout the frame, due to plastic hinge rotations. This redistribution 'relieves' the highly stressed regions and allows the capacity of under-utilised parts of the frame to be mobilised.

Structural Integrity and Failure

Structural integrity and failure is an aspect of engineering which deals with the ability of a structure to support a designed load (weight, force, etc...) without breaking, tearing apart, or collapsing, and includes the study of breakage that has previously occurred in order to prevent failures in future designs.

Structural integrity is a performance characteristic which is applied to a component, a single structure, or a structure consisting of different components. Structural integrity is the quality of an item to hold together under a load, including its own weight, resisting breakage or bending. It assures that the construction will perform its designed function, during reasonable use, for as long as the designed life of the structure. Items are constructed with structural integrity to ensure that catastrophic failure does not occur, which can result in injuries, severe damage, death, or monetary losses.

Structural failure refers to the loss of structural integrity, which is the loss of the load-carrying capacity of a component or member within a structure, or of the structure itself. Structural failure is initiated when the material is stressed to its strength limit, thus causing fracture or excessive deformations. In a well-designed system, a localized failure should not cause immediate or even progressive collapse of the entire structure. Ultimate failure strength is one of the limit states that must be accounted for in structural engineering and structural design.

Introduction

Structural integrity is the ability of a structure or a component to withstand a designed service load, resisting structural failure due to fracture, deformation, or fatigue. Structural integrity is a concept often used in engineering, to produce items that will not only function adequately for their designed purposes, but also to function for a desired service life.

To construct an item with structural integrity, an engineer must first consider the mechanical properties of a material, such as toughness, strength, weight, hardness, and elasticity, and then determine a suitable size, thickness, or shape that will withstand the desired load for a long life.

A material with high strength may resist bending, but, without adequate toughness, it may have to be very large to support a load and prevent breaking.

However, a material with low strength will likely bend under a load even though its high toughness prevents fracture. A material with low elasticity may be able to support a load with minimum deflection (flexing), but can be prone to fracture from fatigue, while a material with high elasticity may be more resistant to fatigue, but may produce too much deflection unless the object is drastically oversized.

Structural integrity must always be considered in engineering when designing buildings, gears or transmissions, support structures, mechanical components, or any other item that may bear a load.

The engineer must carefully balance the properties of a material with its size and the load it is intended to support. Bridge supports, for instance, need good yield strength, whereas the bolts that hold them need good shear and tensile strength. Springs need good elasticity, but lathe tooling needs high rigidity and minimal deflection.

When applied to a structure, the integrity of each component must be carefully matched to its individual application, so that the entire structure can support its load without failure due to weak links. When a weak link breaks, it can put more stress on other parts of the structure, leading to cascading failures.

History

The need to build structure with integrity goes back as far as recorded history. Houses needed to be able to support their own weight, plus the weight of the inhabitants. Castles needed to be fortified to withstand assaults from invaders. Tools needed to be strong and tough enough to do their jobs. However, it was not until the 1920s that the science of fracture mechanics, namely the brittleness of glass, was described by Alan Arnold Griffith. Even so, a real need for the science did not present itself until World War II, when over 200 welded-steel ships broke in half due to brittle fracture, caused by a combination of the stresses created from the welding process, temperature changes, and the stress points created at the square corners of the bulkheads. The squared windows in the De Havilland Comet aircraft of the 1950s caused stress points which allowed cracks to form, causing the pressurized cabins to explode in mid-flight. Failures in pressurized boiler tanks were a common problem during this era, causing severe damage. The growing sizes of bridges and buildings began to lead to even greater catastrophes and loss of life. The need to build constructions with structural integrity led to great advances in the fields of material sciences and fracture mechanics.

Types of Failure

Failure of a structure can occur from many types of problems. Most of these problems are unique to the type of structure or to the various industries. However, most can be traced to one of five main causes.

- The first, whether due to size, shape, or the choice of material, is that the structure is not strong and tough enough to support the load. If the structure or component is not strong enough, catastrophic failure can occur when the overstressed construction reaches a critical stress level.
- The second is instability, whether due to geometry, design or material choice, causing the structure to fail from fatigue or corrosion. These types of failure often occur at stress points, such as squared corners or from bolt holes being too close to the material's edge, causing cracks to slowly form and then progress through cyclic loading. Failure general occurs when the cracks reach a critical length, causing breakage to happen suddenly under normal loading conditions.
- The third type of failure is caused by manufacturing errors. This may be due to improper selection of materials, incorrect

sizing, improper heat treating, failing to adhere to the design, or shoddy workmanship. These types of failure can occur at any time, and are usually unpredictable.

- The fourth is also unpredictable, from the use of defective materials. The material may have been improperly manufactured, or may have been damaged from prior use.
- The fifth cause of failure is from lack of consideration of unexpected problems. Vandalism, sabotage, and natural disasters can all overstress a structure to the point of failure. Improper training of those who use and maintain the construction can also overstress it, leading to potential failures.

Notable Failures

On 24 May 1847 the new railway bridge over the river Dee collapsed as a train passed over it, with the loss of 5 lives. It was designed by Robert Stephenson, using cast iron girders reinforced with wrought iron struts. The bridge collapse was the subject of one of the first formal inquiries into a structural failure. The result of the inquiry was that the design of the structure was fundamentally flawed, as the wrought iron did not reinforce the cast iron at all, and that, owing to repeated flexing, the casting had suffered a brittle failure due to fatigue.

First Tay Rail Bridge

The Dee bridge disaster was followed by a number of cast iron bridge collapses, including the collapse of the first Tay Rail Bridge on 28 December 1879. Like the Dee bridge, the Tay collapsed when a train passed over it causing 75 people to lose their lives. The bridge failed because of poorly made cast iron, and the failure of the designer Thomas Bouch to consider wind loading on the bridge. The collapse resulted in cast iron largely being replaced by steel construction, and a complete redesign in 1890 of the Forth Railway Bridge. As a result, the Forth Bridge was the first entirely steel bridge in the world.

First Tacoma Narrows Bridge

The 1940 collapse of the original Tacoma Narrows Bridge is sometimes characterized in physics textbooks as a classical example of resonance; although, this description is misleading. The catastrophic vibrations that destroyed the bridge were not due to simple mechanical resonance, but to a more complicated oscillation between the bridge and winds passing through it, known as aeroelastic flutter. Robert H. Scanlan, father of the field of bridge aerodynamics, wrote an article about this misunderstanding. This collapse, and the research that followed, led to an increased understanding of wind/structure interactions. Several bridges were altered following the collapse to prevent a similar event occurring again. The only fatality was 'Tubby' the dog.

I-35W Bridge

The I-35W Mississippi River bridge (officially known simply as Bridge 9340) was an eight-lane steel truss arch bridge that carried Interstate 35W across the Mississippi River in Minneapolis, Minnesota, United States. The bridge was completed in 1967, and its maintenance was performed by the Minnesota Department of Transportation. The bridge was Minnesota's fifth–busiest, carrying 140,000 vehicles daily. The bridge catastrophically failed during the evening rush hour on 1 August 2007, collapsing to the river and riverbanks beneath. Thirteen people were killed and 145 were injured. Following the collapse, the Federal Highway Administration (FHWA) advised states to inspect the 700 U.S. bridges of similar construction after a possible design flaw in the bridge was discovered, related to large steel sheets called gusset plates which were used to connect girders together in the truss structure. Officials expressed concern about many other bridges in the United States sharing the same design and raised questions as to why such a flaw would not have been discovered in over 40 years of inspections.

Buildings

Thane Building Collapse: On 4 April 2013, a building collapsed on tribal land in Mumbra, a suburb of Thane in Maharashtra, India. It has been called the worst building collapse in the area. 74 people died, including 18 children, 23 women, and 33 men, while more than 100 people survived. The search for additional survivors ended on 6 April 2013.

The building was under construction and did not have an occupancy certificate for its 100 to 150 low- to middle-income residents. Living in the building were the site construction workers and families. It was reported that the building was illegally constructed because standard practices were not followed for safe, lawful construction; land acquisition and resident occupancy.

By 11 April, a total of 15 suspects were arrested including builders, engineers, municipal officials and other responsible parties. Governmental records indicate that there were two orders to manage the number of illegal buildings in the area: a 2005 Maharashtra state order to use remote sensing and a 2010 Bombay High Court order. There were also complaints made to state and municipal officials.

On 9 April, a campaign began by the Thane Municipal Corporation to demolish area illegal buildings, focusing first on "dangerous" buildings. The forest department said that it will address encroachment of forest land in the Thane district. A call centre was established by the Thane Municipal Corporation to accept and track resolution of caller complaints about illegal buildings.

Savar Building Collapse

On 24 April 2013, *Rana Plaza*, an eight-story commercial building, collapsed in Savar, a sub-district in the Greater Dhaka Area, the capital of Bangladesh. The search for the dead ended on 13 May with the death toll of 1,129. Approximately 2,515 injured people were rescued from the building alive.

It is considered to be the deadliest garment-factory accident in history, as well as the deadliest accidental structural failure in modern human history.

The building contained clothing factories, a bank, apartments, and several other shops. The shops and the bank on the lower floors immediately closed after cracks were discovered in the building. Warnings to avoid using the building after cracks appeared the day before had been ignored. Garment workers were ordered to return the following day and the building collapsed during the morning rush-hour.

Sampoong Department Store Collapse

On 29 June 1995, the 5-story Sampoong Department Store in the Seocho District of Seoul, South Korea collapsed resulting in the deaths of 502 people. In April 1995, cracks began to appear in the ceiling of the fifth floor of the store's south wing due to the presence of an air-conditioning unit on the weakened roof of the poorly built structure. On the morning of 29 June, as the number of cracks in the ceiling increased dramatically, the top floor was closed and managers shut the air conditioning off. The store management failed to shut the building down or issue formal evacuation orders; however, the executives themselves left the premises as a precaution. Five hours before the collapse, the first of several loud bangs was heard emanating from the top floors, as the vibration of the air conditioning caused the cracks in the slabs to widen further. Amid customer reports of vibration, the air conditioning was turned off, but the cracks in the floors had already grown to 10 cm. At about 5:00 p.m. local time, the fifth floor ceiling began to sink; by 5:57 p.m., the roof gave way, and the air conditioning unit crashed through into the already-overloaded fifth floor, trapping more than 1,500 people and killing 502.

Ronan Point

On 16 May 1968, the 22-storey residential tower Ronan Point in the London Borough of Newham collapsed when a relatively small gas explosion on the 18th floor caused a structural wall panel to be blown away from the building. The tower was constructed of precast concrete, and the failure of the single panel caused one entire corner of the building to collapse.

The panel was able to be blown out because there was insufficient reinforcement steel passing between the panels. This also meant that the loads carried by the panel could not be redistributed to other adjacent panels, because there was no route for the forces to follow. As a result of the collapse, building regulations were overhauled to prevent disproportionate collapse and the understanding of precast concrete detailing was greatly advanced. Many similar buildings were altered or demolished as a result of the collapse.

Oklahoma City Bombing

On 19 April 1995, the nine-story concrete framed Alfred P. Murrah Federal Building in Oklahoma was struck by a huge car bomb causing partial collapse, resulting in the deaths of 168 people. The bomb, though large, caused a significantly disproportionate collapse of the structure. The bomb blew all the glass off the front of the building and completely shattered a ground floor reinforced concrete column. At second story level a wider column spacing existed, and loads from upper story columns were transferred into fewer columns below by girders at second floor level. The removal of one of the lower story columns caused neighbouring columns to fail due to the extra load, eventually leading to the complete collapse of the central portion of the building. The bombing was one of the first to highlight the extreme forces that blast loading from terrorism can exert on buildings, and led to increased consideration of terrorism in structural design of buildings.

Versailles Wedding Hall

The Versailles wedding hall (Hebrew: àåìîé åøñàé), located in Talpiot, Jerusalem, is the site of the worst civil disaster in Israel's history. At 22:43 on Thursday night, May 24, 2001 during the wedding of Keren and AsafDror, a large portion of the third floor of the four-story building collapsed.

World Trade Centre Towers 1, 2, and 7

In the September 11 attacks, two commercial airliners were deliberately crashed into the Twin Towers of the World Trade Centre in New York City. The impact and resulting fires caused both towers to collapse within two hours. After the impacts had severed exterior columns and damaged core columns, the loads on these columns were redistributed. The hat trusses at the top of each building played a significant role in this redistribution of the loads in the structure. The impacts dislodged some of the fireproofing from the steel, increasing its exposure to the heat of the fires. Temperatures became high enough to weaken the core columns to the point of creep and plastic deformation under the weight of higher floors. Perimeter columns and floors were also weakened by the heat of the fires, causing the floors to sag and exerting an inward force on exterior walls of the building. WTC Building 7 also collapsed later that day. According the official report, the 47 story skyscraper collapsed within seconds due to a combination of a large fire inside the building and heavy structural damage from the collapse of the north tower.

Aircraft

Repeated structural failures of aircraft types occurred in 1954, when 2 de Havilland Comet C1 jet airliners crashed due to decompression caused by metal fatigue, and in 1963-4, when the vertical stabilizer on 4 Boeing B-52 bombers broke off in mid-air.

Other

Warsaw Radio Mast

On 8 August 1991 at 16:00 UTC Warsaw radio mast, the tallest man-made object ever built before the erection of BurjKhalifa collapsed as consequence of an error in exchanging the guy-wires on the highest stock. The mast first bent and then snapped at roughly half its height. It destroyed at its collapse a small mobile crane of Mostostal Zabrze. As all workers left the mast before the exchange procedures, there were no fatalities, in contrast to the similar collapse of WLBT Tower in 1997.

Hyatt Regency Walkway

On 17 July 1981, two suspended walkways through the lobby of the Hyatt Regency in Kansas City, Missouri, collapsed, killing 114 and injuring 200 people at a tea dance. The collapse was due to a late change in design, altering the method in which the rods supporting the walkways were connected to them, and inadvertently doubling the forces on the connection. The failure highlighted the need for good communication between design engineers and contractors, and rigorous checks on designs and especially on contractor-proposed design changes. The failure is a standard case study on engineering courses around the world, and is used to teach the importance of ethics in engineering.

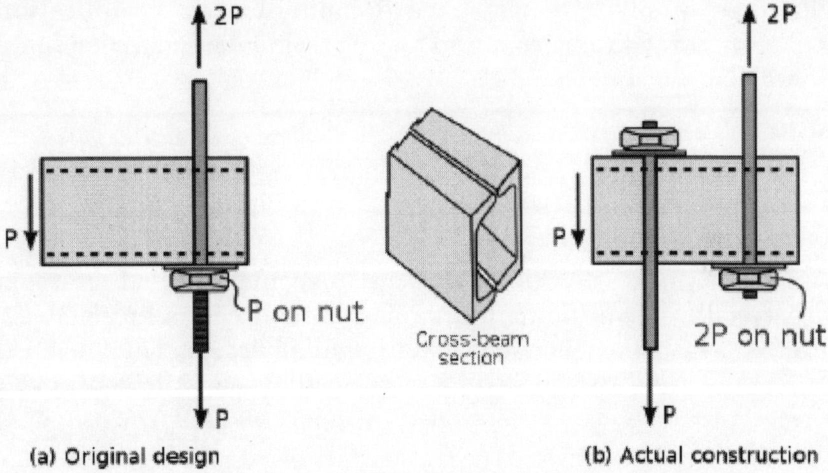

Figure: *Design change on the Hyatt Regency walkways.*

Stress–Strain Analysis

Stress–strain analysis (or just stress analysis, for short) is an engineering discipline covering methods to determine the stresses and strains in materials and structures subjected to forces or loads.

Stress analysis is a primary task for civil, mechanical and aerospace engineers involved in the design of structures of all sizes, such as tunnels, bridges and dams, aircraft and rocket bodies, mechanical parts, and even plastic cutlery and staples. Stress analysis is also used in the maintenance of such structures, and to investigate the causes of structural failures.

Typically, the input data for stress analysis are a geometrical description of the structure, the properties of the materials used for its parts, how the parts are joined, and the maximum or typical forces that are expected to be applied to each point of the structure. The output data is typically a quantitative description of the stress over all those parts and joints, and the deformation caused by those stresses. The analysis may consider forces that vary with time, such as engine vibrations or the load of moving vehicles. In that case, the stresses and deformations will also be functions of time.

In engineering, stress analysis is often a tool rather than a goal in itself; the ultimate goal being the design of structures and artifacts that can withstand a specified load, using the minimum amount of material (or satisfying some other optimality criterion).

Stress analysis may be performed through classical mathematical techniques, analytic mathematical modelling or computational simulation, through experimental testing techniques, or a combination of methods.

Scope

General Principles: Stress analysis is specifically concerned with solid objects. The study of stresses in liquids and gases is the subject of fluid mechanics.

Stress analysis adopts the macroscopic view of materials characteristic of continuum mechanics, namely that all properties of materials are homogeneous at small enough scales. Thus, even the smallest particle considered in stress analysis still contains an enormous number of atoms, and its properties are averages of the properties of those atoms.

In stress analysis one normally disregards the physical causes of forces or the precise nature of the materials. Instead, one assumed that the stresses are related to strain of the material by known constitutive equations.

By Newton's laws of motion, any external forces that act on a system must be balanced by internal reaction forces, or cause the particles in the affected part to accelerate. In a solid object, all particles must move substantially in concert in order to maintain the object's overall shape. It follows that any force applied to one part of a solid object must give rise to internal reaction forces that propagate from particle to particle throughout an extended part of the system. With very rare exceptions (such as ferromagnetic materials or planet-scale bodies), internal forces are due to very short range intermolecular interactions, and are therefore manifested as surface contact forces between adjacent particles — that is, as stress.

Fundamental Problem

The fundamental problem in stress analysis is to determine the distribution of internal stresses throughout the system, given the external forces that are acting on it. In principle, that means determining, implicitly or explicitly, the Cauchy stress tensor at every point.

The external forces may be body forces (such as gravity or magnetic attraction), that act throughout the volume of a material; or concentrated loads (such as friction between an axle and a bearing, or the weight of a train wheel on a rail), that are imagined to act over a two-dimensional area, or along a line, or at single point. The same net external force will have a different effect on the local stress depending on whether it is concentrated or spread out.

Types of Structures

In civil engineering applications, one typically considers structures that can practically static equilibrium: that is, are either unchanging with time, or are changing slowly enough for viscous stresses to be unimportant. By Newton's laws of motion, any external forces that act on such a system must be balanced by internal reaction forces. It follows that any force applied to one part of a solid object must give rise to internal reaction forces that propagate from particle to particle throughout an extended part of the system. With very rare exceptions (such as ferromagnetic materials or planet-scale bodies), internal forces are due to very short range intermolecular interactions, and are therefore manifested as surface contact forces between adjacent particles — that is, as stress.

In mechanical and aerospace engineering, however, stress analysis must often be performed on parts that are far from equilibrium, such as vibrating plates or rapidly spinning wheels and axles. In those cases, the equations of motion must include terms that account for acceleration of the particles.

In structural design applications, one usually tries to ensure the stress is everywhere well below the yield strength of the material. In the case of dynamic loads, the material fatigue must also be taken into account. However, these concerns lie outside the scope of stress analysis proper, being covered in materials science under the names strength of materials, fatigue analysis, stress corrosion, creepmodelling, and other.

Experimental Methods

Stress analysis can be performed experimentally by applying forces to a test element or structure and then determining the resulting stress using sensors. In this case the process would more properly be known as *testing* (destructive or non-destructive). Experimental methods may be used in cases where mathematical approaches are cumbersome or inaccurate. Special equipment appropriate to the experimental method is used to apply the static or dynamic loading.

There are a number of experimental methods which may be used:

- Tensile testing is a fundamental materials science test in which a sample is subjected to uniaxial tension until failure. The results from the test are commonly used to select a material for an application, for quality control, and to predict how a material will react under other types of forces. Properties that are directly measured via a tensile test are ultimate tensile strength, maximum elongation and reduction in area. From these measurements properties such as Young's modulus, Poisson's ratio, yield strength, and strain-hardening characteristics can be determined.
- Strain gauges can be used to experimentally determine the deformation of a physical part. A commonly used type of strain gauge is a thin flat resistor that is affixed to the surface of a part, and which measures the strain in a given direction. From the measurement of strain on a surface in three directions the stress state in the part can be calculated.
- Neutron diffraction is a technique that can be used to determine the subsurface strain in a part.
- The photoelastic method relies on the fact that some materials exhibit birefringence on the application of stress, and the magnitude of the refractive indices at each point in the material is directly related to the state of stress at that point. The stresses in a structure can be determined by making a model of it in a photoelastic material.
- Dynamic mechanical analysis is a technique used to study and characterize viscoelastic materials, particularly polymers. The viscoelastic property of polymer is studied by dynamic mechanical analysis where a sinusoidal force (stress) is applied to a material and the resulting displacement (strain) is measured. For a perfectly elastic solid, the resulting strain and the stress will be perfectly in phase. For a purely viscous fluid, there will be a 90 degree phase lag of strain with respect to stress. Viscoelastic polymers have the characteristics in between where some phase lag will occur during DMA tests.

Mathematical Methods

While experimental techniques are widely used, most stress analysis is done by mathematical methods, especially during design.

Differential Formulation

The basic stress analysis problem can be formulated by Euler's equations of motion for continuous bodies (which are consequences of Newton's laws for conservation of linear momentum and angular momentum) and the Euler-Cauchy stress principle, together with the appropriate constitutive equations.

These laws yield a system of partial differential equations involving the stress tensor field and the strain tensor field as unknown functions to be determined. Solving for either one will yield the other through the constitutive equations. Both fields will normally be continuous within each extended part of the system that does not contain any concentrated loads and that can be regarded as a continuous medium with smoothly varying constitutive equations.

The external body forces will appear as the independent ("right-hand side") term in the differential equations, while the concentrated forces appear as boundary conditions. An external surface force, such as ambient pressure or friction, can be incorporated as an imposed value of the stress tensor across that surface. External forces that are specified as line loads (such as traction on a concrete block plastic applied by an embedded rebar) or point loads (such as the weight of a person standing on a roof) introduce singularities in the stress field, and may be handled by assuming that they are spread over a small arbitrary volume or surface patch. The basic stress analysis problem is therefore a boundary-value problem.

Elastic and Linear Cases

A system is said to be elastic if any deformations caused by external forces will spontaneously and completely disappear once the external forces are removed. Stress analysis for such systems is based on the theory of elasticity and infinitesimal strain theory. When the applied loads cause permanent deformation, one must use more complicated constitutive equations, that can account for the physical processes involved (plastic flow, fracture, phase change, etc.)

Engineered structures are usually designed so that the maximum expected stresses are well within the range of linear elasticity (the generalization of Hooke's law for continuous media); that is, the deformations caused by internal stresses are linearly related to them. In this case the differential equations that define the stress tensor are linear, too. Linear equations are much better understood than non-linear ones; for one thing, their solution (the stress at any given point) will also be a linear function of the external forces.

For small enough stresses, even non-linear systems can usually be assumed to be linear.

Built-in Stress

The mathematical problem represented is typically ill-posed because it has an infinitude of solutions. In fact, in any three-dimensional solid body one may have infinitely many and infinitely complicated) non-zero stress tensor fields that are in stable equilibrium even in the absence of external forces.

Such built-in stress may occur due to many physical causes, either during manufacture (in processes like extrusion, casting or cold working), or after it (for example because of uneven heating, or changes in moisture content or chemical composition). However, if the system can be assumed to be linear, then the built-in stress can be ignored in the analysis, since it will merely add to the solution without interfering with it.

If linearity cannot be assumed, however, any built-in stress may affect the distribution of externally-induced stress (for example, by changing the effective stiffness of the material) or even cause the unexpected material failure. For these reasons, a number of techniques have been developed to avoid or reduce built-in stress, such as annealing of cold-worked glass and metal parts, expansion joints in buildings, and roller joints for bridges.

Simplifications

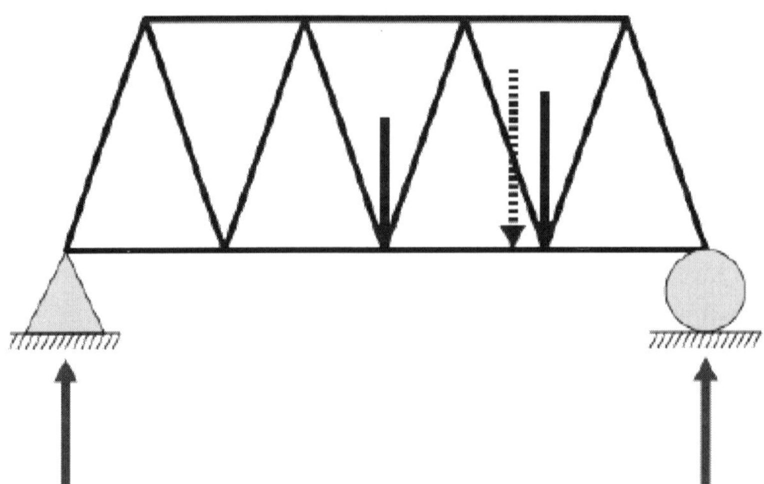

Figure: *Simplified modelling of a truss by unidimensional elements under uniaxial uniform stress.*

Stress analysis is simplified when the physical dimensions and the distribution of loads allow the structure to be treated as one- or two-dimensional. In the analysis of trusses, for example, the stress field may be assumed to be uniform and uniaxial over each member. Then the differential equations reduce to a finite set of equations (usually linear) with finitely many unknowns.

If the stress distribution can be assumed to be uniform (or predictable, or unimportant) in one direction, then one may use a plane stress and plane strain formulation where the stress field is a function of two coordinates only, instead of three.

Even under the assumption of linear elasticity, the relation between the stress and strain tensors is generally expressed by a fourth-order stiffness tensor with 21 independent coefficients. This complexity may be required for general anisotropic materials, but for many common materials it can be simplified. For orthotropic materials like wood, whose stiffness is symmetric about three orthogonal planes, nine coefficients suffice to express the stress-strain relationship. For isotropic materials, these reduce to only two.

One may be able to determine a priori that, in some parts of the system, the stress will be of a certain type, such as uniaxial tension or compression, simple shear, isotropic compression or tension, torsion, bending, etc. In those parts the stress field may then be represented by fewer than six numbers, possibly just one.

Solving the Equations

In any case, for two- or three-dimensional domains one must solve a system of partial differential equations with boundary conditions. Analytical or closed-form solutions to the differential equations can be obtained when the geometry, constitutive relations, and boundary conditions are simple enough. Otherwise one must generally resort to numerical approximations such as the finite element method, the finite difference method, and the boundary element method.

Factor of Safety

The factor of safety (also "margin of safety" or "design factor") is a design requirement for the structure based on the uncertainty in loads, material strength, and consequences of failure. In design of structures, calculated stresses are restricted to be less than a specified allowable stress, also known as working, designed or limit stresses, that is chosen as some fraction of the yield strength or of the ultimate strength of the material which the structure is made of. The ratio of the ultimate stress to the allowable stress is defined as the factor of safety.

Laboratory test are usually performed on material samples in order to determine the yield strength and the ultimate strength that the material can withstand before failure. Often a separate factor of safety is applied to the yield strength and to the ultimate strength. The factor of safety on yield strength is to prevent detrimental deformations and the factor of safety on ultimate strength is to prevent collapse.

The factor of safety is used to calculate a maximum allowable stress:

$$\text{Factor of safety} = \frac{\text{Ultimate tensile strength}}{\text{Maximum stress}}$$

Load Transfer

The evaluation of loads and stresses within structures is directed to finding the load transfer path. Loads will be transferred by physical contact between the various component parts and within structures. The load transfer may be identified visually or by simple logic for simple structures. For more complex structures more complex methods, such as theoretical solid mechanics or numerical methods may be required. Numerical methods include direct stiffness method which is also referred to as the finite element method.

The object is to determine the critical stresses in each part, and compare them to the strength of the material.

For parts that have broken in service, a forensic engineering or failure analysis is performed to identify weakness, where broken parts are analysed for the cause or causes of failure. The method seeks to identify the weakest component in the load path. If this is the part which actually failed, then it may corroborate independent evidence of the failure. If not, then another explanation has to be sought, such as a defective part with a lower tensile strength than it should for example.

Uniaxial Stress

If two of the dimensions of the object are very large or very small compared to the others, the object may be modelled as one-dimensional. In this case the stress tensor has only one component and is indistinguishable from a scalar. One-dimensional objects include a piece of wire loaded at the ends and a metal sheet loaded on the face and viewed up close and through the cross section.

When a structural element is subjected to tension or compression its length will tend to elongate or shorten, and its cross-sectional area changes by an amount that depends on the Poisson's ratio of the material. In engineering applications, structural members experience small deformations and the reduction in cross-sectional area is very small and can be neglected, i.e., the cross-sectional area is assumed constant during deformation. For this case, the stress is called *engineering stress* or *nominal stress.* In some other cases, e.g., elastomers and plastic materials, the change in cross-sectional area is significant. In this case, engineers sometimes still use the original cross section when calculating "engineering stress" as

$$\sigma_e = P / Ao,$$

whereAo is the original cross-sectional area, and P is the applied load.

If the "true stress" is desired, it must be calculated using the current cross-sectional area instead of the initial cross-sectional area, as

$$\sigma_{true} = (1 + \varepsilon_e)(\sigma_e),$$

where

ε_e is the nominal (engineering) strain, and

σ_e is nominal (engineering) stress.

The relationship between true strain and engineering strain is given by

$$\varepsilon_{true} = \ln(1 + \varepsilon_e).$$

In uniaxial tension, true stress is then greater than nominal stress. The converse holds in compression.

Graphical Representation of Stress at a Point

Mohr's circle, Lame's stress ellipsoid (together with the *stress director surface*), and *Cauchy's stress quadric* are two-dimensional graphical representations of the state of stress at a point. They allow for the graphical determination of the magnitude of the stress tensor at a given point for all planes passing through that point. Mohr's circle is the most common graphical method.

Mohr's Circle

Mohr's circle, named after Christian Otto Mohr, is the locus of points that represent the state of stress on individual planes at all their orientations. The abscissa, σ_n, and ordinate, τ_n, of each point on the circle are the normal stress and shear stress components, respectively, acting on a particular cut plane with a unit vector *n* with components (n_1, n_2, n_3).

Lame's Stress Ellipsoid

The surface of the ellipsoid represents the locus of the endpoints of all stress vectors acting on all planes passing through a given point in the continuum body. In other words, the endpoints of all stress vectors at a given point in the continuum body lie on the stress ellipsoid surface, i.e., the radius-vector from the centre of the ellipsoid, located at the material point in consideration, to a point on the surface of the ellipsoid is equal to the stress vector on some plane passing through the point. In two dimensions, the surface is represented by an ellipse.

Cauchy's Stress Quadric

The Cauchy's stress quadric, also called the *stress surface,* is a surface of the second order that traces the variation of the normal stress vector σ_n as the orientation of the planes passing through a given point is changed.

Graphical Representation of the Stress Field

The complete state of stress in a body at a particular deformed configuration, i.e., at a particular time during the motion of the body, implies knowing the six independent components of the stress tensor $(\sigma_{11}, \sigma_{22}, \sigma_{33}, \sigma_{12}, \sigma_{23}, \sigma_{13})$, or the three principal stresses $(\sigma_1, \sigma_2, \sigma_3)$, at each material point in the body at that time. However, numerical analysis and analytical methods allow only for the calculation of the stress tensor at a certain number of discrete material points. To graphically represent in two dimensions this partial picture of the stress field different sets of contour lines can be used:

- *Isobars* are curves along which the principal stress, e.g., σ_1 is constant.
- *Isochromatics* are curves along which the maximum shear stress is constant. This curves are directly determined using photoelasticity methods.
- *Isopachs* are curves along which the mean normal stress is constant
- *Isostatics* or *stress trajectories* are a system of curves which are at each material point tangent to the principal axes of stress.
- *Isoclinics* are curves on which the principal axes make a constant angle with a given fixed reference direction. These curves can also be obtained directly by photoelasticity methods.
- *Slip lines* are curves on which the shear stress is a maximum.

Rockwell Scale

The Rockwell scale is a hardness scale based on indentation hardness of a material. The Rockwell test determines the hardness by measuring the depth of penetration of an indenter under a large load compared to the penetration made by a preload. There are different scales, denoted by a single letter, that use different loads or indenters. The result is a dimensionless number noted as HRA, where A is the scale letter.

When testing metals, indentation hardness correlates linearly with tensile strength. This important relation permits economically important nondestructive testing of bulk metal deliveries with lightweight, even portable equipment, such as hand-held Rockwell hardness testers.

History

The differential depth hardness measurement was conceived in 1908 by a Viennese professor Paul Ludwik in his book *Die Kegelprobe* (crudely, "the cone trial"). The differential-depth method subtracted out the errors associated with the mechanical imperfections of the system, such as backlash and surface imperfections. The Brinell hardness test, invented in Sweden, was developed earlier—in 1900—but it was slow, not useful on fully hardened steel, and left too large an impression to be considered nondestructive.

Hugh M. Rockwell (1890–1957) and Stanley P. Rockwell (1886–1940) from Connecticut in the United States co-invented the "Rockwell hardness tester," a differential-depth machine. They applied for a patent on July 15, 1914. The requirement for this tester was to quickly determine the effects of heat treatment on steel bearing races. The application was subsequently approved on February 11, 1919, and holds U.S. Patent 1,294,171. At the time of invention, both Hugh and Stanley Rockwell worked for the New Departure Manufacturing Co. of Bristol, CT. New Departure was a major ball bearing manufacturer which in 1916 became part of United Motors and, shortly thereafter, General Motors Corp.

After leaving the Connecticut company, Stanley Rockwell, then in Syracuse, NY, applied for an improvement to the original invention on September 11, 1919, which was approved on November 18, 1924. The new tester holds U.S. Patent 1,516,207. Rockwell moved to West Hartford, CT, and made an additional improvement in 1921. Stanley collaborated with instrument manufacturer Charles H. Wilson of the Wilson-Mauelen Company in 1920 to commercialize his invention and develop standardized testing machines. Stanley started a heat-treating firm circa 1923, the Stanley P. Rockwell Company, which still exists in Hartford, CT. The later-named Wilson Mechanical Instrument Company has changed ownership over the years, and was most recently acquired by Instron Corp. in 1993.

Operation

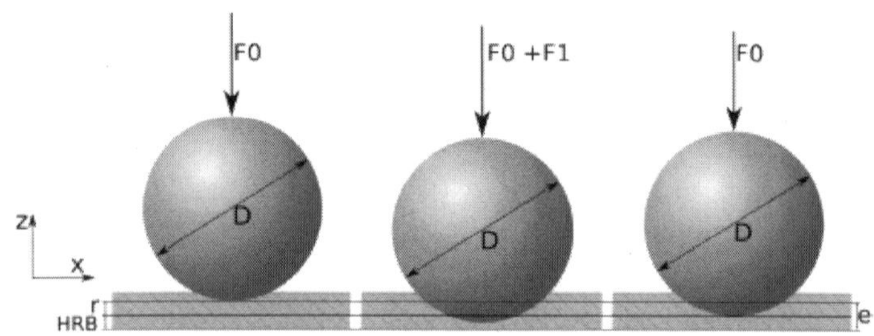

Figure: *Force diagram of Rockwell test*

The determination of the Rockwell hardness of a material involves the application of a minor load followed by a major load, and then noting the depth of penetration from a dial, on which a harder material gives a higher number. The chief advantage of Rockwell hardness is its ability to display hardness values directly, thus obviating tedious calculations involved in other hardness measurement techniques.

It is typically used in engineering and metallurgy. Its commercial popularity arises from its speed, reliability, robustness, resolution and small area of indentation.

In order to get a reliable reading the thickness of the test-piece should be at least 10 times the depth of the indentation. Also, readings should be taken from a flat perpendicular surface, because convex surfaces give lower readings. A correction factor can be used if the hardness of a convex surface is to be measured.

Scales and Values

There are several alternative scales, the most commonly used being the "B" and "C" scales. Both express hardness as an arbitrary dimensionless number.

- Except for testing thin materials in accordance with A623 the steel indenter balls have been replaced by tungsten carbide balls of the varying diameters. When a ball indenter is used, the letter "W is used to indicate a tungsten/carbide ball was used, and the letter "S" indicates the use of a steel ball. E.g.: 70 HRCW indicates the reading was 70 in the Rockwell C scale using a Tungsten/Carbide indenter.

Various Rockwell scales				
Scale	***Abbreviation***	***Load***	***Indenter***	***Use***
A	HRA	60 kgf	120° diamond cone†	Tungsten carbide
B	HRB	100 kgf	$^1/_{16}$-inch-diameter (1.588 mm) steel sphere	Aluminium, brass, and soft steels
C	HRC	150 kgf	120° diamond cone	Harder steels >B100
D	HRD	100 kgf	120° diamond cone	
E	HRE	100 kgf	$^1/_8$-inch-diameter (3.175 mm) steel sphere	
F	HRF	60 kgf	$^1/_{16}$-inch-diameter (1.588 mm) steel sphere	
G	HRG	150 kgf	$^1/_{16}$-inch-diameter (1.588 mm) steel sphere	
†Also called a *brale indenter*				

The *superficial* Rockwell scales use lower loads and shallower impressions on brittle and very thin materials. The 45N scale employs a 45-kgf load on a diamond cone-shaped Brale indenter, and can be used on dense ceramics. The 15T scale employs a 15-kgf load on a ^1D $_{16}$-inch-diameter (1.588 mm) hardened steel ball, and can be used on sheet metal.

Readings below HRC 20 are generally considered unreliable, as are readings much above HRB 100.

Typical Values

- Very hard steel (e.g. chisels, quality knife blades): HRC 55–66 (Hardened High Speed Carbon and Tool Steels such as M2, W2, O1, CPM-M4, and D2, as well as many of the newer powder metallurgy Stainless Steels such as S30V, 154CM, ZDP-189, etc.)
- Axes: about HRC 45-55
- Brass: HRB 55 (Low brass, UNS C24000, H01 Temper) to HRB 93 (Cartridge Brass, UNS C26000 (260 Brass), H10 Temper)

Several other scales, including the extensive A-scale, are used for specialised applications. There are special scales for measuring case-hardened specimens.

Standards

- International (ISO)
- ISO 6508-1: Metallic materials — Rockwell hardness test — Part 1: Test method (scales A, B, C, D, E, F, G, H, K, N, T)
- ISO 2039-2: Plastics — Determination of hardness — Part 2: Rockwell hardness
- US standard (ASTM International)

- ASTM E18 : Standard methods for Rockwell hardness and Rockwell superficial hardness of metallic materials

Structural Engineering

Structural engineering is a field of engineering dealing with the analysis and design of structures that support or resist loads. Structural engineering is usually considered a speciality within civil engineering, but it can also be studied in its own right.

Structural engineers are most commonly involved in the design of buildings and large nonbuilding structures but they can also be involved in the design of machinery, medical equipment, vehicles or any item where structural integrity affects the item's function or safety. Structural engineers must ensure their designs satisfy given design criteria, predicated on safety (e.g. structures must not collapse without due warning) or serviceability and performance (e.g. building sway must not cause discomfort to the occupants).

Structural engineering theory is based upon physical laws and empirical knowledge of the structural performance of different materials and geometries. Structural engineering design utilises a number of simple structural elements to build complex structural systems. Structural engineers are responsible for making creative and efficient use of funds, structural elements and materials to achieve these goals.

Buckling

In science, buckling is a mathematical instability, leading to a failure mode. Theoretically, buckling is caused by a bifurcation in the solution to the equations of static equilibrium. At a certain stage under an increasing load, further load is able to be sustained in one of two states of equilibrium: an undeformed state or a laterally-deformed state.

In practice, buckling is characterized by a sudden failure of a structural member subjected to high compressive stress, where the actual compressive stress at the point of failure is less than the ultimate compressive stresses that the material is capable of withstanding. Mathematical analysis of buckling often makes use of an axial load eccentricity that introduces a secondary bending moment, which is not a part of the primary applied forces to which the member is subjected. As an applied load is increased on a member, such as column, it will ultimately become large enough to cause the member to become unstable and is said to have buckled. Further load will cause significant and somewhat unpredictable deformations, possibly leading to complete loss of the member's load-carrying capacity. If the deformations that follow buckling are not catastrophic the member will continue to carry the load that caused it to buckle. If the buckled member is part of a larger assemblage of components such as a building, any load applied to the structure beyond that which caused the member to buckle will be redistributed within the structure.

Columns

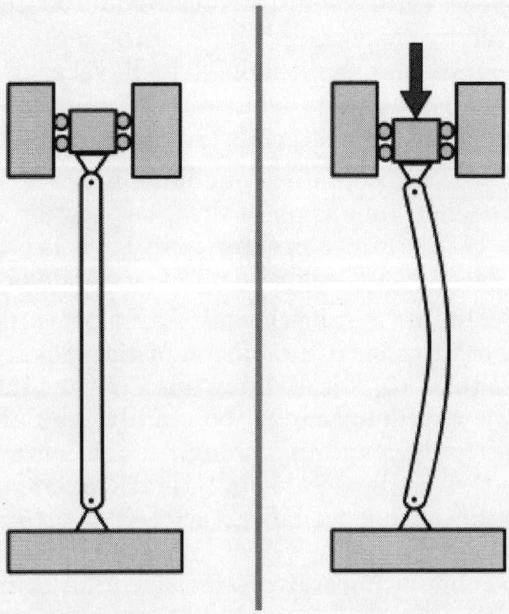

Figure: *A column under a concentric axial load exhibiting the characteristic deformation of buckling*

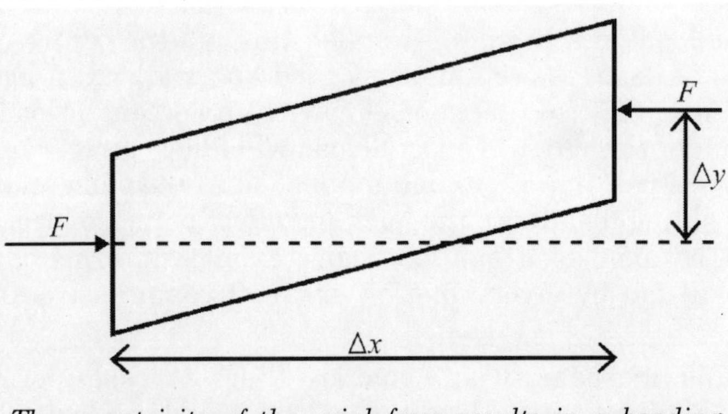

Figure: *The eccentricity of the axial force results in a bending moment acting on the beam element.*

The ratio of the effective length of a column to the least radius of gyration of its cross section is called the *slenderness ratio* (sometimes expressed with the Greek letter lambda, ë). This ratio affords a means of classifying columns. Slenderness ratio is important for design considerations. All the following are approximate values used for convenience.

- A short steel column is one whose slenderness ratio does not exceed 50; an intermediate length steel column has a slenderness ratio ranging from about 50 to 200, and are dominated by the strength limit of the material, while a long steel column may be assumed to have a slenderness ratio greater than 200 and its behaviour is dominated by the modulus of elasticity of the material.
- A short concrete column is one having a ratio of unsupported length to least dimension of the cross section equal to or less than 10. If the ratio is greater than 10, it is considered a long column (sometimes referred to as a slender column).
- Timber columns may be classified as short columns if the ratio of the length to least dimension of the cross section is equal to or less than 10. The dividing line between intermediate and long timber columns cannot be readily evaluated. One way of defining the lower limit of long timber columns would be to set it as the smallest value of the ratio of length to least cross sectional area that would just exceed a certain constant K of the material. Since K depends on the modulus of elasticity and the allowable compressive stress parallel to the grain, it can be seen that this arbitrary limit would vary with the species of the timber. The value of K is given in most structural handbooks.

If the load on a column is applied through the centre of gravity (centroid) of its cross section, it is called an axialload. A load at any other point in the cross section is known as an eccentric load. A short column under the action of an axial load will fail by direct compression before it buckles, but a long column loaded in the same manner will fail by buckling (bending), the buckling effect being so large that the effect of the axial load may be neglected. The intermediate-length column will fail by a combination of direct compressive stress and bending.

In 1757, mathematician Leonhard Euler derived a formula that gives the maximum axial load that a long, slender, ideal column can carry without buckling. An ideal column is one that is perfectly straight, homogeneous, and free from initial stress. The maximum load, sometimes called the critical load, causes the column to be in a state of unstable equilibrium; that is, the introduction of the slightest lateral force will cause the column to fail by buckling. The formula derived by Euler for columns with no consideration for lateral forces is given below. However, if lateral forces are taken into consideration the value of critical load remains approximately the same.

$$F = \frac{\pi^2 EI}{(KL)^2}$$

where

F= maximum or critical force (vertical load on column),

E= modulus of elasticity,

I= area moment of inertia,

L= unsupported length of column,

K= column effective length factor, whose value depends on the conditions of end support of the column, as follows.

For both ends pinned (hinged, free to rotate), K= 1.0.

For both ends fixed, K= 0.50.

For one end fixed and the other end pinned, K= 0.699....

For one end fixed and the other end free to move laterally, K= 2.0.

KL is the effective length of the column.

Examination of this formula reveals the following interesting facts with regard to the load-bearing ability of slender columns.

1. Elasticity and not the compressive strength of the materials of the column determines the critical load.
2. The critical load is directly proportional to the second moment of area of the cross section.
3. The boundary conditions have a considerable effect on the critical load of slender columns. The boundary conditions determine the mode of bending and the distance between inflection points on the deflected column. The inflection points in the deflection shape of the column are the points at which the curvature of the column change sign and are also the points at which the internal bending moments are zero. The closer together the inflection points are, the higher the resulting capacity of the column.

The strength of a column may therefore be increased by distributing the material so as to increase the moment of inertia. This can be done without increasing the weight of the column by distributing the material as far from the principal axis of the cross section as possible, while keeping the material thick enough to prevent local buckling. This bears out the well-known fact that a tubular section is much more efficient than a solid section for column service.

Another bit of information that may be gleaned from this equation is the effect of length on critical load. For a given size column, doubling the unsupported length quarters the allowable load. The restraint offered by the end connections of a column also affects the critical load. If the connections are perfectly rigid, the critical load will be four times that for a similar column where there is no resistance to rotation (in which case the column is idealized as having hinges at the ends).

Since the radius of gyration is defined as the square root of the ratio of the column's cross section to its area, the above formula may be rearranged as follows. Using the Euler formula for hinged ends, and substituting $A \cdot r^2$ for I, the following formula results.

$$\sigma = \frac{F}{A} = \frac{\pi^2 E}{(\ell / r)^2}$$

where F / A is the allowable stress of the column, and l / r is the slenderness ratio.

Since structural columns are commonly of intermediate length, and it is impossible to obtain an ideal column, the Euler formula on its own has little practical application for ordinary design. Issues that cause deviation from the pure Euler column behaviour include imperfections in geometry in combination with plasticity/non-linear stress strain behaviour of the column's material. Consequently, a number of empirical column formulae have been developed to agree with test data, all of which embody the slenderness ratio. For design, appropriate safety factors are introduced into these formulae. One such formula is the Perry Robertson formula which estimates the critical buckling load based on an initial (small) curvature. The Rankine Gordon formula (Named for William John Macquorn Rankine and Perry Hugesworth Gordon (1899 – 1966)) is also based on experimental results and suggests that a column will buckle at a load F_{max} given by:

$$\frac{1}{F_{max}} = \frac{1}{F_e} + \frac{1}{F_c}$$

where F_e is the Euler maximum load and F_c is the maximum compressive load. This formula typically produces a conservative estimate of F_{max}.

Self-Buckling

A free-standing, vertical column, with density ρ, Young's modulus *E*, and radius *r*, will buckle under its own weight if its height exceeds a certain critical height:

$$h_{crit} = \left(\frac{9B^2}{4} \frac{EI}{\rho g \pi r^2} \right)^{1/3}$$

where*g* is the acceleration due to gravity, *I* is the second moment of area of the beam cross section, and *B* is the first zero of the Bessel function of the first kind of order -1/3, which is equal to 1.86635086...

Buckling Under Tensile Dead Loading

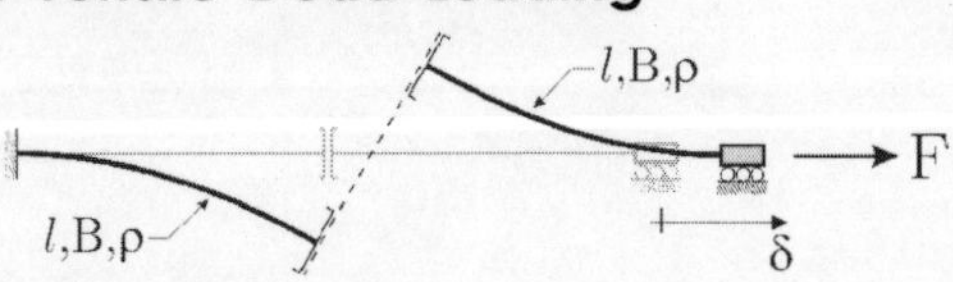

$$F_{cr} = 0.145 \frac{\pi^2 B}{l^2}$$

Figure: *Elastic beam system showing buckling under tensile dead loading.*

Usually buckling and instability are associated to compression, but recently Zaccaria, Bigoni, Noselli and Misseroni (2011) have shown that buckling and instability can also occur in elastic structures subject to dead tensile load.

An example of a single-degree-of-freedom structure is shown, where the critical load is also indicated. Another example involving flexure of a structure made up of beam elements governed by the equation of the Euler's elastica is shown. In both cases, there are no elements subject to compression.

The instability and buckling in tension are related to the presence of the slider, the junction between the two rods, allowing only relative sliding between the connected pieces.

Constraints, Curvature and Multiple Buckling

Buckling of an elastic structure strongly depends on the curvature of the constraints against which the ends of the structure are prescribed to move. In fact, even a single-degree-of-freedom system may exhibit a tensile (or a compressive) buckling load as related to the fact that one end has to move along the circular profile labelled 'Ct' (labelled 'Cc').

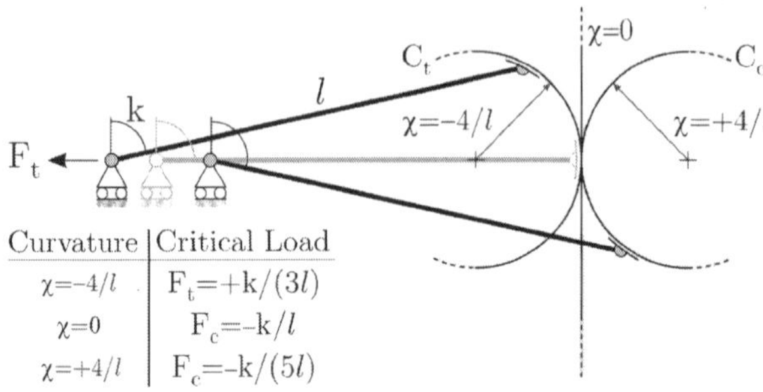

Figure: *A one-degree-of-freedom structure exhibiting a tensile (compressive) buckling load as related to the fact that the right end has to move along the circular profile labelled 'Ct' (labelled 'Cc').*

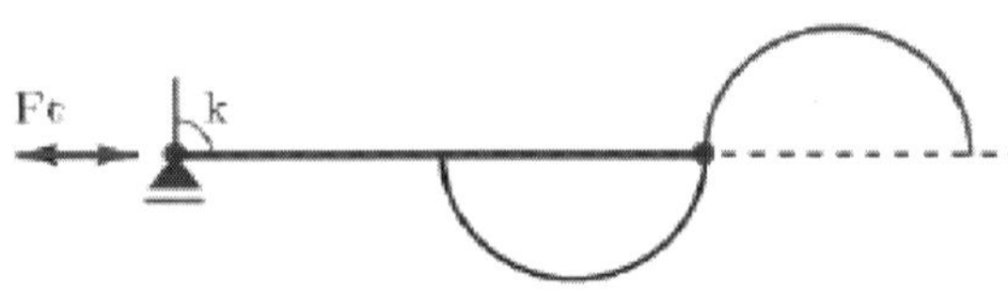

Figure: *4: A one-degree-of-freedom structure with a 'S'-shaped bicircular profile exhibiting multiple bifurcations (both tensile and compressive).*

The two circular profiles can be arranged in a 'S'-shaped profile; in that case a discontinuity of the constraint's curvature is introduced, leading to multiple bifurcations. Note that the single-degree-of-freedom structure shown in Figure depicted earlier, has two buckling loads (one tensile and one compressive). Watch a movie for more details.

Flutter Instability

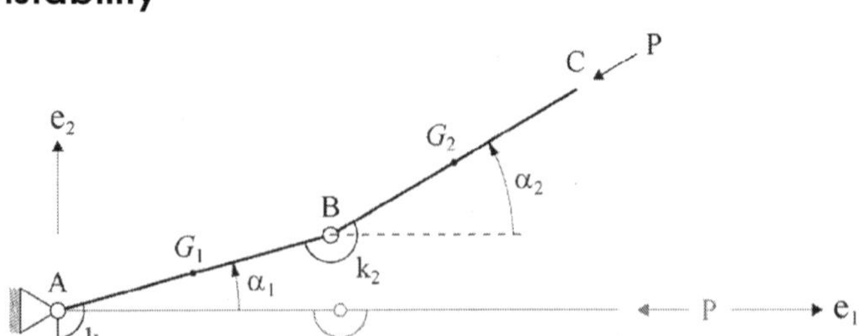

Figure: *5: A sketch of the 'Ziegler column', a two-degree-of-freedom system subject to a follower load (the force P remains always parallel to the rod BC), exhibiting flutter and divergence instability. The two rods, of linear mass density ρ, are rigid and connected through two rotational springs of stiffness k1 and k2.*

Structures subject to a follower (nonconservative) load may suffer instabilities which are not of the buckling type and therefore are not detectable with a static approach. For instance, the so-called 'Ziegler column' is shown.

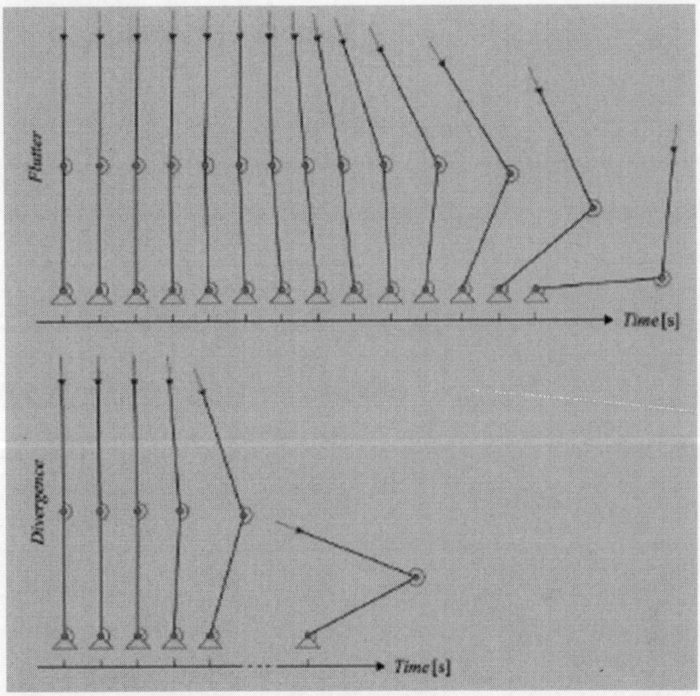

Figure: *A sequence of deformed shapes at consecutive times intervals of the structure sketched in Figure given earlier, and exhibiting flutter (upper part) and divergence (lower part) instability.*

This two-degree-of-freedom system does not display a quasi-static buckling, but becomes dynamically unstable. To see this, we note that the equations of motion are

$$\begin{cases} \frac{1}{3}\rho l_1^2(l_1+3l_2)\ddot{\alpha}_1+\frac{1}{2}\rho l_1 l_2^2\cos(\alpha_1-\alpha_2)\ddot{\alpha}_2+\frac{1}{2}\rho l_1 l_2^2\sin(\alpha_1-\alpha_2)\dot{\alpha}_2^2+(k_1+k_2)\alpha_1-k_2\alpha_2+ \\ [5mm]+(\beta_1+\beta_2)\dot{\alpha}_1-\beta_2\dot{\alpha}_2-l_1P\sin(\alpha_1-\alpha_2)=0, \\ [5mm]\frac{1}{2}\rho l_1 l_2^2\cos(\alpha_1-\alpha_2)\ddot{\alpha}_1+\frac{1}{3}\rho l_2^3\ddot{\alpha}_2-\frac{1}{2}\rho l_1 l_2^2\sin(\alpha_1-\alpha_2)\dot{\alpha}_1^2-k_2(\alpha_1-\alpha_2)-\beta_2(\dot{\alpha}_1-\dot{\alpha}_2)=0, \end{cases}$$

and their linearized version is

$$\begin{cases} \frac{1}{3}\rho l_1^2(l_1+3l_2)\ddot{\alpha}_1+\frac{1}{2}\rho l_1 l_2^2\ddot{\alpha}_2+(k_1+k_2)\alpha_1-k_2\alpha_2-l_1P(\alpha_1-\alpha_2)=0, \\ [5mm]\frac{1}{2}\rho l_1 l_2^2\ddot{\alpha}_1+\frac{1}{3}\rho l_2^3\ddot{\alpha}_2-k_2(\alpha_1-\alpha_2)=0. \end{cases}$$

Assuming a time-harmonic solution in the form

$$\alpha_j = A_j e^{-i\Omega t}, \quad j = 1, 2,$$

we find the critical loads for flutter (P_f) and divergence (P_d),

$$P_{f,d} = \frac{k_2}{l_1} \cdot \frac{k + (1+\lambda)^3 \mp \lambda\sqrt{k(3+4\lambda)}}{1 + 3\lambda/2}$$

where $\lambda = l_1 / l_2$ and $k = k_1 / k_2$.

Flutter instability corresponds to a vibrational motion of increasing amplitude and is shown in Figure (upper part) together with the divergence instability (lower part) consisting in an exponential growth.

Recently, Bigoni and Noselli (2011) have experimentally shown that flutter and divergence instabilities can be directly related to dry friction, watch the movie for more details.

Various Forms of Buckling

Buckling is a state which defines a point where an equilibrium configuration becomes unstable and can manifest itself in several different phenomena. All can be classified as forms of bifurcation.

There are four basic forms of bifurcation associated with loss of structural stability or buckling. These comprise two types of pitchfork bifurcation, one saddle-node bifurcation (often referred to as a limit point) and one transcritical bifurcation. The pitchfork bifurcations are the most commonly studied forms and include the buckling of columns and struts, sometimes known as Euler buckling; the buckling of plates, sometimes known as local buckling, which is well known to be relatively safe and the buckling of shells, which is well-known to be a highly dangerous.

In Euler buckling, the applied load is increased by a small amount beyond the critical load, the structure deforms into a buckled configuration which is adjacent to the original configuration. For example, the Euler column pictured will start to bow when loaded slightly above its critical load, but will not suddenly collapse.

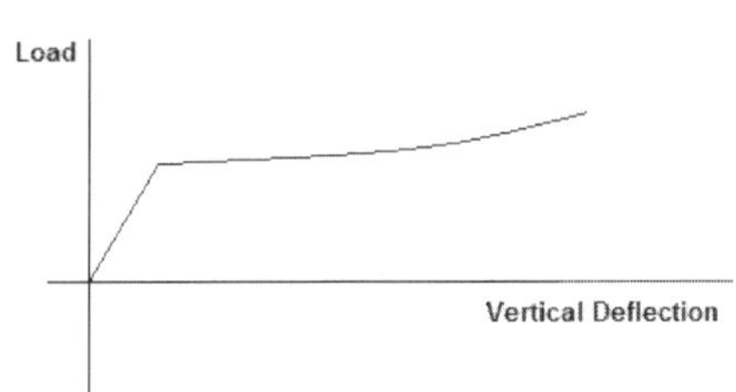

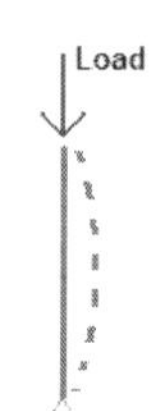

In structures experiencing limit point instability, if the load is increased infinitesimally beyond the critical load, the structure undergoes a large deformation into a different stable configuration which is not adjacent to the original configuration. An example of this type of buckling is a toggle frame (pictured) which 'snaps' into its buckled configuration.

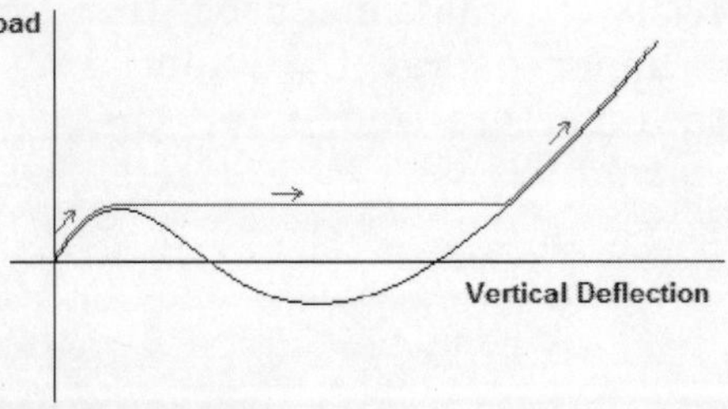

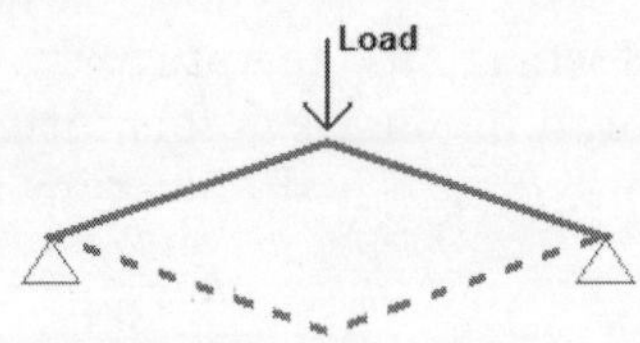

Bicycle Wheels

A conventional bicycle wheel consists of a thin rim kept under high compressive stress by the (roughly normal) inward pull of a large number of spokes. It can be considered as a loaded column that has been bent into a circle. If spoke tension is increased beyond a safe level, the wheel spontaneously fails into a characteristic saddle shape (sometimes called a "taco" or a "pringle") like a three-dimensional Euler column. This is normally a purely elastic deformation and the rim will resume its proper plane shape if spoke tension is reduced slightly.

Surface Materials

Buckling is also a failure mode in pavement materials, primarily with concrete, since asphalt is more flexible. Radiant heat from the sun is absorbed in the road surface, causing it to expand, forcing adjacent pieces to push against each other. If the stress is great enough, the pavement can lift up and crack without warning. Going over a buckled section can be very jarring to automobile drivers, described as running over a speed hump at highway speeds.

Similarly, rail tracks also expand when heated, and can fail by buckling, a phenomenon called sun kink. It is more common for rails to move laterally, often pulling the underlain railroad ties (sleepers) along.

Energy Method

Often it is very difficult to determine the exact buckling load in complex structures using the Euler formula, due to the difficulty in deciding the constant K. Therefore, maximum buckling load often is approximated using energy conservation. This way of calculating the maximum buckling load is often referred to as the energy method in structural analysis. The first step in this method is to suggest a displacement function. This function must satisfy the most important boundary conditions, such as displacement and rotation. The more accurate the displacement function, the more accurate the result.

In this method, there are two equations used (for small deformations) to approximate the "inner" energy (the potential energy stored in elastic deformation of the structure) and "outer" energy (the work done on the system by external forces).

$$U_{\text{inner}} = \frac{E}{2} \int I(x)(w_{xx}(x))^2 \, dx$$

$$U_{\text{outer}} = \frac{P_{\text{Crit}}}{2} \int (w_x(x))^2 \, dx$$

where $w(x)$ is the displacement function and the subscripts x and xx refer to the first and second derivatives of the displacement. Energy conservation yields:

$$U_{\text{Inner}} = U_{\text{Outer}}$$

Flexural-Torsional Buckling

Occurs in compression members only and it can be described as a combination of bending and twisting of a member. And it must be considered for design purposes, since the shape and cross sections are very critical. This mostly occurs in channels, structural tees, double-angle shapes, and equal-leg single angles.

Lateral-Torsional Buckling

When a simply supported beam is loaded in flexure, the top side is in compression, and the bottom side is in tension. When a slender member is subjected to an axial force, failure takes place due to bending or torsion rather than direct compression of the material.

If the beam is not supported in the lateral direction (i.e., perpendicular to the plane of bending), and the flexural load increases to a critical limit, the beam will fail due to lateral buckling of the compression flange. In wide-flange sections, if the compression flange buckles laterally, the cross section will also twist in torsion, resulting in a failure mode known as *lateral-torsional buckling*.

The Modification Factor (C_b)

C_b is a modification factor used in the equation for nominal flexural strength when determining lateral-torsional buckling. The reason for this factor is to allow for non-uniform moment diagrams when the ends of a beam segment are braced. The conservative value for C_b can be taken as 1, regardless of beam configuration or loading, but in some cases it may be excessively conservative. C_b is always equal to or greater than 1, never less. For cantilevers or overhangs where the free end is unbraced, C_b is equal to 1.

If an appropriate value of C_b is not given in the table, it can be obtained via the following formula:

$$C_b = \frac{12.5M_{\max}}{2.5M_{\max} + 3M_A + 4M_B + 3M_C}$$

where

$M_{\max}$ = absolute value of maximum moment in the unbraced segment, (kip-in)

M_A = absolute value of maximum moment at quarter point of the unbraced segment, (kip-in)

M_B = absolute value of maximum moment at centreline of the unbraced segment, (kip-in)

M_C = absolute value of maximum moment at three-quarter point of the unbraced segment, (kip-in)

Plastic Buckling

Buckling will generally occur slightly before the calculated elastic buckling strength of a structure, due to non-linear behaviour of the material. When the compressive load is near the buckling load, the structure will bow significantly and the material of the column will diverge from a linear stress-strain behaviour.

The stress-strain behaviour of materials is not strictly linear even below yield, and the modulus of elasticity decreases as stress increases, and significantly so as the stresses approach the yield strength. This lower rigidity reduces the buckling strength of the structure and causes at a load less than that predicted by the assumption of lineal elastic behaviour.

A more accurate approximation of the buckling load can be had by the use of the tangent modulus of elasticity, E_t, in place of the elastic modulus of elasticity.

The tangent modulus is a line drawn tangent to the stress-strain curve at a particular value of strain. Plots of the tangent modulus of elasticity for a variety of materials are available in standard references.

Dynamic Buckling

If a column is loaded suddenly and then the load released, the column can sustain a much higher load than its static (slowly applied) buckling load. This can happen in a long, unsupported column (rod) used as a drop hammer. The duration of compression at the impact end is the time required for a stress wave to travel up the rod to the other (free) end and back down as a relief wave. Maximum buckling occurs near the impact end at a wavelength much shorter than the length of the rod, and at a stress many times the buckling stress of a statically-loaded column. The critical condition for buckling amplitude to remain less than about 25 times the effective rod straightness imperfection at the buckle wavelength is

$$\sigma L = \rho c^2 h$$

where σ is the impact stress, L is the length of the rod, c is the elastic wave speed, and h is the smaller lateral dimension of a rectangular rod. Because the buckle wavelength depends only on σ and h, this same formula holds for thin cylindrical shells of thickness h.

Buckling of Thin Cylindrical Shells Subject to Axial Loads

Solutions of Donnell's eight order differential equation gives the various buckling modes of a thin cylinder under compression. But this analysis, which is in accordance with the small deflection theory gives much higher values than shown from experiments. So it is customary to find the critical buckling load for various structures which are cylindrical in shape from pre-existing design curves where critical buckling load F_{cr} is plotted against the ratio R/t, where R is the radius and t is the thickness of the cylinder for various values of L/R, L the length of the cylinder. If cut-outs are present in the cylinder, critical buckling loads as well as pre-buckling modes will be affected. Presence or absence of reinforcements of cut-outs will also affect the buckling load.

Buckling of Pipes and Pressure Vessels Subject to External Overpressure

Pipes and pressure vessels subject to external overpressure, caused for example by steam cooling within the pipe and condensing into water with subsequent massive pressure drop, risk buckling due to compressive hoop stresses. Design rules for calculation of the required wall thickness or reinforcement rings are given in various piping and pressure vessel codes.

Corrosion

Corrosion is the gradual destruction of materials, (usually metals), by chemical reaction with its environment. In the most common use of the word, this means electrochemical oxidation of metals in reaction with an oxidant such as oxygen. Rusting, the formation of ironoxides, is a well-known example of electrochemical corrosion. This type of damage typically produces oxide(s) or salt(s) of the original metal. Corrosion can also occur in materials other than metals, such as ceramics or polymers, although in this context, the term degradation is more common. Corrosion degrades the useful properties of materials and structures including strength, appearance and permeability to liquids and gases.

Many structural alloys corrode merely from exposure to moisture in air, but the process can be strongly affected by exposure to certain substances. Corrosion can be concentrated locally to form a pit or crack, or it can extend across a wide area more or less uniformly corroding the surface. Because corrosion is a diffusion-controlled process, it occurs on exposed surfaces. As a result, methods to reduce the activity of the exposed surface, such as passivation and chromate conversion, can increase a material's corrosion resistance. However, some corrosion mechanisms are less visible and less predictable.

Galvanic Corrosion

Figure: *Galvanic corrosion of aluminium. A 5-mm-thick Al alloy plate is physically (and hence, electrically) connected to a 10-mm-thick mild steel structural support. Galvanic corrosion occurred on the Al plate along the joint with the steel. Perforation of Al plate occurred within 2 years.*

Galvanic corrosion occurs when two different metals have physical or electrical contact with each other and are immersed in a common electrolyte, or when the same metal is exposed to electrolyte with different concentrations. In a galvanic couple, the more active metal (the anode) corrodes at an accelerated rate and the more noble metal (the cathode) corrodes at a retarded rate.

When immersed separately, each metal corrodes at its own rate. What type of metal(s) to use is readily determined by following the galvanic series. For example, zinc is often used as a sacrificial anode for steel structures. Galvanic corrosion is of major interest to the marine industry and also anywhere water (containing salts) contacts pipes or metal structures.

Factors such as relative size of anode, types of metal, and operating conditions (temperature, humidity, salinity, etc.) affect galvanic corrosion. The surface area ratio of the anode and cathode directly affects the corrosion rates of the materials. Galvanic corrosion is often prevented by the use of sacrificial anodes.

Galvanic Series

In a given environment (one standard medium is aerated, room-temperature seawater), one metal will be either more *noble* or more *active* than others, based on how strongly its ions are bound to the surface. Two metals in electrical contact share the same electrons, so that the "tug-of-war" at each surface is analogous to competition for free electrons between the two materials. Using the electrolyte as a host for the flow of ions in the same direction, the noble metal will take electrons from the active one.

The resulting mass flow or electrical current can be measured to establish a hierarchy of materials in the medium of interest. This hierarchy is called a *galvanic series* and is useful in predicting and understanding corrosion. This method is expensive but offers maximum protection against corrosion.

Corrosion Removal

Often it is possible to chemically remove the products of corrosion. For example phosphoric acid in the form of naval jelly is often applied to ferrous tools or surfaces to remove rust. Corrosion removal should not be confused with electropolishing, which removes some layers of the underlying metal to make a smooth surface. For example, phosphoric acid may also be used to electropolish copper but it does this by removing copper, not the products of copper corrosion.

Resistance to Corrosion

Some metals are more intrinsically resistant to corrosion than others. There are various ways of protecting metals from corrosion (oxidation) including painting, hot dip galvanizing, and combinations of these.

Intrinsic Chemistry

The materials most resistant to corrosion are those for which corrosion is thermodynamically unfavourable. Any corrosion products of gold or platinum tend to decompose spontaneously into pure metal, which is why these elements can be found in metallic form on Earth and have long been valued. More common "base" metals can only be protected by more temporary means.

Some metals have naturally slow reaction kinetics, even though their corrosion is thermodynamically favourable. These include such metals as zinc, magnesium, and cadmium. While corrosion of these metals is continuous and ongoing, it happens at an acceptably slow rate. An extreme example is graphite, which releases large amounts of energy upon oxidation, but has such slow kinetics that it is effectively immune to electrochemical corrosion under normal conditions.

Passivation

Passivation refers to the spontaneous formation of an ultrathin film of corrosion products known as passive film, on the metal's surface that act as a barrier to further oxidation. The chemical composition and microstructure of a passive film are different from the underlying metal. Typical passive film thickness on aluminium, stainless steels and alloys is within 10 nanometres. The passive film is different from oxide layers that are formed upon heating and are in the micrometre thickness range – the passive film recovers if removed or damaged whereas the oxide layer does not. Passivation in natural environments such as air, water and soil at moderate pH is seen in such materials as aluminium, stainless steel, titanium, and silicon.

Passivation is primarily determined by metallurgical and environmental factors. The effect of pH is summarized using Pourbaix diagrams, but many other factors are influential. Some conditions that inhibit passivation include high pH for aluminium and zinc, low pH or the presence of chloride ions for stainless steel, high temperature for titanium (in which case the oxide dissolves into the metal, rather than the electrolyte) and fluoride ions for silicon. On the other hand, unusual conditions may result in passivation of materials that are normally unprotected, as the alkaline environment of concrete does for steelrebar. Exposure to a liquid metal such as mercury or hot solder can often circumvent passivation mechanisms. Passivation is primarily determined by metallurgical and environmental factors.

Corrosion in Passivated Materials

Passivation is extremely useful in mitigating corrosion damage, however even a high-quality alloy will corrode if its ability to form a passivating film is hindered. Proper selection of the right grade of material for the specific environment is important for the long-lasting performance of this group of materials.

If breakdown occurs in the passive film due to chemical or mechanical factors, the resulting major modes of corrosion may include pitting corrosion, crevice corrosion and stress corrosion cracking.

Pitting Corrosion

Certain conditions, such as low concentrations of oxygen or high concentrations of species such as chloride which complete as anions, can interfere with a given alloy's ability to re-form a passivating film.

In the worst case, almost all of the surface will remain protected, but tiny local fluctuations will degrade the oxide film in a few critical points. Corrosion at these points will be greatly amplified, and can cause *corrosion pits* of several types, depending upon conditions.

While the corrosion pits only nucleate under fairly extreme circumstances, they can continue to grow even when conditions return to normal, since the interior of a pit is naturally deprived of oxygen and locally the pH decreases to very low values and the corrosion rate increases due to an autocatalytic process.

In extreme cases, the sharp tips of extremely long and narrow corrosion pits can cause stress concentration to the point that otherwise tough alloys can shatter; a thin film pierced by an invisibly small hole can hide a thumb sized pit from view.

These problems are especially dangerous because they are difficult to detect before a part or structure fails. Pitting remains among the most common and damaging forms of corrosion in passivate dalloys, but it can be prevented by control of the alloy's environment.

Weld Decay and Knifeline Attack

Stainless steel can pose special corrosion challenges, since its passivating behaviour relies on the presence of a major alloying component (chromium, at least 11.5%). Because of the elevated temperatures of welding and heat treatment, chromium carbides can form in the grain boundaries of stainless alloys. This chemical reaction robs the material of chromium in the zone near the grain boundary, making those areas much less resistant to corrosion. This creates a galvanic couple with the well-protected alloy nearby, which leads to *weld decay* (corrosion of the grain boundaries in the heat affected zones) in highly corrosive environments.

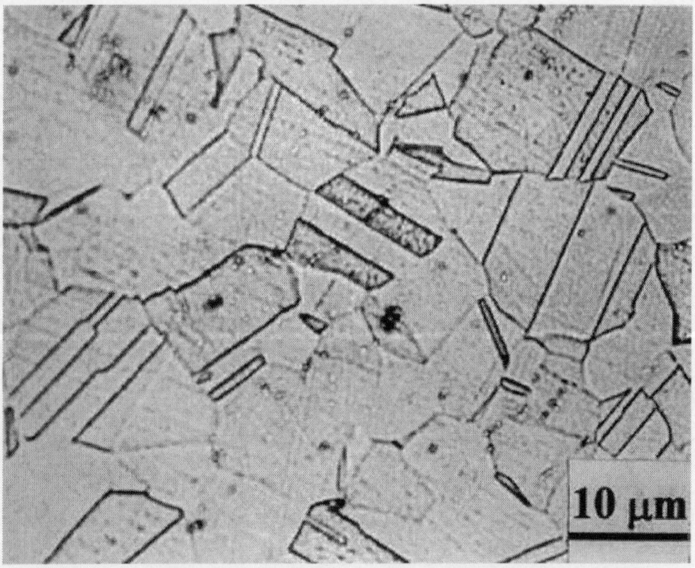

Figure: *Normal microstructure*

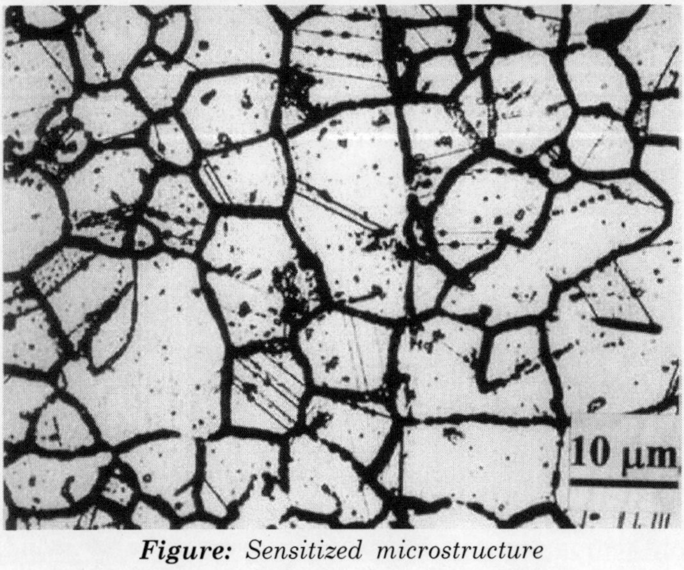

Figure: *Sensitized microstructure*

A stainless steel is said to be sensitized if chromium carbides are formed in the microstructure. A typical microstructure of a normalized type-304 stainless steel shows no signs of sensitization while a heavily sensitized steel shows the presence of grain boundary precipitates. The dark lines in the sensitized microstructure are networks of chromium carbides formed along the grain boundaries.

Special alloys, either with low carbon content or with added carbon "getters" such as titanium and niobium (in types 321 and 347, respectively), can prevent this effect, but the latter require special heat treatment after welding to prevent the similar phenomenon of *knifeline attack*. As its name implies, corrosion is limited to a very narrow zone adjacent to the weld, often only a few micrometres across, making it even less noticeable.

Crevice Corrosion

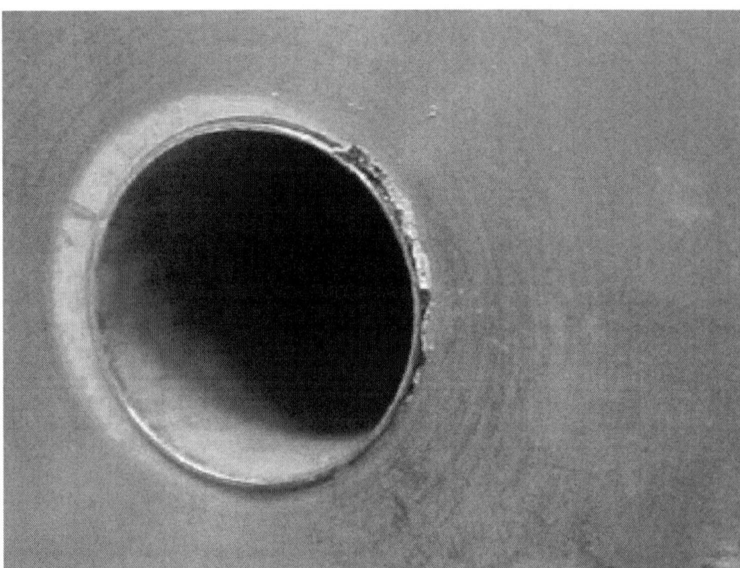

Figure: *Corrosion in the crevice between the tube and tube sheet (both made of type-316 stainless steel) of a heat exchanger in a seawater desalination plant.*

Crevice corrosion is a localized form of corrosion occurring in confined spaces (crevices), to which the access of the working fluid from the environment is limited. Formation of a differential aeration cell leads to corrosion inside the crevices. Examples of crevices are gaps and contact areas between parts, under gaskets or seals, inside cracks and seams, spaces filled with deposits and under sludge piles.

Crevice corrosion is influenced by the crevice type (metal-metal, metal-nonmetal), crevice geometry (size, surface finish), and metallurgical and environmental factors. The susceptibility to crevice corrosion can be evaluated with ASTM standard procedures. A critical crevice corrosion temperature is commonly used to rank a material's resistance to crevice corrosion.

Microbial Corrosion

Microbial corrosion, or commonly known as microbiologically influenced corrosion (MIC), is a corrosion caused or promoted by microorganisms, usually chemoautotrophs. It can apply to both metallic and non-metallic materials, in the presence or absence of oxygen. Sulfate-reducing bacteria are active in the absence of oxygen (anaerobic); they produce hydrogen sulfide, causing sulfide stress cracking. In the presence of oxygen (aerobic), some bacteria may directly oxidize iron to iron oxides and hydroxides, other bacteria oxidize sulfur and produce sulfuric acid causing biogenic sulfide corrosion. Concentration cells can form in the deposits of corrosion products, leading to localized corrosion.

Accelerated low-water corrosion (ALWC) is a particularly aggressive form of MIC that affects steel piles in seawater near the low water tide mark. It is characterized by an orange sludge, which smells of hydrogen sulfide when treated with acid. Corrosion rates can be very high and design corrosion allowances can soon be exceeded leading to premature failure of the steel pile. Piles that have been coating and have cathodic protection installed at the time of construction are not susceptible to ALWC. For unprotected piles, sacrificial anodes can be installed local to the affected areas to inhibit the corrosion or a complete retrofitted sacrificial anode system can be installed. Affected areas can also be treated electrochemically by using an electrode to first produce chlorine to kill the bacteria, and then to produced a calcareous deposit, which will help shield the metal from further attack.

High-Temperature Corrosion

High-temperature corrosion is chemical deterioration of a material (typically a metal) as a result of heating. This non-galvanic form of corrosion can occur when a metal is subjected to a hot atmosphere containing oxygen, sulfur or other compounds capable of oxidizing (or assisting the oxidation of) the material concerned. For example, materials used in aerospace, power generation and even in car engines have to resist sustained periods at high temperature in which they may be exposed to an atmosphere containing potentially highly corrosive products of combustion. The products of high-temperature corrosion can potentially be turned to the advantage of the engineer. The formation of oxides on stainless steels, for example, can provide a protective layer preventing further atmospheric attack, allowing for a material to be used for sustained periods at both room and high temperatures in hostile conditions. Such high-temperature corrosion products, in the form of compacted oxide layer glazes, prevent or reduce wear during high-temperature sliding contact of metallic (or metallic and ceramic) surfaces.

Metal Dusting

Metal dusting is a catastrophic form of corrosion that occurs when susceptible materials are exposed to environments with high carbon activities, such as synthesis gas and other high-CO environments. The corrosion manifests itself as a break-up of bulk metal to metal powder. The suspected mechanism is firstly the deposition of a graphite layer on the surface of the metal, usually from carbon monoxide (CO) in the vapour phase. This graphite layer is then thought to form metastable M_3C species (where M is the metal), which migrate away from the metal surface. However, in some regimes no M_3C species is observed indicating a direct transfer of metal atoms into the graphite layer.

Protection from Corrosion

Surface Treatments:

Applied Coatings: Plating, painting, and the application of enamel are the most common anti-corrosion treatments. They work by providing a barrier of corrosion-resistant material between the damaging environment and the structural material. Aside from cosmetic and manufacturing issues, there may be tradeoffs in mechanical flexibility versus resistance to abrasion and high temperature. Platings usually fail only in small sections, but if the plating is more noble than the substrate (for example, chromium on steel), a galvanic couple will cause any exposed area to corrode much more rapidly than an unplated surface would. For this reason, it is often wise to plate with active metal such as zinc or cadmium.

Painting either by roller or brush is more desirable for tight spaces; spray would be better for larger coating areas such as steel decks and waterfront applications. Flexible polyurethane coatings, like Durabak-M26 for example, can provide an anti-corrosive seal with a highly durable slip resistant membrane. Painted coatings are relatively easy to apply and have fast drying times although temperature and humidity may cause dry times to vary.

Reactive Coatings

If the environment is controlled (especially in recirculating systems), corrosion inhibitors can often be added to it. These chemicals form an electrically insulating or chemically impermeable coating on exposed metal surfaces, to suppress electrochemical reactions. Such methods make the system less sensitive to scratches or defects in the coating, since extra inhibitors can be made available wherever metal becomes exposed. Chemicals that inhibit corrosion include some of the salts in hard water (Roman water systems are famous for their mineral deposits), chromates, phosphates, polyaniline, other conducting polymers and a wide range of specially-designed chemicals that resemble surfactants (i.e. long-chain organic molecules with ionic end groups).

Anodization

Aluminium alloys often undergo a surface treatment. Electrochemical conditions in the bath are carefully adjusted so that uniform pores, several nanometres wide, appear in the metal's oxide film. These pores allow the oxide to grow much thicker than passivating conditions would allow. At the end of the treatment, the pores are allowed to seal, forming a harder-than-usual surface layer. If this coating is scratched, normal passivation processes take over to protect the damaged area.

Anodizing is very resilient to weathering and corrosion, so it is commonly used for building facades and other areas that the surface will come into regular contact with the elements. Whilst being resilient, it must be cleaned frequently. If left without cleaning, panel edge staining will naturally occur.

Biofilm Coatings

A new form of protection has been developed by applying certain species of bacterial films to the surface of metals in highly corrosive environments. This process increases the corrosion resistance substantially. Alternatively, antimicrobial-producing biofilms can be used to inhibit mild steel corrosion from sulfate-reducing bacteria.

Controlled Permeability Formwork

Controlled permeability formwork (CPF) is a method of preventing the corrosion of reinforcement by naturally enhancing the durability of the cover during concrete placement. CPF has been used in environments to combat the effects of carbonation, chlorides, frost and abrasion.

Cathodic Protection

Cathodic protection (CP) is a technique to control the corrosion of a metal surface by making that surface the cathode of an electrochemical cell. Cathodic protection systems are most commonly used to protect steel, water, and fuel pipelines and tanks; steel pier piles, ships, and offshore oil platforms.

Sacrificial Anode Protection

For effective CP, the potential of the steel surface is polarized (pushed) more negative until the metal surface has a uniform potential. With a uniform potential, the driving force for the corrosion reaction is halted. For galvanic CP systems, the anode material corrodes under the influence of the steel, and eventually it must be replaced. The polarization is caused by the current flow from the anode to the cathode, driven by the difference in electrochemical potential between the anode and the cathode.

Impressed Current Cathodic Protection

For larger structures, galvanic anodes cannot economically deliver enough current to provide complete protection. Impressed current cathodic protection (ICCP) systems use anodes connected to a DC power source (such as a cathodic protection rectifier). Anodes for ICCP systems are tubular and solid rod shapes of various specialised materials. These include high silicon cast iron, graphite, mixed metal oxide or platinum coated titanium or niobium coated rod and wires.

Anodic Protection

Anodic protection impresses anodic current on the structure to be protected (opposite to the cathodic protection). It is appropriate for metals that exhibit passivity (e.g., stainless steel) and suitably small passive current over a wide range of potentials. It is used in aggressive environments, e.g., solutions of sulfuric acid.

Rate of Corrosion

A simple test for measuring corrosion is the weight loss method. The method involves exposing a clean weighed piece of the metal or alloy to the corrosive environment for a specified time followed by cleaning to remove corrosion products and weighing the piece to determine the loss of weight.

The rate of corrosion (R) is calculated as

$$R = KW/(\rho At)$$

where k is a constant, W is the weight loss of the metal in time t, A is the surface area of the metal exposed, and ρ is the density of the metal (in g/cm^3).

Economic Impact

In 2002, the US Federal Highway Administration released a study titled *Corrosion Costs and Preventive Strategies in the United States* on the direct costs associated with metallic corrosion in the U.S. industry. In 1998, the total annual direct cost of corrosion in the U.S. was ca. $276 billion (ca. 3.2% of the US gross domestic product).

Rust is one of the most common causes of bridge accidents. As rust has a much higher volume than the originating mass of iron, its build-up can also cause failure by forcing apart adjacent parts.

It was the cause of the collapse of the Mianus river bridge in 1983, when the bearings rusted internally and pushed one corner of the road slab off its support. Three drivers on the roadway at the time died as the slab fell into the river below.

The following NTSB investigation showed that a drain in the road had been blocked for road re-surfacing, and had not been unblocked; as a result, runoff water penetrated the support hangers. Rust was also an important factor in the Silver Bridge disaster of 1967 in West Virginia, when a steel suspension bridge collapsed within a minute, killing 46 drivers and passengers on the bridge at the time.

Similarly, corrosion of concrete-covered steel and iron can cause the concrete to spall, creating severe structural problems. It is one of the most common failure modes of reinforced concretebridges. Measuring instruments based on the half-cell potential can detect the potential corrosion spots before total failure of the concrete structure is reached.

Until 20–30 years ago; galvanized steel pipe was used extensively in the potable water systems for single and multi-family residents as well as commercial and public construction.

Today, these systems have long ago consumed the protective zinc and are corroding internally resulting in poor water quality and pipe failures.

The economic impact on homeowners, condo dwellers, and the public infrastructure is estimated at 22 billion dollars as insurance industry braces for a wave of claims due to pipe failures.

Corrosion in Nonmetals

Most ceramic materials are almost entirely immune to corrosion. The strong chemical bonds that hold them together leave very little free chemical energy in the structure; they can be thought of as already corroded. When corrosion does occur, it is almost always a simple dissolution of the material or chemical reaction, rather than an electrochemical process. A common example of corrosion protection in ceramics is the lime added to soda-lime glass to reduce its solubility in water; though it is not nearly as soluble as pure sodium silicate, normal glass does form sub-microscopic flaws when exposed to moisture. Due to its brittleness, such flaws cause a dramatic reduction in the strength of a glass object during its first few hours at room temperature.

Corrosion of Polymers

Polymer degradation involves several complex and often poorly understood physiochemical processes. These are strikingly different from the other processes discussed here, and so the term "corrosion" is only applied to them in a loose sense of the word. Because of their large molecular weight, very little entropy can be gained by mixing a given mass of polymer with another substance, making them generally quite difficult to dissolve. While dissolution is a problem in some polymer applications, it is relatively simple to design against. A more common and related problem is *swelling*, where small molecules infiltrate the structure, reducing strength and stiffness and causing a volume change. Conversely, many polymers (notably flexible vinyl) are intentionally swelled with plasticizers, which can be leached out of the structure, causing brittleness or other undesirable changes. The most common form of degradation, however, is a decrease in polymer chain length. Mechanisms which break polymer chains are familiar to biologists because of their effect on DNA: ionizing radiation (most commonly ultraviolet light), free radicals, and oxidizers such as oxygen, ozone, and chlorine. Ozone cracking is a well-known problem affecting natural rubber for example. Additives can slow these process very effectively, and can be as simple as a UV-absorbing pigment (i.e., titanium dioxide or carbon black). Plastic shopping bags often do not include these additives so that they break down more easily as litter.

Corrosion of Glasses

Glass disease is the corrosion of silicate glasses in aqueous solutions. It is governed by two mechanisms: diffusion-controlled leaching (ion exchange) and hydrolytic dissolution of the glass network. Both mechanisms strongly depend on the pH of contacting solution:

the rate of ion exchange decreases with pH as $10^{-0.5pH}$ whereas the rate of hydrolytic dissolution increases with pH as $10^{0.5pH}$.

Mathematically, corrosion rates of glasses are characterized by normalized corrosion rates of elements NR_i ($g/cm^2 \cdot d$) which are determined as the ratio of total amount of released species into the water M_i (g) to the water-contacting surface area S (cm^2), time of contact t (days) and weight fraction content of the element in the glass f_i:

$$NR_i = \frac{M_i}{Sf_i t}.$$

The overall corrosion rate is a sum of contributions from both mechanisms (leaching + dissolution) $NR_i = Nrx_i + NRh$. Diffusion-controlled leaching (ion exchange) is characteristic of the initial phase of corrosion and involves replacement of alkali ions in the glass by a hydronium (H_3O^+) ion from the solution. It causes an ion-selective depletion of near surface layers of glasses and gives an inverse square root dependence of corrosion rate with exposure time. The diffusion-controlled normalized leaching rate of cations from glasses ($g/cm^2 \cdot d$) is given by:

$$NRx_i = 2\rho\sqrt{\frac{D_i}{\pi t}},$$

wheret is time, D_i is the i-thcation effective diffusion coefficient (cm^2/d), which depends on pH of contacting water as $D_i = D_{i0} \cdot 10^{-pH}$, and ñ is the density of the glass (g/cm^3).

Glass network dissolution is characteristic of the later phases of corrosion and causes a congruent release of ions into the water solution at a time-independent rate in dilute solutions ($g/cm^2 \cdot d$):

$$NRh = \rho r_h,$$

wherer $_h$ is the stationary hydrolysis (dissolution) rate of the glass (cm/d). In closed systems the consumption of protons from the aqueous phase increases the pH and causes a fast transition to hydrolysis. However, a further saturation of solution with silica impedes hydrolysis and causes the glass to return to an ion-exchange, e.g. diffusion-controlled regime of corrosion.

In typical natural conditions normalized corrosion rates of silicate glasses are very low and are of the order of 10^{-7}–10^{-5} g/($cm^2 \cdot d$). The very high durability of silicate glasses in water makes them suitable for hazardous and nuclear waste immobilisation.

Glass Corrosion Tests

There exist numerous standardized procedures for measuring the corrosion (also called chemical durability) of glasses in neutral, basic, and acidic environments, under simulated environmental conditions, in simulated body fluid, at high temperature and pressure, and under other conditions.

The standard procedure ISO 719 describes a test of the extraction of water-soluble basic compounds under neutral conditions: 2 g of glass, particle size 300–500 μm, is kept for 60 min in 50 ml de-ionized water of grade 2 at 98 °C; 25 ml of the obtained solution is titrated against 0.01 mol/l HCl solution. The volume of HCl required for neutralization is classified according to the table below.

Amount of 0.01M HCl needed to Neutralize Extracted Basic oxides, ml	*Extracted Na_2 Oequivalent, μg*	*Hydrolyticclass*
< 0.1	< 31	1
0.1-0.2	31-62	2
0.2-0.85	62-264	3
0.85-2.0	264-620	4
2.0-3.5	620-1085	5
> 3.5	> 1085	> 5

Creep (Deformation)

In materials science, creep (sometimes called cold flow) is the tendency of a solid material to move slowly or deform permanently under the influence of mechanical stresses. It can occur as a result of long-term exposure to high levels of stress that are still below the yield strength of the material. Creep is more severe in materials that are subjected to heat for long periods, and generally increases as they near their melting point. Creep always increases with temperature.

The rate of deformation is a function of the material properties, exposure time, exposure temperature and the applied structural load. Depending on the magnitude of the applied stress and its duration, the deformation may become so large that a component can no longer perform its function — for example creep of a turbine blade will cause the blade to contact the casing, resulting in the failure of the blade. Creep is usually of concern to engineers and metallurgists when evaluating components that operate under high stresses or high temperatures. Creep is a deformation mechanism that may or may not constitute a failure mode. For example, moderate creep in concrete is sometimes welcomed because it relieves tensile stresses that might otherwise lead to cracking.

Unlike brittle fracture, creep deformation does not occur suddenly upon the application of stress. Instead, strain accumulates as a result of long-term stress. Therefore, creep is a "time-dependent" deformation.

The temperature range in which creep deformation may occur differs in various materials. For example, tungsten requires a temperature in the thousands of degrees before creep deformation can occur, while ice will creep at temperatures near 0 °C (32 °F). As a general guideline, the effects of creep deformation generally become noticeable at approximately 30% of the melting point (as measured on a thermodynamic temperature scale such as kelvin or rankine) for metals, and at 40–50% of melting point for ceramics. Virtually any material will creep upon approaching its melting temperature. Since the creep minimum temperature is related to the melting point, creep can be seen at relatively low temperatures for some materials. Plastics and low-melting-temperature metals, including many solders, can begin to creep at room temperature, as can be seen markedly in old lead hot-water pipes. Glacier flow is an example of creep processes in ice.

Stages of Creep

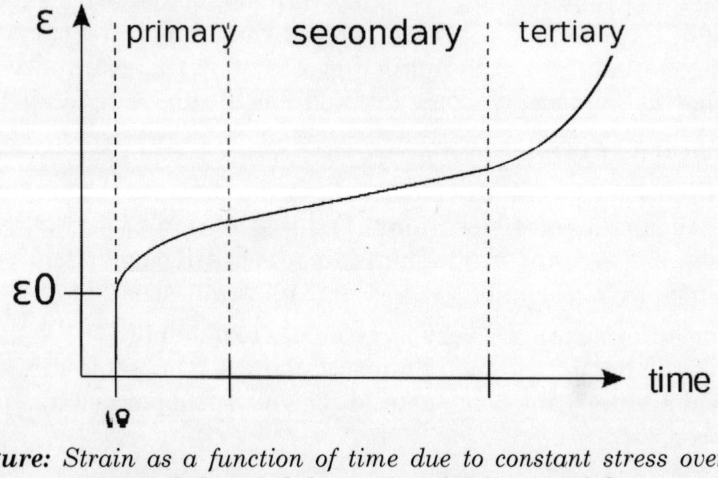

Figure: *Strain as a function of time due to constant stress over an extended period for a viscoelastic material.*

In the initial stage, or primary creep, the strain rate is relatively high, but slows with increasing time. This is due to work hardening. The strain rate eventually reaches a minimum and becomes near constant. This is due to the balance between work hardening and annealing (thermal softening). This stage is known as secondary or steady-state creep. This stage is the most understood. The characterized "creep strain rate" typically refers to the rate in this secondary stage. Stress dependence of this rate depends on the creep mechanism. In tertiary creep, the strain rate exponentially increases with stress because of necking phenomena.

Mechanisms of Creep

The mechanism of creep depends on temperature and stress. Various mechanisms are:

- Bulk diffusion (Nabarro-Herring creep)
- Climb — here the strain is actually accomplished by climb
- Climb-assisted glide — here the climb is an *enabling* mechanism, allowing dislocations to get around obstacles
- Grain boundary diffusion (Coble creep)
- Thermally activated glide — e.g., via cross-slip

General Creep Equation

$$\frac{d\varepsilon}{dt} = \frac{C\sigma^m}{d^b} e^{\frac{-Q}{kT}}$$

where ε is the creep strain, C is a constant dependent on the material and the particular creep mechanism, m and b are exponents dependent on the creep mechanism, Q is the activation energy of the creep mechanism, σ is the applied stress, d is the grain size of the material, k is Boltzmann's constant, and T is the absolute temperature.

Dislocation Creep

At high stresses (relative to the shear modulus), creep is controlled by the movement of dislocations. For dislocation creep, Q = *Q(self diffusion)*, m = 4-6, and b = 0. Therefore, dislocation creep has a strong dependence on the applied stress and no grain size dependence.

Some alloys exhibit a very large stress exponent ($n > 10$), and this has typically been explained by introducing a "threshold stress," σ_{th}, below which creep can't be measured. The modified power law equation then becomes:

$$\frac{d\epsilon}{dt} = A\left(\sigma - \sigma_{th}\right)^n e^{\frac{-Q}{\bar{R}T}}$$

where A, Q and n can all be explained by conventional mechanisms (so $3 \leq n \leq 10$).

Nabarro-Herring Creep

Nabarro-Herring creep is a form of diffusion creep. In Nabarro-Herring creep, atoms diffuse through the lattice causing grains to elongate along the stress axis; k is related to the diffusion coefficient of atoms through the lattice, $Q = Q(self\ diffusion)$, $m = 1$, and $b = 2$. Therefore Nabarro-Herring creep has a weak stress dependence and a moderate grain size dependence, with the creep rate decreasing as grain size is increased.

Nabarro-Herring creep is strongly temperature dependent. For lattice diffusion of atoms to occur in a material, neighbouring lattice sites or interstitial sites in the crystal structure must be free. A given atom must also overcome the energy barrier to move from its current site (it lies in an energetically favourable potential well) to the nearby vacant site (another potential well).

The general form of the diffusion equation is $D = D_0 exp(E/KT)$ where D_0 has a dependence on both the attempted jump frequency and the number of nearest neighbour sites and the probability of the sites being vacant. Thus there is a double dependence upon temperature. At higher temperatures the diffusivity increases due to the direct temperature dependence of the equation, the increase in vacancies through Schottky defect formation, and an increase in the average energy of atoms in the material. Nabarro-Herring creep dominates at very high temperatures relative to a material's melting temperature.

Coble Creep

Coble creep is a second form of diffusion controlled creep. In Coble creep the atoms diffuse along grain boundaries to elongate the grains along the stress axis. This causes Coble creep to have a stronger grain size dependence than Nabarro-Herring creep. For Coble creep k is related to the diffusion coefficient of atoms along the grain boundary, $Q = Q(grain\ boundary\ diffusion)$, $m = 1$, and $b = 3$. Because $Q(grain\ boundary\ diffusion) < Q(self\ diffusion)$, Coble creep occurs at lower temperatures than Nabarro-Herring creep. Coble creep is still temperature dependent, as the temperature increases so does the grain boundary diffusion.

However, since the number of nearest neighbours is effectively limited along the interface of the grains, and thermal generation of vacancies along the boundaries is less prevalent, the temperature dependence is not as strong as in Nabarro-Herring creep. It also exhibits the same linear dependence on stress as Nabarro-Herring creep.

Creep of Polymers

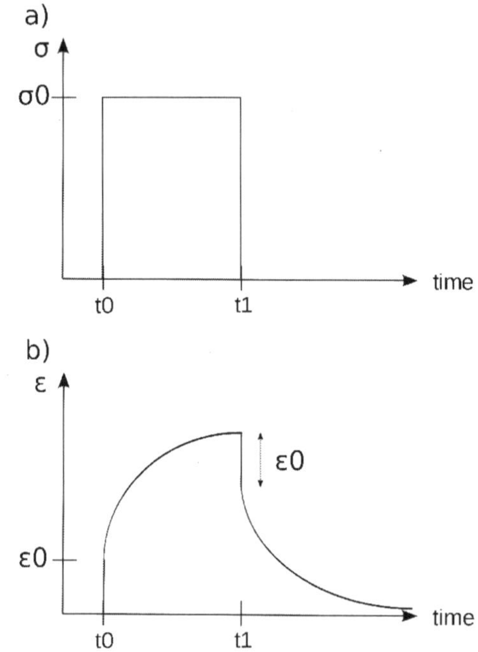

Figure: *a) Applied stress and b) induced strain as functions of time over a short period for a viscoelastic material.*

Creep can occur in polymers and metals which are considered viscoelastic materials. When a polymeric material is subjected to an abrupt force, the response can be modelled using the Kelvin-Voigt model. In this model, the material is represented by a Hookean spring and a Newtonian dashpot in parallel. The creep strain is given by the following convolution integral:

$$\varepsilon(t) = \sigma C_0 + \sigma C \int_0^{\infty} f(\tau)(1 - \exp[-t/\tau])d\tau$$

where:

- σ = applied stress
- C_0 = *instantaneous creep compliance*
- C = creep compliance coefficient
- τ= retardation time
- $f(\tau)$= distribution of retardation times

When subjected to a step constant stress, viscoelastic materials experience a time-dependent increase in strain. This phenomenon is known as viscoelastic creep.

At a time t_0, a viscoelastic material is loaded with a constant stress that is maintained for a sufficiently long time period. The material responds to the stress with a strain that increases until the material ultimately fails. When the stress is maintained for a shorter time period, the material undergoes an initial strain until a time t_1 at which the stress is relieved, at which time the strain immediately decreases (discontinuity) then continues decreasing gradually to a residual strain.

Viscoelastic creep data can be presented in one of two ways. Total strain can be plotted as a function of time for a given temperature or temperatures. Below a critical value of applied stress, a material may exhibit linear viscoelasticity. Above this critical stress, the creep rate grows disproportionately faster. The second way of graphically presenting viscoelastic creep in a material is by plotting the creep modulus (constant applied stress divided by total strain at a particular time) as a function of time. Below its critical stress, the viscoelastic creep modulus is independent of stress applied. A family of curves describing strain versus time response to various applied stress may be represented by a single viscoelastic creep modulus versus time curve if the applied stresses are below the material's critical stress value.

Additionally, the molecular weight of the polymer of interest is known to affect its creep behaviour. The effect of increasing molecular weight tends to promote secondary bonding between polymer chains and thus make the polymer more creep resistant. Similarly, aromatic polymers are even more creep resistant due to the added stiffness from the rings. Both molecular weight and aromatic rings add to polymers' thermal stability, increasing the creep resistance of a polymer.

Both polymers and metals can creep. Polymers experience significant creep at temperatures above ca. –200°C; however, there are three main differences between polymeric and metallic creep.

Polymers show creep basically in two different ways. At typical work loads (5 up to 50%) ultra high molecular weight polyethylene (Spectra, Dyneema) will show time-linear creep, whereas polyester or aramids (Twaron, Kevlar) will show a time-logarithmic creep.

Creep of Concrete

The creep of concrete, which originates from the calcium silicate hydrates (C-S-H) in the hardened Portland cement paste (which is the binder of mineral aggregates), is fundamentally different from the creep of metals as well as polymers. Unlike the creep of metals, it occurs at all stress levels and, within the service stress range, is linearly dependent on the stress if the pore water content is constant. Unlike the creep of polymers and metals, it exhibits multi-months aging, caused by chemical hardening due to hydration which stiffens the microstructure, and multi-year aging, caused by long-term relaxation of self-equilibrated micro-stresses in the nano-porous microstructure of the C-S-H.

If concrete is fully dried, it does not creep, but it is next to impossible to dry concrete fully without severe cracking.

Fatigue (Material)

In materials science, fatigue is the weakening of a material caused by repeatedly applied loads. It is the progressive and localized structural damage that occurs when a material is subjected to cyclic loading.

The nominal maximum stress values that cause such damage may be much less than the strength of the material typically quoted as the ultimate tensile stress limit, or the yield stress limit.

Fatigue occurs when a material is subjected to repeated loading and unloading. If the loads are above a certain threshold, microscopic cracks will begin to form at the stress concentrators such as the surface, persistent slip bands (PSBs), and grain interfaces. Eventually a crack will reach a critical size, the crack will propagate suddenly, and the structure will fracture.

The shape of the structure will significantly affect the fatigue life; square holes or sharp corners will lead to elevated local stresses where fatigue cracks can initiate. Round holes and smooth transitions or fillets will therefore increase the fatigue strength of the structure.

Fatigue Life

ASTM defines *fatigue life*, N_f, as the number of stress cycles of a specified character that a specimen sustains before failure of a specified nature occurs.

One method to predict fatigue life of materials is the Uniform Material Law (UML). UML was developed for fatigue life prediction of aluminum and titanium alloys by the end of 20th century and extended to high-strength steels and cast iron.

For some materials, there is a theoretical value for stress amplitude below which the material will not fail for any number of cycles, called a fatigue limit, endurance limit, or fatigue strength.

Characteristics of Fatigue

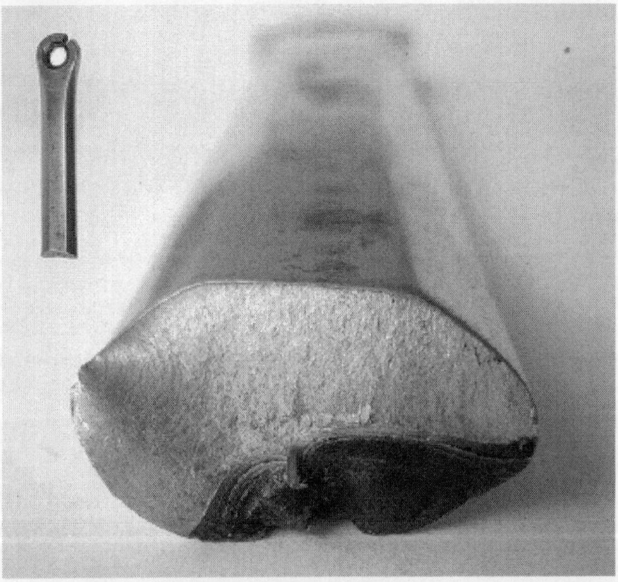

Figure: *Fracture of an aluminium crank arm. Dark area of striations: slow crack growth. Bright granular area: sudden fracture.*

- In metal alloys, when there are no macroscopic or microscopic discontinuities, the process starts with dislocation movements, which eventually form persistent slip bands that become the nucleus of short cracks.
- Macroscopic and microscopic discontinuities as well as component design features which cause stress concentrations (holes, keyways, sharp changes of direction etc.) are common locations at which the fatigue process begins.
- Fatigue is a process that has a degree of randomness (stochastic), often showing considerable scatter even in well controlled environments.
- Fatigue is usually associated with tensile stresses but fatigue cracks have been reported due to compressive loads.
- The greater the applied stress range, the shorter the life.
- Fatigue life scatter tends to increase for longer fatigue lives.
- Damage is cumulative. Materials do not recover when rested.
- Fatigue life is influenced by a variety of factors, such as temperature, surface finish, metallurgical microstructure, presence of oxidizing or inert chemicals, residual stresses, scuffing contact (fretting), etc.

- Some materials (e.g., some steel and titanium alloys) exhibit a theoretical fatigue limit below which continued loading does not lead to fatigue failure.
- In recent years, researchers (see, for example, the work of Bathias, Murakami, and Stanzl-Tschegg) have found that failures can occur below the theoretical fatigue limit at very high fatigue lives (10^9 to 10^{10} cycles). An ultrasonic resonance technique is used in these experiments with frequencies around 10–20 kHz.
- High cycle fatigue strength (about 10^4 to 10^8 cycles) can be described by stress-based parameters. A load-controlled servo-hydraulic test rig is commonly used in these tests, with frequencies of around 20–50 Hz. Other sorts of machines—like resonant magnetic machines—can also be used, to achieve frequencies up to 250 Hz.
- Low cycle fatigue (loading that typically causes failure in less than 10^4 cycles) is associated with localized plastic behaviour in metals; thus, a strain-based parameter should be used for fatigue life prediction in metals. Testing is conducted with constant strain amplitudes typically at 0.01–5 Hz.

Timeline of Early Fatigue Research History

- 1837: Wilhelm Albert publishes the first article on fatigue. He devised a test machine for conveyor chains used in the Clausthalmines.
- 1839: Jean-Victor Poncelet describes metals as being *tired* in his lectures at the military school at Metz.
- 1842: William John Macquorn Rankine recognises the importance of stress concentrations in his investigation of railroad axle failures. The Versailles train crash was caused by axle fatigue.
- 1843: Joseph Glynn reports on fatigue of axle on a locomotive tender. He identifies the keyway as the crack origin.
- 1848: The Railway Inspectorate reports one of the first tyre failures, probably from a rivet hole in tread of railway carriage wheel. It was likely a fatigue failure.
- 1849: Eaton Hodgkinson is granted a *small sum of money* to report to the UK Parliament on his work in *ascertaining by direct experiment, the effects of continued changes of load upon iron structures and to what extent they could be loaded without danger to their ultimate security*.

- 1854: Braithwaite reports on common service fatigue failures and coins the term *fatigue*.
- 1860: Systematic fatigue testing undertaken by Sir William Fairbairn and August Wöhler.
- 1870: Wöhlersummarises his work on railroad axles. He concludes that cyclic stress range is more important than peak stress and introduces the concept of *endurance limit*.
- 1903: Sir James Alfred Ewing demonstrates the origin of fatigue failure in microscopic cracks.
- 1910: O. H. Basquin proposes a log-log relationship for SN curves, using Wöhler's test data.
- 1945: A. M. MinerpopularisesA. Palmgren's (1924) linear damage hypothesis as a practical design tool.
- 1954: L. F. Coffin and S. S. Manson explain fatigue crack-growth in terms of plasticstrain in the tip of cracks.
- 1961: P. C. Paris proposes methods for predicting the rate of growth of individual fatigue cracks in the face of initial scepticism and popular defence of Miner's phenomenological approach.
- 1968: Tatsuo Endo and M. Matsuishi devise the rainflow-counting algorithm and enable the reliable application of Miner's rule to random loadings.
- 1970: W. Elber elucidates the mechanisms and importance of crack closure in slowing the growth of a fatigue crack due to the wedging effect of plastic deformation left behind the tip of the crack.

High-Cycle Fatigue

Historically, most attention has focused on situations that require more than 10^4 cycles to failure where stress is low and deformation is primarily elastic.

S-N Curve

In high-cycle fatigue situations, materials performance is commonly characterized by an *S-N curve*, also known as a *Wöhlercurve* . This is a graph of the magnitude of a cyclic stress (*S*) against the logarithmic scale of cycles to failure (*N*).

S-N curves are derived from tests on samples of the material to be characterized (often called *coupons*) where a regular sinusoidal stress is applied by a testing machine which also counts the number of cycles to failure. This process is sometimes known as *coupon testing.* Each coupon test generates a point on the plot though in some cases there is a *runout* where the time to failure exceeds that available for the test. Analysis of fatigue data requires techniques from statistics, especially survival analysis and linear regression.

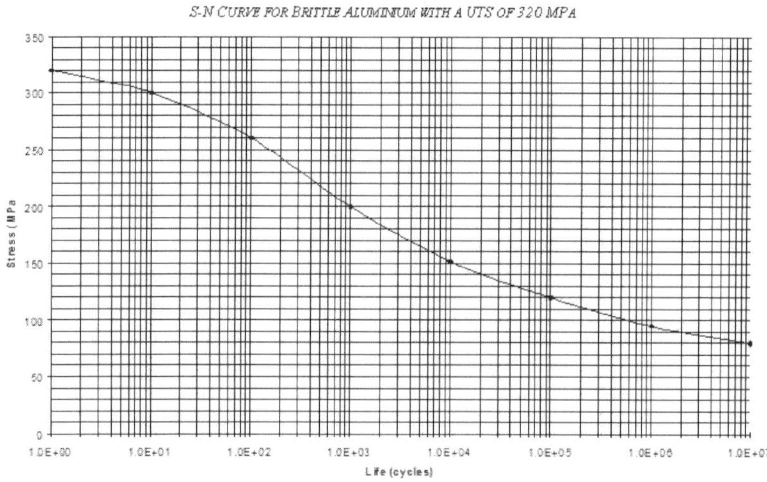

The progression of the *S-N curve* can be influenced by many factors such as corrosion, temperature, residual stresses, and the presence of notches. The Goodman-Line is a method to estimate the influence of the mean stress on the fatigue strength.

Probabilistic Nature of Fatigue

As coupons sampled from a homogeneous frame will display a variation in their number of cycles to failure, the S-N curve should more properly be an *S-N-P curve* capturing the probability of failure after a given number of cycles of a certain stress. Probability distributions that are common in data analysis and in design against fatigue include the log-normal distribution, extreme value distribution, Birnbaum–Saunders distribution, and Weibull distribution.

Complex Loadings

In practice, a mechanical part is exposed to a complex, often random, sequence of loads, large and small. In order to assess the safe life of such a part:

1. Reduce the complex loading to a series of simple cyclic loadings using a technique such as rainflow analysis;
2. Create a histogram of cyclic stress from the rainflow analysis to form a fatigue damage spectrum;

3. For each stress level, calculate the degree of cumulative damage incurred from the S-N curve; and
4. Combine the individual contributions using an algorithm such as *Miner's rule.*

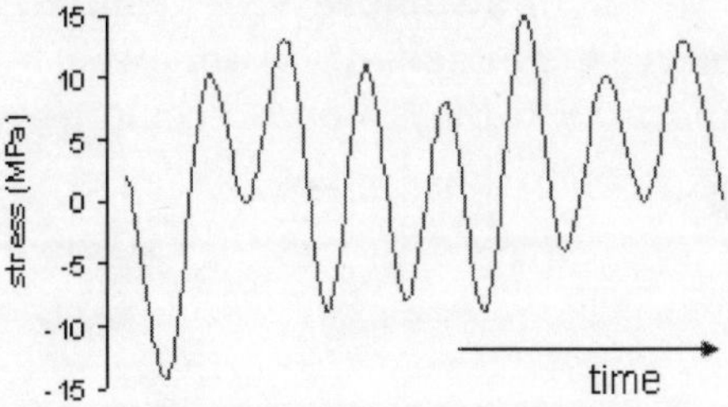

Figure: *Spectrum loading*

For Multiaxial Loading

Since S-N curves are typically generated for uniaxial loading, some equivalence rule is needed whenever the loading is multiaxial. For simple, proportional loading histories, Sines rule may be applied. For more complex situations, such as nonproportional loading, Critical plane analysis must be applied.

Miner's Rule

In 1945, M. A. Miner popularised a rule that had first been proposed by A. Palmgren in 1924. The rule, variously called *Miner's rule* or the *Palmgren-Miner linear damage hypothesis*, states that where there are k different stress magnitudes in a spectrum, S_i ($1 \leq i \leq k$), each contributing $n_i(S_i)$ cycles, then if $N_i(S_i)$ is the number of cycles to failure of a constant stress reversal S_i, failure occurs when:

$$\sum_{i=1}^{k} \frac{n_i}{N_i} = C$$

C is experimentally found to be between 0.7 and 2.2. Usually for design purposes, C is assumed to be 1. This can be thought of as assessing what proportion of life is consumed by stress reversal at each magnitude then forming a linear combination of their aggregate.

Though Miner's rule is a useful approximation in many circumstances, it has several major limitations:

1. It fails to recognise the probabilistic nature of fatigue and there is no simple way to relate life predicted by the rule with the characteristics of a probability distribution. Industry analysts often use design curves, adjusted to account for scatter, to calculate $N_i(S_i)$.

2. There is sometimes an effect in the order in which the reversals occur. In some circumstances, cycles of low stress followed by high stress cause more damage than would be predicted by the rule. It does not consider the effect of an overload or high stress which may result in a compressive residual stress that may retard crack growth. High stress followed by low stress may have less damage due to the presence of compressive residual stress.

Paris' Relationship

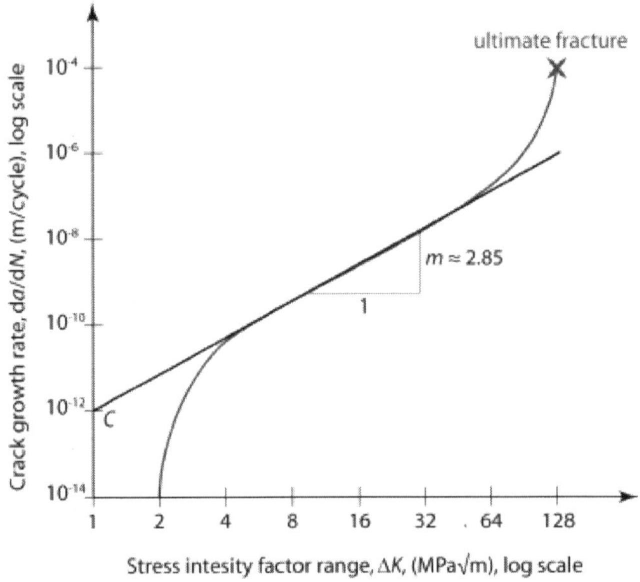

Figure: *Typical fatigue crack growth rate graph*

In Fracture mechanics, Anderson, Gomez and Paris derived relationships for the stage II crack growth with cycles N, in terms of the cyclical component ΔK of the Stress Intensity FactorK

$$\frac{da}{dN} = C(\Delta K)^m$$

where a is the crack length and m is typically in the range 3 to 5 (for metals).

This relationship was later modified (by Forman, 1967) to make better allowance for the mean stress, by introducing a factor depending on (1-R) where R = min stress/max stress, in the denominator.

Goodman Relation

In the presence of a steady stress superimposed on the cyclic loading, the Goodman relation can be used to estimate a failure condition. It plots stress amplitude against mean stress with the fatigue limit and the ultimate tensile strength of the material as the two extremes. Alternative failure criteria include Soderberg and Gerber.

Low-Cycle Fatigue

Where the stress is high enough for plastic deformation to occur, the accounting of the loading in terms of stress is less useful and the strain in the material offers a simpler and more accurate description. Low-cycle fatigue is usually characterised by the *Coffin-Manson relation* (published independently by L. F. Coffin in 1954 and S. S. Manson 1953):

$$\frac{\Delta\epsilon_p}{2} = \epsilon_{f'}(2N)^c$$

-where:

- $\Delta\varepsilon_p$ /2 is the plastic strain amplitude;
- ε_f' is an empirical constant known as the *fatigue ductility coefficient*, the failure strain for a single reversal;
- $2N$ is the number of reversals to failure (N cycles);
- c is an empirical constant known as the *fatigue ductility exponent*, commonly ranging from -0.5 to -0.7 for metals in time independent fatigue. Slopes can be considerably steeper in the presence of creep or environmental interactions.

A similar relationship for materials such as Zirconium, is used in the nuclear industry.

Fatigue and Fracture Mechanics

The account above is purely empirical and, though it allows life prediction and design assurance, life improvement or design optimisation can be enhanced using Fracture mechanics. It can be developed in four stages.

1. Crack nucleation;
2. Stage I crack-growth;
3. Stage II crack-growth; and
4. Ultimate ductile failure.

Factors that Affect Fatigue-Life

- Cyclic stress state: Depending on the complexity of the geometry and the loading, one or more properties of the stress state need to be considered, such as stress amplitude, mean stress, biaxiality, in-phase or out-of-phase shear stress, and load sequence,

- Geometry: Notches and variation in cross section throughout a part lead to stress concentrations where fatigue cracks initiate.
- Surface quality: Surface roughness can cause microscopic stress concentrations that lower the fatigue strength. Compressive residual stresses can be introduced in the surface by e.g. shot peening to increase fatigue life. Such techniques for producing surface stress are often referred to as *peening*, whatever the mechanism used to produce the stress. Low plasticity burnishing, laser peening, and ultrasonic impact treatment can also produce this surface compressive stress and can increase the fatigue life of the component. This improvement is normally observed only for high-cycle fatigue.
- Material Type: Fatigue life, as well as the behaviour during cyclic loading, varies widely for different materials, e.g. composites and polymers differ markedly from metals.
- Residual stresses: Welding, cutting, casting, grinding, and other manufacturing processes involving heat or deformation can produce high levels of tensile residual stress, which decreases the fatigue strength.
- Size and distribution of internal defects: Casting defects such as gas porosity voids, non-metallic inclusions and shrinkage voids can significantly reduce fatigue strength.
- Air or Vacuum: Certain materials like Metals are more prone to fatigue in air than in a vacuum. Depending upon the level of humidity and temperature, the lifetime for metals such as aluminum or iron might be as much as 5 to 10 times greater in a vacuum. This is mostly due to the effect of the oxygen and water vapour in the air which will aggressively attack the material and so encourage the propagation of cracks. Other environments such as oil or seawater may reduce the fatigue life at an even greater rate.
- Direction of loading: For non-isotropic materials, fatigue strength depends on the direction of the principal stress.
- Grain size: For most metals, smaller grains yield longer fatigue lives, however, the presence of surface defects or scratches will have a greater influence than in a coarse grained alloy.
- Environment: Environmental conditions can cause erosion, corrosion, or gas-phase embrittlement, which all affect fatigue life. Corrosion fatigue is a problem encountered in many aggressive environments.

- Temperature: Extreme high or low temperatures can decrease fatigue strength.
- Crack Closure:Crack closure is a phenomenon in fatigue loading, during which the crack will tend to remain in a closed position even though some external tensile force is acting on the material. During this process the crack will open only at a nominal stress above a particular crack opening stress. This is due to several factors such as plastic deformation or phase transformation during crack propagation, corrosion of crack surfaces, presence of fluids in the crack, or roughness at cracked surfaces etc. this will provide a longer fatigue life for the material than expected, by slowing the crack growth rate.

Design against Fatigue

Dependable design against fatigue-failure requires thorough education and supervised experience in structural engineering, mechanical engineering, or materials science. There are four principal approaches to life assurance for mechanical parts that display increasing degrees of sophistication:

1. Design to keep stress below threshold of fatigue limit (infinite lifetime concept);
2. fail-safe, graceful degradation, and fault-tolerant design: Instruct the user to replace parts when they fail. Design in such a way that there is no single point of failure, and so that when any one part completely fails, it does not lead to catastrophic failure of the entire system.
3. Safe-life design: Design (conservatively) for a fixed life after which the user is instructed to replace the part with a new one (a so-called *lifed* part, finite lifetime concept, or "safe-life" design practice); planned obsolescence and disposable product are variants that design for a fixed life after which the user is instructed to replace the entire device;
4. damage tolerant design: Instruct the user to inspect the part periodically for cracks and to replace the part once a crack exceeds a critical length. This approach usually uses the technologies of nondestructive testing and requires an accurate prediction of the rate of crack-growth between inspections. The designer sets some aircraft maintenance checks schedule frequent enough that parts are replaced while the crack is still in the "slow growth" phase. This is often referred to as damage tolerant design or "retirement-for-cause".

Stopping Fatigue

Fatigue cracks that have begun to propagate can sometimes be stopped by drilling holes, called *drill stops*, in the path of the fatigue crack. This is not recommended as a general practice because the hole represents a stress concentration factor which depends on the size of the hole and geometry, though the hole is typically less of a stress concentration than the removed tip of the crack. The possibility remains of a new crack starting in the side of the hole. It is always far better to replace the cracked part entirely.

Material Change

Changes in the materials used in parts can also improve fatigue life. For example, parts can be made from better fatigue rated metals. Complete replacement and redesign of parts can also reduce if not eliminate fatigue problems. Thus helicopter rotor blades and propellers in metal are being replaced by composite equivalents. They are not only lighter, but also much more resistant to fatigue. They are more expensive, but the extra cost is amply repaid by their greater integrity, since loss of a rotor blade usually leads to total loss of the aircraft. A similar argument has been made for replacement of metal fuselages, wings and tails of aircraft.

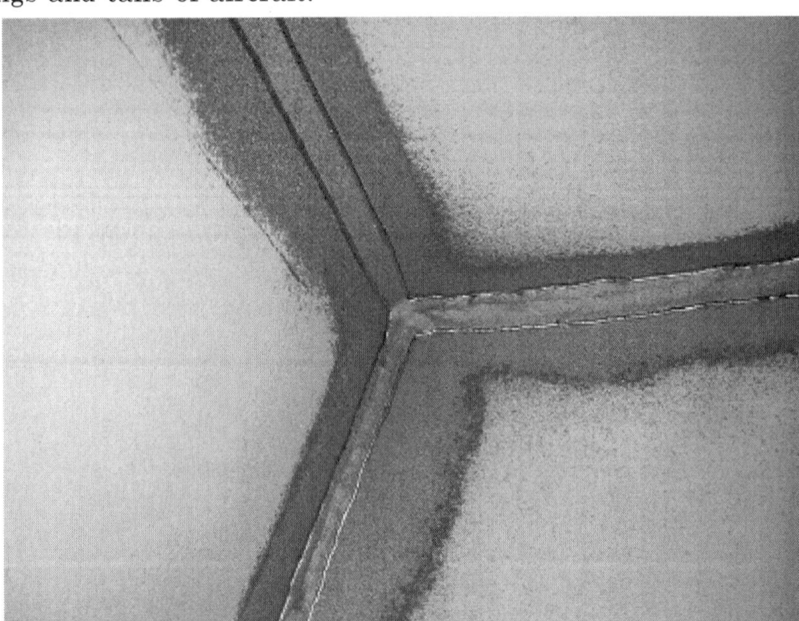

Figure: *Example of a HFMI treated steel highway bridge to avoid fatigue along the weld transition.*

High Frequency Mechanical Impact (HFMI) Treatment of Welds

The durability and life of dynamically loaded, welded steel structures are determined often by the welds, particular by the weld transitions. By selective treatment of weld transitions with the High Frequency Mechanical Impact (HFMI) treatment method, the durability of many designs can be increased significantly.

This method is universally applicable, requires only technical equipment and offers high reproducibility and a high grade of quality control.

Fracture

A fracture is the separation of an object or material into two, or more, pieces under the action of stress.

The fracture of a solid almost always occurs due to the development of certain displacement discontinuity surfaces within the solid. If a displacement develops in this case perpendicular to the surface of displacement, it is called a normal tensile crack or simply a crack; if a displacement develops tangentially to the surface of displacement, it is called a shear crack, slip band, or dislocation.

The word *fracture* is often applied to bones of living creatures (that is, a bone fracture), or to crystals or crystalline materials, such as gemstones or metal. Sometimes, in crystalline materials, individual crystals fracture without the body actually separating into two or more pieces. Depending on the substance which is fractured, a fracture reduces strength (most substances) or inhibits transmission of light (optical crystals).

A detailed understanding of how fracture occurs in materials may be assisted by the study of fracture mechanics.

A fracture is also the term used for a particular mask data preparation procedure within the realm of integrated circuit design that involves transposing complex polygons into simpler shapes such as trapezoids and rectangles.

Fracture Strength

1. Ultimate tensile strength
2. Yield strength
3. Proportional limit stress
4. Fracture
5. Offset strain (typically 0.2%)

Fracture strength, also known as breaking strength, is the stress at which a specimen fails via fracture. This is usually determined for a given specimen by a tensile test, which charts the stress-strain curve. The final recorded point is the fracture strength.

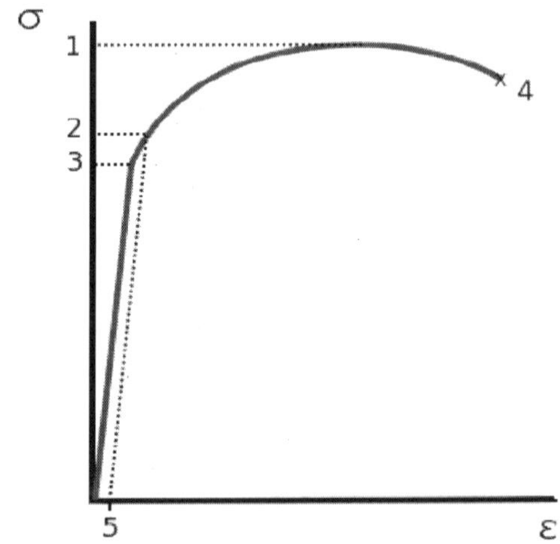

Figure: *Stress vs. strain curve typical of aluminum*

Ductile materials have a fracture strength lower than the ultimate tensile strength (UTS), whereas in brittle materials the fracture strength is equivalent to the UTS. If a ductile material reaches its ultimate tensile strength in a load-controlled situation, it will continue to deform, with no additional load application, until it ruptures. However, if the loading is displacement-controlled, the deformation of the material may relieve the load, preventing rupture.

If the stress-strain curve is plotted in terms of *true stress* and *true strain* the curve will always slope upwards and never reverse, as true stress is corrected for the decrease in cross-sectional area. The true stress on the material at the time of rupture is known as the breaking strength. This is the maximum stress on the true stress-strain curve, given by point 1 on curve B.

Types

Brittle Fracture

In *brittle fracture*, no apparent plastic deformation takes place before fracture. In brittle crystalline materials, fracture can occur by *cleavage* as the result of tensile stress acting normal to crystallographic planes with low bonding (cleavage planes). In amorphous solids, by contrast, the lack of a crystalline structure results in a conchoidal fracture, with cracks proceeding normal to the applied tension.

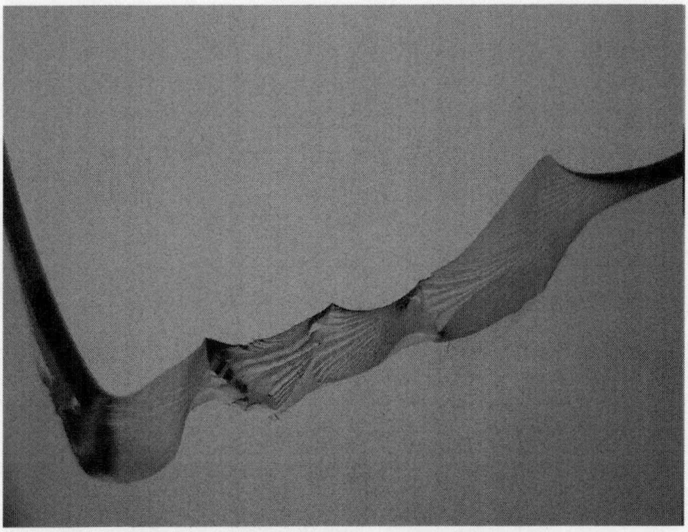

Figure: *Brittle fracture in glass.*

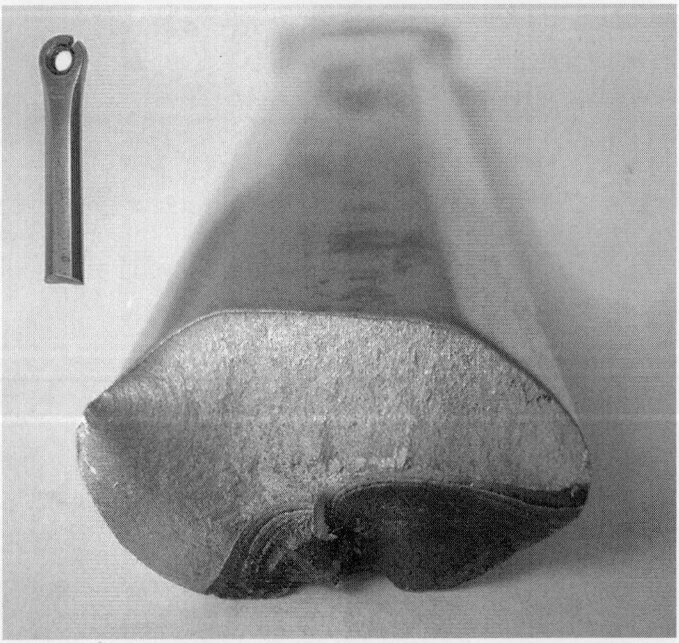

Figure: *Fracture of an aluminum crank arm. Bright: brittle fracture. Dark: fatigue fracture.*

The theoretical strength of a crystalline material is (roughly)

$$\sigma_{\text{theoretical}} = \sqrt{\frac{E\gamma}{r_o}}$$

where: -

E is the Young's modulus of the material,

γ is the surface energy, and

r_o is the equilibrium distance between atomic centres.

On the other hand, a crack introduces a stress concentration modelled by

$$\sigma_{\text{elliptical crack}} = \sigma_{\text{applied}}\left(1+2\sqrt{\frac{a}{\rho}}\right) = 2\sigma_{\text{applied}}\sqrt{\frac{a}{\rho}} \quad \text{(For sharp cracks)}$$

where:

σ_{applied} is the loading stress,

a is half the length of the crack, and

ρ is the radius of curvature at the crack tip.

Putting these two equations together, we get

$$\sigma_{\text{fracture}} = \sqrt{\frac{E\gamma\rho}{4ar_o}}.$$

Looking closely, we can see that sharp cracks (small ρ) and large defects (large) both lower the fracture strength of the material.

Recently, scientists have discovered supersonic fracture, the phenomenon of crack motion faster than the speed of sound in a material. This phenomenon was recently also verified by experiment of fracture in rubber-like materials.

Ductile Fracture

Figure: *Ductile failure of a specimen strained axially.*

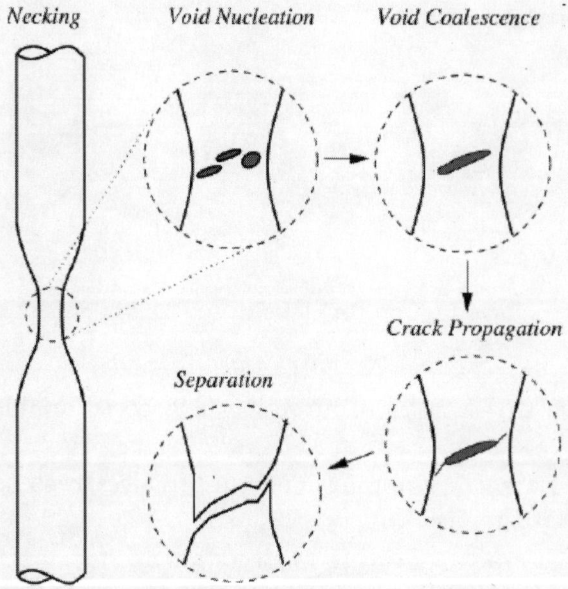

Figure: *Schematic representation of the steps in ductile fracture (in pure tension).*

In *ductile fracture*, extensive plastic deformation (necking) takes place before fracture. The terms *rupture* or *ductile rupture* describe the ultimate failure of tough ductile materials loaded in tension. Rather than cracking, the material "pulls apart," generally leaving a rough surface. In this case there is slow propagation and an absorption of a large amount energy before fracture.

Many ductile metals, especially materials with high purity, can sustain very large deformation of 50–100% or more strain before fracture under favourable loading condition and environmental condition. The strain at which the fracture happens is controlled by the purity of the materials. At room temperature, pure iron can undergo deformation up to 100% strain before breaking, while cast iron or high-carbon steels can barely sustain 3% of strain.

Because ductile rupture involves a high degree of plastic deformation, the fracture behaviour of a propagating crack as modelled above changes fundamentally. Some of the energy from stress concentrations at the crack tips is dissipated by plastic deformation before the crack actually propagates.

The basic steps are: void formation, void coalescence (also known as crack formation), crack propagation, and failure, often resulting in a cup-and-cone shaped failure surface.

Crack Separation Modes

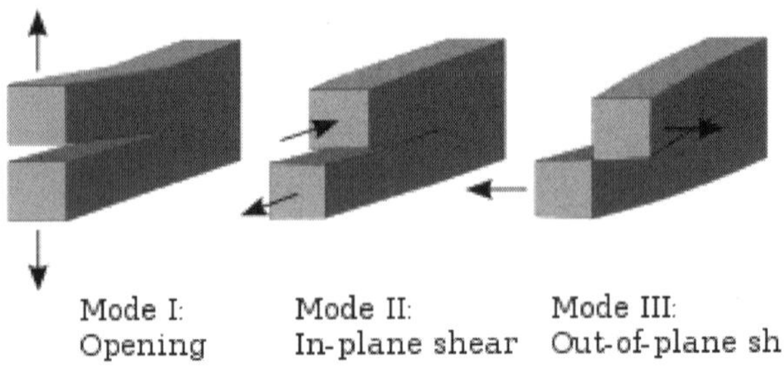

Figure: *The three fracture modes.*

There are three ways of applying a force to enable a crack to propagate:

- Mode I crack – Opening mode (a tensile stress normal to the plane of the crack)
- Mode II crack – Sliding mode (a shear stress acting parallel to the plane of the crack and perpendicular to the crack front)
- Mode III crack – Tearing mode (a shear stress acting parallel to the plane of the crack and parallel to the crack front)

Crack initiation and propagation accompany fracture. The manner through which the crack propagates through the material gives great insight into the mode of fracture. In ductile materials (ductile fracture), the crack moves slowly and is accompanied by a large amount of plastic deformation. The crack will usually not extend unless an increased stress is applied. On the other hand, in dealing with brittle fracture, cracks spread very rapidly with little or no plastic deformation. The cracks that propagate in a brittle material will continue to grow and increase in magnitude once they are initiated. Another important mannerism of crack propagation is the way in which the advancing crack travels through the material. A crack that passes through the grains within the material is undergoing transgranular fracture. However, a crack that propagates along the grain boundaries is termed an intergranular fracture.

Mechanical Overload (Engineering)

The failure or fracture of a product or component as a result of a single event is known as mechanical overload. It is a common failure mode, and may be contrasted with fatigue, creep, rupture, or stress relaxation. The terms are used in forensic engineering and structural engineering when analysing product failure. Failure may occur because either the product is weaker than expected owing to a stress concentration, or the applied load is greater than expected and exceeds the normal tensile strength, shear strength or compressive strength of the product.

Examples

Examples include the many components which fail in car crashes, train crashes, and airplane crashes as a result of impact loading. The problem for the investigator is to determine which failures have been caused by the crash, and which may have caused the crash. It usually involves examining the broken parts for signs of fatigue crack growth or other damage to the part which cannot be attributed to the crash itself. For very large structural failures such as the collapse of bridges, it is necessarily a long and tedious process of sifting through the broken parts.

Stress Corrosion Cracking

Stress corrosion cracking (SCC) is the growth of crack formation in a corrosive environment. It can lead to unexpected sudden failure of normally ductile metals subjected to a tensile stress, especially at elevated temperature in the case of metals. SCC is highly chemically specific in that certain alloys are likely to undergo SCC only when exposed to a small number of chemical environments. The chemical environment that causes SCC for a given alloy is often one which is only mildly corrosive to the metal otherwise. Hence, metal parts with severe SCC can appear bright and shiny, while being filled with microscopic cracks. This factor makes it common for SCC to go undetected prior to failure. SCC often progresses rapidly, and is more common among alloys than pure metals. The specific environment is of crucial importance, and only very small concentrations of certain highly active chemicals are needed to produce catastrophic cracking, often leading to devastating and unexpected failure.

The stresses can be the result of the crevice loads due to stress concentration, or can be caused by the type of assembly or residual stresses from fabrication (e.g. cold working); the residual stresses can be relieved by annealing.

Metals Attacked

Certain austeniticstainless steels and aluminiumalloys crack in the presence of chlorides, mild steel cracks in the presence of alkali (boiler cracking) and nitrates, copper alloys crack in ammoniacal solutions (season cracking). This limits the usefulness of austenitic stainless steel for containing water with higher than few ppm content of chlorides at temperatures above 50 °C. Worse still, high-tensile structural steels crack in an unexpectedly brittle manner in a whole variety of aqueous environments, especially containing chlorides. With the possible exception of the latter, which is a special example of *hydrogen cracking*, all the others display the phenomenon of subcritical crack growth, i.e. small surface flaws propagate (usually smoothly) under conditions where fracture mechanics predicts that failure should not occur. That is, in the presence of a corrodent, cracks develop and propagate well below K_{Ic}. In fact, the subcritical value of the stress intensity, designated as K_{Iscc}, may be less than 1% of K_{Ic}, as the following table shows:

A close-up of the surface of a steel pipeline showing indications of stress corrosion cracking (two clusters of small black lines) revealed by magnetic particle inspection. Cracks which would normally have been invisible are detectable due to the magnetic particles clustering at the crack openings. The scale at the bottom is in millimetres.

Alloy	K_{Ic} *MN/m*$^{3/2}$	*SCC environment*	K_{Iscc} *MN/m*$^{3/2}$
13Cr steel	60	3% NaCl	12
18Cr-8Ni	200	42% $MgCl_2$	10
Cu-30Zn	200	NH_4OH, pH7	1
Al-3Mg-7Zn	25	Aqueous halides	5
Ti-6Al-1V	60	0.6M KCl	20

Polymers Attacked

A similar process occurs in polymers, when products are exposed to aggressive chemicals such as acids and alkalis. As with metals, attack is confined to specific polymers and particular chemicals. Thus polycarbonate is sensitive to attack by alkalis, but not by acids. On the other hand, polyesters are readily degraded by acids, and SCC is a likely failure mechanism. Polymers are also susceptible to environmental stress cracking where attacking agents do not necessarily degrade the materials chemically. Nylon is sensitive to degradation by acids, a process known as hydrolysis, and nylon mouldings will crack when attacked by strong acids.

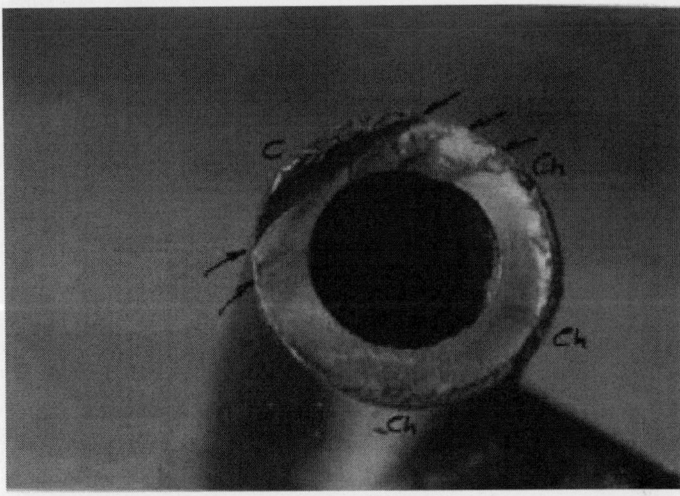

Figure: *Close-up of broken nylon fuel pipe connector caused by SCC*

For example, the fracture surface of a fuel connector showed the progressive growth of the crack from acid attack to the final cusp (C) of polymer. In this case the failure was caused by hydrolysis of the polymer by contact with sulfuric acid leaking from a car battery. The degradation reaction is the reverse of the synthesis reaction of the polymer:

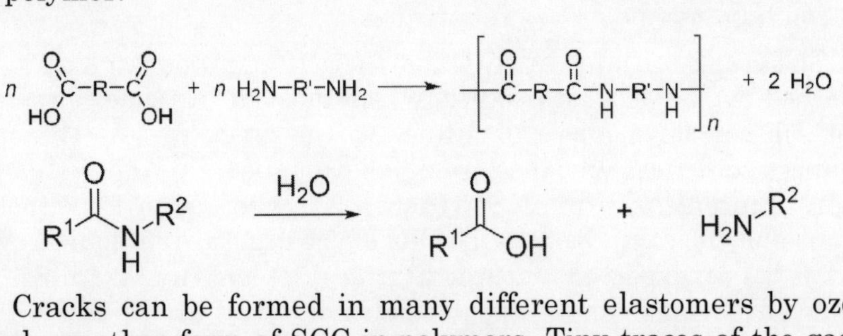

Cracks can be formed in many different elastomers by ozone attack, another form of SCC in polymers. Tiny traces of the gas in the air will attack double bonds in rubber chains, with Natural rubber, Styrene-butadiene rubber and NBR being most sensitive to degradation.

Ozone cracks form in products under tension, but the critical strain is very small. The cracks are always oriented at right angles to the strain axis, so will form around the circumference in a rubber tube bent over. Such cracks are very dangerous when they occur in fuel pipes because the cracks will grow from the outside exposed surfaces into the bore of the pipe, so fuel leakage and fire may follow.

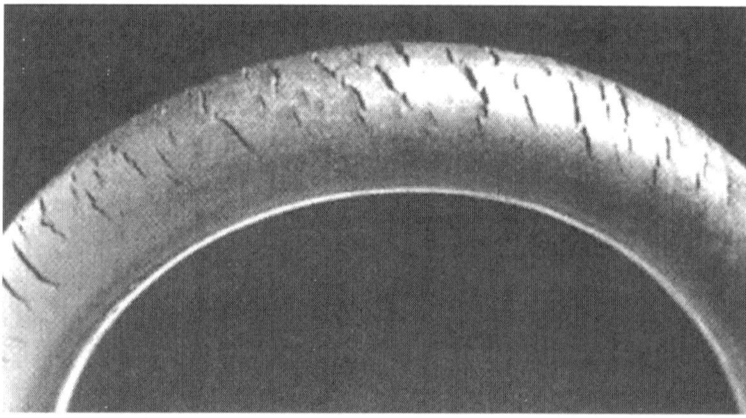

Figure: *Ozone cracking in Natural rubber tubing*

The problem of ozone cracking can be prevented by adding anti-ozonants to the rubber before vulcanization. Ozone cracks were commonly seen in automobile tire sidewalls, but are now seen rarely thanks to the use of these additives. On the other hand, the problem does recur in unprotected products such as rubber tubing and seals.

Crack Growth

The subcritical nature of propagation may be attributed to the chemical energy released as the crack propagates. That is,

elastic energy released + chemical energy = surface energy + deformation energy

The crack initiates at K_{Iscc} and thereafter propagates at a rate governed by the slowest process, which most of the time is the rate at which corrosive ions can diffuse to the crack tip. As the crack advances so K rises (because crack length appears in the calculation of stress intensity). Finally it reaches K_{Ic}, whereupon fast fracture ensues and the component fails. One of the practical difficulties with SCC is its unexpected nature. Stainless steels, for example, are employed because under most conditions they are "passive", i.e. effectively inert. Very often one finds a single crack has propagated while the rest of the metal surface stays apparently unaffected. The crack propagates perpendicular to the applied stress.

Prevention

SCC is the result of a combination of three factors – a susceptible material, exposure to a corrosive environment, and tensile stresses above a threshold. If you eliminate any one of these factors SCC initiation becomes impossible. There are, consequently, a number of approaches that we can use to prevent SCC, or at least to give an acceptable lifetime. In an ideal world a stress corrosion cracking control strategy will start operating at the design stage, and will focus on the selection of material, the limitation of stress and the control of the environment.

The skill of the engineer then lies in selecting the strategy that delivers the required performance at minimum cost. In this context we should appreciate that a part of the performance requirement relates to the acceptability of failure. For the primary containment pressure vessel in a nuclear reactor we obviously require a very low risk of failure.

For the pressed brass decorative trim on a light switch, the occasional stress corrosion crack is not going to be a serious problem, although frequent failures would have an undesirable impact on product returns and the corporate image. The conventional approach to controlling the problem has been to develop new alloys that are more resistant to SCC. This is a costly proposition and can require a massive time investment to achieve only marginal success.

Selection and Control of Material

The first line of defence in controlling stress corrosion cracking is to be aware of the possibility at the design and construction stages. By choosing a material that is not susceptible to SCC in the service environment, and by processing and fabricating it correctly, subsequent SCC problems can be avoided. Unfortunately, it is not always quite that simple. Some environments, such as high temperature water, are very aggressive, and will cause SCC of most materials. Mechanical requirements, such as a high yield strength, can be very difficult to reconcile with SCC resistance (especially where hydrogen embrittlement is involved).

Control of Stress

As one of the requirements for stress corrosion cracking is the presence of stress in the components, one method of control is to eliminate that stress, or at least reduce it below the threshold stress for SCC. This is not usually feasible for working stresses (the stress that the component is intended to support), but it may be possible where the stress causing cracking is a residual stress introduced during welding or forming. Residual stresses can be relieved by stress-relief annealing, and this is widely used for carbon steels. These have the advantage of a relatively high threshold stress for most environments, consequently it is relatively easy to reduce the residual stresses to a low enough level. In contrast austenitic stainless steels have a very low threshold stress for chloride SCC.

This, combined with the high annealing temperatures that are necessary to avoid other problems, such as sensitization and sigma phase embrittlement, means that stress relief is rarely successful as a method of controlling SCC for this system. For large structures, for which full stress-relief annealing is difficult or impossible, partial stress relief around welds and other critical areas may be of value. However, this must be done in a controlled way to avoid creating new regions of high residual stress, and expert advice is advisable if this approach is adopted. Stresses can also be relieved mechanically.

For example, hydrostatic testing beyond yield will tend to 'even-out' the stresses and thereby reduce the peak residual stress. Similarly shot-peening or grit-blasting tend to introduce a surface compressive stress, and are beneficial for the control of SCC. The uniformity with which these processes are applied is important. If, for example, only the weld region is shot-peened, damaging tensile stresses may be created at the border of the peened area.

Control of Environment

The most direct way of controlling SCC through control of the environment is to remove or replace the component of the environment that is responsible for the problem. Unfortunately, it is relatively rare for this approach to be applicable. If the active species is present in an environment over which we have some control, then it may be feasible to remove the active species, although even then it may be difficult.

For example, chloride stress corrosion cracking of austenitic stainless steels has been experienced in hot-water jackets around chocolate pipes (that is to say, pipes carrying molten chocolate) in the food industry. In this situation we can't easily change the material or the temperature, and it is virtually impossible to eliminate the residual stresses associated with welding and forming of the stainless steel. However, we can remove the chloride from the water by an ion exchange process, and, with proper control and monitoring, this approach could be successful. In the latter situation, where the species responsible for cracking are a required component of the environment, the environmental control options consist of adding inhibitors, modifying the electrode potential of the metal, or isolating the metal from the environment with coatings.

Examples

A 32 inch diameter gas transmission pipeline, north of Natchitoches, Louisiana, belonging to the Tennessee Gas Pipeline exploded and burned from SCC on March 4, 1965, killing 17 people. At least 9 others were injured, and 7 homes 450 feet from the rupture were destroyed.

SCC caused the catastrophic collapse of the Silver Bridge in December 1967, when an eyebarsuspension bridge across the Ohio river at Point Pleasant, West Virginia, suddenly failed. The main chain joint failed and the whole structure fell into the river, killing 46 people in vehicles on the bridge at the time. Rust in the eyebar joint had caused a stress corrosion crack, which went critical as a result of high bridge loading and low temperature. The failure was exacerbated by a high level of residual stress in the eyebar. The disaster led to a nationwide reappraisal of bridges.

Suspended ceilings in indoor swimming pools are safety-relevant components. As was demonstrated by the collapses of the ceiling of the Uster (Switzerland) indoor swimming pool (1985) and again at Steenwijk (Netherlands, 2001), attention must be paid to selecting suitable materials and inspecting the state of such components. The reason for the failures was stress corrosion cracking of metal fastening components made of stainless steel. The active chemical was chlorine added to the water as a disinfectant.

A classic example of SCC is season cracking of brass cartridge cases, a problem experienced by the British army in India in the early 19th century. It was initiated by ammonia from dung and horse manure decomposing at the higher temperatures of the spring and summer. There was substantial residual stress in the cartridge shells as a result of cold forming. The problem was solved by annealing the shells to ameliorate the stress.

Yield (Engineering)

A yield strength or yield point of a material is defined in engineering and materials science as the stress at which a material begins to deform plastically. Prior to the yield point the material will deform elastically and will return to its original shape when the applied stress is removed. Once the yield point is passed, some fraction of the deformation will be permanent and non-reversible.

In the three-dimensional space of the principal stresses ($\sigma_1, \sigma_2, \sigma_3$), an infinite number of yield points form together a yield surface.

Knowledge of the yield point is vital when designing a component since it generally represents an upper limit to the load that can be applied. It is also important for the control of many materials production techniques such as forging, rolling, or pressing. In structural engineering, this is a soft failure mode which does not normally cause catastrophic failure or ultimate failure unless it accelerates buckling.

Definition

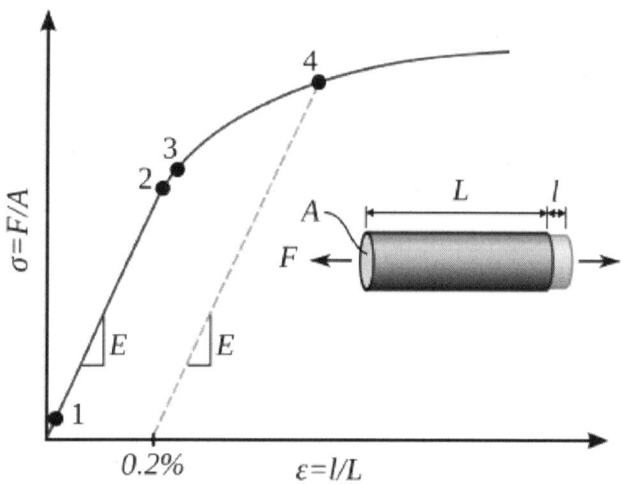

Figure: *Typical yield behaviour for non-ferrous alloys.*

1: True elastic limit

2: Proportionality limit

3: Elastic limit

4: Offset yield strength

It is often difficult to precisely define yielding due to the wide variety of stress–strain curves exhibited by real materials. In addition, there are several possible ways to define yielding:

True Elastic Limit: The lowest stress at which dislocations move. This definition is rarely used, since dislocations move at very low stresses, and detecting such movement is very difficult.

Proportionality Limit: Up to this amount of stress, stress is proportional to strain (Hooke's law), so the stress-strain graph is a straight line, and the gradient will be equal to the elastic modulus of the material.

Elastic Limit (Yield Strength): Beyond the elastic limit, permanent deformation will occur. The lowest stress at which permanent deformation can be measured. This requires a manual load-unload procedure, and the accuracy is critically dependent on equipment and operator skill. For elastomers, such as rubber, the elastic limit is much larger than the proportionality limit. Also, precise strain measurements have shown that plastic strain begins at low stresses.

Yield Point : The point in the stress-strain curve at which the curve levels off and plastic deformation begins to occur.

Offset Yield Point (Proof Stress): When a yield point is not easily defined based on the shape of the stress-strain curve an *offset yield point* is arbitrarily defined.

The value for this is commonly set at 0.1 or 0.2% strain. The offset value is given as a subscript, e.g., $R_{p0.2}$=310 MPa. High strength steel and aluminum alloys do not exhibit a yield point, so this offset yield point is used on these materials.

Upper Yield Point and Lower Yield Point: Some metals, such as mild steel, reach an upper yield point before dropping rapidly to a lower yield point. The material response is linear up until the upper yield point, but the lower yield point is used in structural engineering as a conservative value. If a metal is only stressed to the upper yield point, and beyond, Lüders bands can develop.

Yield Criterion

A yield criterion, often expressed as yield surface, or yield locus, is a hypothesis concerning the limit of elasticity under any combination of stresses. There are two interpretations of yield criterion: one is purely mathematical in taking a statistical approach while other models attempt to provide a justification based on established physical principles. Since stress and strain are tensor qualities they can be described on the basis of three principal directions, in the case of stress these are denoted by σ_1, σ_2, and σ_3.

The following represent the most common yield criterion as applied to an isotropic material (uniform properties in all directions). Other equations have been proposed or are used in specialist situations.

Isotropic Yield Criteria

Maximum Principal Stress Theory – by W.J.M Rankine(1850). Yield occurs when the largest principal stress exceeds the uniaxial tensile yield strength. Although this criterion allows for a quick and easy comparison with experimental data it is rarely suitable for design purposes. This theory gives good predictions for brittle materials.

$$\sigma_1 \le \sigma_y$$

Maximum Principal Strain Theory – by St.Venant. Yield occurs when the maximum principal strain reaches the strain corresponding to the yield point during a simple tensile test. In terms of the principal stresses this is determined by the equation:

$$\sigma_1 - \nu(\sigma_2 + \sigma_3) \le \sigma_y.$$

Maximum Shear Stress Theory – Also known as the Tresca yield criterion, after the French scientist Henri Tresca. This assumes that yield occurs when the shear stress τ exceeds the shear yield strength τ_y :

$$\tau = \frac{\sigma_1 - \sigma_3}{2} \le \tau_y.$$

Total Strain Energy Theory – This theory assumes that the stored energy associated with elastic deformation at the point of yield is independent of the specific stress tensor. Thus yield occurs when the strain energy per unit volume is greater than the strain energy at the elastic limit in simple tension. For a 3-dimensional stress state this is given by:

$$\sigma_1^2 + \sigma_2^2 + \sigma_3^2 - 2\nu(\sigma_1\sigma_2 + \sigma_2\sigma_3 + \sigma_1\sigma_3) \le \sigma_y^2.$$

Distortion Energy Theory – This theory proposes that the total strain energy can be separated into two components: the *volumetric* (hydrostatic) strain energy and the *shape* (distortion or shear) strain energy. It is proposed that yield occurs when the distortion component exceeds that at the yield point for a simple tensile test.

Based on a different theoretical underpinning this expression is also referred to as octahedral shear stress theory.

Other commonly used isotropic yield criteria are the

- Mohr-Coulomb yield criterion
- Drucker-Prager yield criterion
- Bresler-Pister yield criterion
- Willam-Warnke yield criterion

The yield surfaces corresponding to these criteria have a range of forms. However, most isotropic yield criteria correspond to convex yield surfaces.

Anisotropic Yield Criteria

When a metal is subjected to large plastic deformations the grain sizes and orientations change in the direction of deformation. As a result the plastic yield behaviour of the material shows directional dependency. Under such circumstances, the isotropic yield criteria such as the von Mises yield criterion are unable to predict the yield behaviour accurately. Several anisotropic yield criteria have been developed to deal with such situations. Some of the more popular anisotropic yield criteria are:

- Hill's quadratic yield criterion.
- Generalized Hill yield criterion.
- Hosford yield criterion.

Factors Influencing Yield Stress

The stress at which yield occurs is dependent on both the rate of deformation (strain rate) and, more significantly, the temperature at which the deformation occurs. In general, the yield strength increases with strain rate and decreases with temperature. When the latter is not the case, the material is said to exhibit yield strength anomaly, which is typical for superalloys and leads to their use in applications requiring high strength at high temperatures. Early work by Alder and Philips in 1954 found that the relationship between yield stress and strain rate (at constant temperature) was best described by a power law relationship of the form

$$\sigma_y = C(\dot{\epsilon})^m$$

where C is a constant and m is the strain rate sensitivity. The latter generally increases with temperature, and materials where m reaches a value greater than ~0.5 tend to exhibit super plastic behaviour.

Later, more complex equations were proposed that simultaneously dealt with both temperature and strain rate:

$$\sigma_y = \frac{1}{\alpha}\sinh^{-1}\left[\frac{Z}{A}\right]^{(1/n)}$$

where α and A are constants and Z is the temperature-compensated strain-rate – often described by the Zener-Hollomon parameter:

$$Z = (\dot{\epsilon})\exp\left(\frac{Q_{HW}}{RT}\right)$$

where Q_{HW} is the activation energy for hot deformation and T is the absolute temperature.

Strengthening Mechanisms

There are several ways in which crystalline and amorphous materials can be engineered to increase their yield strength. By altering dislocation density, impurity levels, grain size (in crystalline materials), the yield strength of the material can be fine tuned. This occurs typically by introducing defects such as impurities dislocations in the material. To move this defect (plastically deforming or yielding the material), a larger stress must be applied. This thus causes a higher yield stress in the material. While many material properties depend only on the composition of the bulk material, yield strength is extremely sensitive to the materials processing as well for this reason. These mechanisms for crystalline materials include:

- Work hardening
- Solid solution strengthening
- Precipitation strengthening
- Grain boundary strengthening

Work Hardening

Where deforming the material will introduce dislocations, which increases their density in the material. This increases the yield strength of the material, since now more stress must be applied to move these dislocations through a crystal lattice. Dislocations can also interact with each other, becoming entangled.

The governing formula for this mechanism is:

$$\Delta\sigma_y = Gb\sqrt{\rho}$$

where σ_y is the yield stress, G is the shear elastic modulus, b is the magnitude of the Burgers vector, and ρ is the dislocation density.

Solid Solution Strengthening

By alloying the material, impurity atoms in low concentrations will occupy a lattice position directly below a dislocation, such as directly below an extra half plane defect. This relieves a tensile strain directly below the dislocation by filling that empty lattice space with the impurity atom.

The relationship of this mechanism goes as:

$$\Delta\tau = Gb\sqrt{C_s}\epsilon^{3/2}$$

where τ is the shear stress, related to the yield stress, G and b are the same as in the above example, C_s is the concentration of solute and ϵ is the strain induced in the lattice due to adding the impurity.

Particle/Precipitate Strengthening

Where the presence of a secondary phase will increase yield strength by blocking the motion of dislocations within the crystal. A line defect that, while moving through the matrix, will be forced against a small particle or precipitate of the material. Dislocations can move through this particle either by shearing the particle, or by a process known as bowing or ringing, in which a new ring of dislocations is created around the particle.

The shearing formula goes as:

$$\Delta\tau = \frac{r_{\text{particle}}}{l_{\text{interparticle}}}\gamma_{\text{particle-matrix}}$$

and the bowing/ringing formula:

$$\Delta\tau = \frac{Gb}{l_{\text{interparticle}} - 2r_{\text{particle}}}$$

In these formulas, r_{particle} is the particle radius, $\gamma_{\text{particle-matrix}}$ is the surface tension between the matrix and the particle, $l_{\text{interparticle}}$ is the distance between the particles.

Grain Boundary Strengthening

Where a buildup of dislocations at a grain boundary causes a repulsive force between dislocations. As grain size decreases, the surface area to volume ratio of the grain increases, allowing more buildup of dislocations at the grain edge. Since it requires a lot of energy to move dislocations to another grain, these dislocations build up along the boundary, and increase the yield stress of the material. Also known as Hall-Petch strengthening, this type of strengthening is governed by the formula:

$$\sigma_y = \sigma_0 + kd^{-1/2}$$

where

σ_0 is the stress required to move dislocations,

k is a material constant, and

d is the grain size.

Testing

Yield strength testing involves taking a small sample with a fixed cross-section area, and then pulling it with a controlled, gradually increasing force until the sample changes shape or breaks. Longitudinal and/or transverse strain is recorded using mechanical or optical extensometres. Indentation hardness correlates linearly with tensile strength for most steels. Hardness testing can therefore be an economical substitute for tensile testing, as well as providing local variations in yield strength due to e.g. welding or forming operations.

Implications for Structural Engineering

Yielded structures have a lower stiffness, leading to increased deflections and decreased buckling strength. The structure will be permanently deformed when the load is removed, and may have residual stresses. Engineering metals display strain hardening, which implies that the yield stress is increased after unloading from a yield state. Highly optimized structures, such as airplane beams and components, rely on yielding as a fail-safe failure mode. No safety factor is therefore needed when comparing limit loads (the highest loads expected during normal operation) to yield criteria.

4

Beams and Girders

Beam (Structure)

A beam is a structural element that is capable of withstanding load primarily by resisting bending. The bending force induced into the material of the beam as a result of the external loads, own weight, span and external reactions to these loads is called a bending moment.

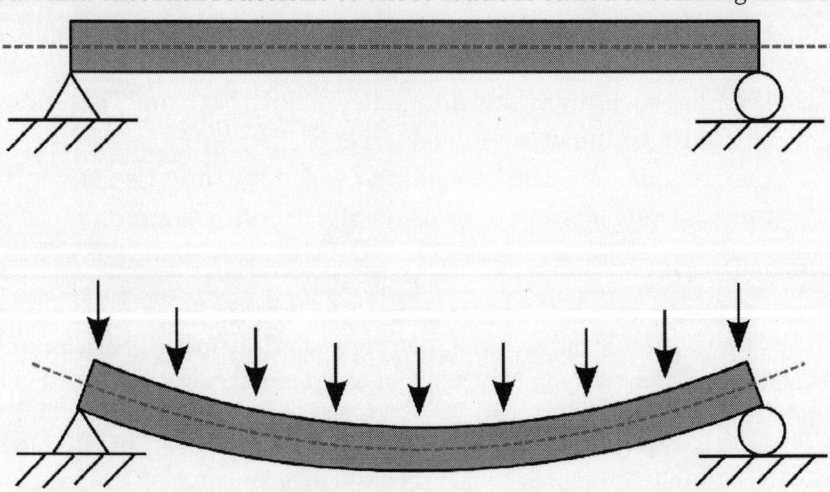

Figure: *A statically determinate beam, bending (sagging) under an evenly distributed load.*

Beams are traditionally descriptions of building or civil engineering structural elements, but smaller structures such as truck or automobile frames, machine frames, and other mechanical or structural systems contain beam structures that are designed and analyzed in a similar fashion.

Overview

Historically beams were squared timbers but are also metal, stone, or combinations of wood and metal such as a flitch beam. Beams generally carry vertical gravitational forces but can also be used to carry horizontal loads (i.e., loads due to an earthquake or wind or in tension to resist rafter thrust as a tie beam or (usually) compression as a collar beam). The loads carried by a beam are transferred to columns, walls, or girders, which then transfer the force to adjacent structural compression members. In light frame construction joists may rest on beams.

In carpentry a beam is called a plate as in a sill plate or wall plate, beam as in a summer beam or dragon beam.

Types of Beams

In engineering, beams are of six types:

1. Simply supported - a beam supported on the ends which are free to rotate and have no moment resistance.
2. Fixed - a beam supported on both ends and restrained from rotation.
3. Over hanging - a simple beam extending beyond its support on one end.
4. Double overhanging - a simple beam with both ends extending beyond its supports on both ends.
5. Continuous - a beam extending over more than two supports.
6. Cantilever - a projecting beam fixed only at one end.

General Shapes

Most beams in reinforced concrete buildings have rectangular cross sections, but a more efficient cross section for a beam is an I or H section which is typically seen in steel construction. Because of the parallel axis theorem and the fact that most of the material is away from the neutral axis, the second moment of area of the beam increases, which in turn increases the stiffness.

An I-beam is only the most efficient shape in one direction of bending: up and down looking at the profile as anI. If the beam is bent side to side, it functions as an H where it is less efficient. The most efficient shape for both directions in 2D is a box (a square shell) however the most efficient shape for bending in any direction is a cylindrical shell or tube. But, for unidirectional bending, the I or wide flange beam is superior.

Efficiency means that for the same cross sectional area (volume of beam per length) subjected to the same loading conditions, the beam deflects less. Other shapes, like L (angles), C (channels) or tubes, are also used in construction when there are special requirements.

Moment of Inertia

Moment of inertia is the mass property of a rigid body that defines the torque needed for a desired angular acceleration about an axis of rotation. Moment of inertia depends on the shape of the body and may be different around different axes of rotation. A larger moment of inertia around a given axis requires more torque to increase the rotation, or to stop the rotation, of a body about that axis. Moment of inertia depends on the amount and distribution of its mass, and can be found through the sum of moments of inertia of the masses making up the whole object, under the same conditions. For example, if $m_a + m_b = m_c$, then $I_a + I_b = I_c$. In classical mechanics, moment of inertia may also be called mass moment of inertia, rotational inertia, polar moment of inertia, or the angular mass.

For planar movement of a body, the trajectories of all of its points lie in parallel planes, and the rotation occurs only about an axis perpendicular to this plane. In this case, the body has a single moment of inertia, which is measured around this axis.

For spatial movement of a body, the moment of inertia is defined by its symmetric 3x3 inertia matrix. The inertia matrix is often described as a symmetric rank two tensor, having six independent components. The inertia matrix includes off-diagonal terms called products of inertia that couple torque around one axis to acceleration about another axis. Each body has a set of mutually perpendicular axes, called principal axes, for which the off-diagonal terms of the inertia matrix are zero, and a torque around a principal axis only affects the acceleration about that axis.

Introduction

When a body is rotating around an axis, a torque must be applied to change its angular momentum. The amount of torque needed for any given change in angular momentum is proportional to the size of that change. Moment of inertia has units of $kg \cdot m^2$ in SI units and $lb_m\ ft^2$ in US units.

In 1673 Christiaan Huygens introduced this parameter in his study of the oscillation of a body hanging from a pivot, known as a compound pendulum. The term *moment of inertia* was introduced by Leonhard Euler in his book *Theoriamo-tuscorporum-solidorum-seurigidorum* in 1765, and it is incorporated into Euler's second law.

The natural frequency of oscillation of a compound pendulum is obtained from the ratio of the torque imposed by gravity on the mass of the pendulum to the resistance to acceleration defined by the moment of inertia. Comparison of this natural frequency to that of a simple pendulum consisting of a single point of mass provides a mathematical formulation for moment of inertia of an extended body.

Moment of inertia also appears in momentum, kinetic energy, and in Newton's laws of motion for a rigid body as a physical parameter that combines its shape and mass. There is an interesting difference in the way moment of inertia appears in planar and spatial movement. Planar movement has a single scalar that defines the moment of inertia, while for spatial movement the same calculations yield a 3x3 matrix of moments of inertia, called the inertia matrix or inertia tensor.

The moment of inertia of a rotating flywheel is used in a machine to resist variations in applied torque in order to smooth its rotational output. The moment of inertia of an airplane about its longitudinal, horizontal and vertical axes determines how steering forces on the control surfaces of its wings, elevators and tail affect the plane in roll, pitch and yaw.

Simple Pendulum

A simple pendulum is a point mass suspended by a string so that its movement is constrained to a circle around a pivot point. The mass of a simple pendulum supported by a light string accelerates due to the force of gravity.

The moment of inertia of the pendulum about the pivot point is its resistance to movement due to the torque due to gravity. Mathematically, it is the ratio of the torque due to gravity about the pivot of a pendulum to its angular acceleration about that pivot point. For a simple pendulum this is found to be the product of the mass of the particle with the square of its distance to the pivot. This is shown as follows:

The force of gravity on the mass of a simple pendulum generates a torque $\tau = r \times F$ around the axis perpendicular to the plane of the pendulum movement. Here F is the tangential component of the net force on the mass. Associated with this torque is an angular acceleration α of the string and mass around this axis. Since the mass is constrained to a circle the tangential acceleration of the mass is $a = \alpha \times r$. Since $F = ma$ the torque equation becomes:

$$\tau = r \times F = r \times (m\alpha \times r) = (mr^2)\alpha = I\alpha e,$$

where e is a unit vector perpendicular to the plane of the pendulum. (The second to the last step occurs because of the BAC-CAB rule using the fact that α is always perpendicular to r.) The quantity $I = mr^2$ is the *moment of inertia* of this single mass around the pivot point.

The quantity $I = mr^2$ also appears in the angular momentum of a simple pendulum, which is calculated from the velocity $v = \omega \times r$ of the pendulum mass around the pivot, where ω is the angular velocity of the mass about the pivot point. This angular momentum is given by

$$L = r \times (mv) = (mr^2)\omega = I\omega e,$$

using math similar to that used to derive the previous equation.

Similarly, the kinetic energy of the pendulum mass is defined by the velocity of the pendulum around the pivot to yield

$$E_K = \frac{1}{2} m v \cdot v = \frac{1}{2}(mr^2)\omega^2 = \frac{1}{2} I\omega^2.$$

This shows that the quantity $I = mr^2$ is how mass combines with the shape of a body to define rotational inertia. The moment of inertia of an arbitrarily shaped body is the sum of the values mr^2 for all of the elements of mass in the body.

Compound Pendulum

A compound pendulum is a body formed from an assembly of particles or continuous shapes that rotates rigidly around a pivot. Its moments of inertia is the sum the moments of inertia of each of the particles that is composed of. The natural frequency (ω_n) of a compound pendulum depends on its moment of inertia, I_P,

$$\omega_n = \sqrt{\frac{mgr}{I_P}},$$

where μ is the mass of the object, g is local acceleration of gravity, and r is the distance from the pivot point to the centre of mass of the object. Measuring this frequency of oscillation over small angular displacements provides an effective way of measuring moment of inertia of a body.

Thus, to determine the moment of inertia of the body, simply suspend it from a convenient pivot point P so that it swings freely in a plane perpendicular to the direction of the desired moment of inertia, then measure its natural frequency or period of oscillation (t), to obtain

$$I_P = \frac{mgr}{\omega_n^2} = \frac{mgrt^2}{4\pi^2},$$

where is the period (duration) of oscillation (usually averaged over multiple periods).

The moment of inertia of the body about its centre of mass, I_C, is then calculated using the parallel axis theorem to be

$$I_C = I_P - mr^2,$$

where m is the mass of the body and r is the distance from the pivot point P to the centre of mass c.

Moment of inertia of a body is often defined in terms of its *radius of gyration*, which is the radius of a ring of equal mass around the centre of mass of a body that has the same moment of inertia. The radius of gyration K is calculated from the body's moment of inertia I_C and mass m as the length,

$$K = \sqrt{\frac{I_C}{m}}.$$

Centre of Oscillation

A simple pendulum that has the same natural frequency as a compound pendulum defines the length L from the pivot to a point called the centre of oscillation of the compound pendulum. This point also corresponds to the centre of percussion. The length L is determined from the formula,

$$\omega_n = \sqrt{\frac{g}{L}} = \sqrt{\frac{mgr}{I_P}},$$

or

$$L = \frac{g}{\omega_n^2} = \frac{I_P}{mr}.$$

The seconds pendulum, which provides the "tick" and "tock" of a grandfather clock, takes one second to swing from side-to-side. This is a period of two seconds, or a natural frequency of π radians/second for the pendulum. In this case, the length L is given by,

$$L = \frac{g}{\omega_n^2} = \frac{9.81\ \text{m/s}^2}{(3.14\ \text{rad/s})^2} = 0.99\ \text{m}.$$

Notice that the centre of oscillation of the seconds pendulum must be adjusted to accommodate use in locations with different values for the local acceleration of gravity. Kater's pendulum is an example of a compound pendulum that is used to measure gravity called a gravimetre.

Measuring Moment of Inertia

The moment of inertia of complex systems such as a vehicle or airplane around its vertical axis can be measured by suspending the system from three points to form a trifilar pendulum. A trifilar pendulum is a platform supported by three wires designed to oscillate in torsion around its vertical centroidal axis. The period of oscillation of the trifilar pendulum yields the moment of inertia of the system.

Definition

The moment of inertia, I, is defined as the ratio of an applied torque to the angular acceleration along a principal axis of the object, where then τ, α and I are scalars, that is

$$\tau = I\alpha.$$

An equivalent definition of I uses the angular momentum L as follows,

$$L = I\omega,$$

where ω is the angular velocity of the object.

Apply this definition to a simple pendulum to see that the moment of inertia of the mass m about the pivot point at a distance r is

$$I = mr^2.$$

This generalizes to define the moment of inertia of a body about an axis S as the sum of all elemental point masses dm each multiplied by the square of its perpendicular distance r to the axis.

Calculating Moment of Inertia about an Axis

The moment of inertia about an axis of a body is calculated by summing mr^2 for every particle in the body, where r is the perpendicular distance to the specified axis. In order to see how moment of inertia arises in the study of the movement of an extended body, it is convenient to consider a rigid assembly of point masses. (This equation can be used for axes that are not principal axes provided that it is understood that this does not fully describe the moment of inertia.)

Consider the kinetic energy of an assembly of N masses m_i that lie at the distances r_i from the pivot point P, which is the nearest point on the axis of rotation. It is the sum of the kinetic energy of the individual masses,

$$E_K = \sum_{i=1}^{N} \frac{1}{2} m_i \mathrm{v}_i \cdot \mathrm{v}_i = \sum_{i=1}^{N} \frac{1}{2} m_i (\omega r_i)^2 = \frac{1}{2} \omega^2 \sum_{i=1}^{N} m_i r_i^2.$$

This shows that the moment of inertia of the body is the sum of each of the mr^2 terms, that is

$$I_P = \sum_{i=1}^{N} m_i r_i^2.$$

Thus, moment of inertia is a physical property that combines the mass and distribution of the particles around the rotation axis. Notice that rotation about different axes of the same body yield different moments of inertia.

The moment of inertia of a continuous body rotating about a specified axis is calculated in the same way, with the summation replaced by the integral,

$$I_P = \int_V \rho(\mathrm{r}) \mathrm{r}^2 \, dV.$$

Again r is the radius vector to a point in the body from the specified axis through the pivot P, and $\rho(\mathrm{r})$ is the mass density at each point r. The integration is evaluated over the volume V of the body. The moment of inertia of a flat surface is similar with the mass density being replaced by its areal mass density with the integral evaluated over its area.

Note on second moment of area: The moment of inertia of a body moving in a plane and the second moment of area of a beam's cross-section are often confused. The moment of inertia of body with the shape of the cross-section is the second moment of this area about the z-axis perpendicular to the cross-section, weighted by its density. This is also called the *polar moment of the area*, and is the sum of the second moments about the x and y axes. The stresses in a beam are calculated using the second moment of the cross-sectional area around either the x-axis or y-axis depending on the load.

Example Calculation of Moment of Inertia

The moment of inertia of a compound pendulum constructed from a thin disc mounted at the end of a thin rod that oscillates around a pivot at the other end of the rod, begins with the calculation of the moment of inertia of the thin rod and thin disc about their respective centres of mass.

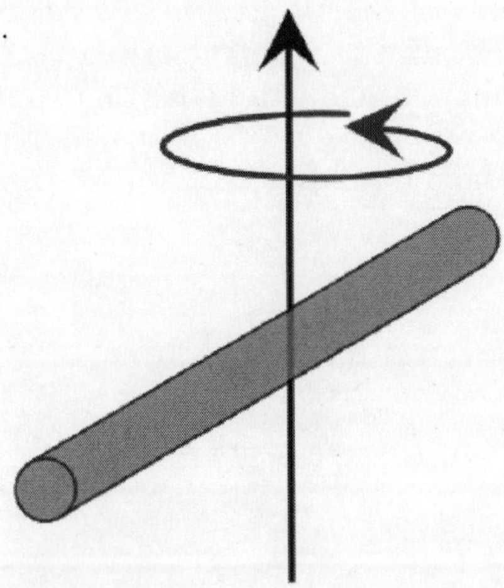

- The moment of inertia of a thin rod with constant cross-section s and density ñ and with length l about a perpendicular axis through its centre of mass is determined by integration. Align the x-axis with the rod and locate the origin its centre of mass at the centre of the rod, then

$$I_{C,\text{rod}} = \int \rho x^2 dV = \int_{-\ell/2}^{\ell/2} \rho x^2 s dx = \rho s \frac{x^3}{3}\Big|_{-\ell/2}^{\ell/2} = \frac{\rho s}{3}(\ell^3/8 + \ell^3/8) = \frac{1}{12} m\ell^2,$$

where $m = \rho s \ell$ is the mass of the rod.

- The moment of an inertia of a thin disc of constant thickness s, radius R, and density ρ about an axis through its centre and perpendicular to its face (parallel to its axis of rotational symmetry) is determined by integration. Align the z-axis with the axis of the disc and define a volume element as $dV = srdrd\theta$, then

$$I_{C,\text{disc}} = \int \rho r^2 dV = \int_0^{2\pi} \int_0^R \rho r^2 (srdrd\theta) = 2\pi \rho s \frac{R^4}{4} = \frac{1}{2} mR^2,$$

where $m = \rho R^2 \rho s$ is its mass.

- The moment of inertia of the compound pendulum is now obtained by adding the moment of inertia of the rod and the disc around the pivot point P as,

$$I_P = I_{C,\text{rod}} + M_{\text{rod}}(L/2)^2 + I_{C,\text{disc}} + M_{\text{disc}}(L+R)^2,$$

where L is the length of the pendulum. Notice that the parallel axis theorem is used to shift the moment of inertia from the centre of mass to the pivot point of the pendulum.

A list of moments of inertia formulas for standard body shapes provides a way to obtain the moment of inertial of a complex body as an assembly of simpler shaped bodies. The parallel axis theorem is used to shift the reference point of the individual bodies to the reference point of the assembly.

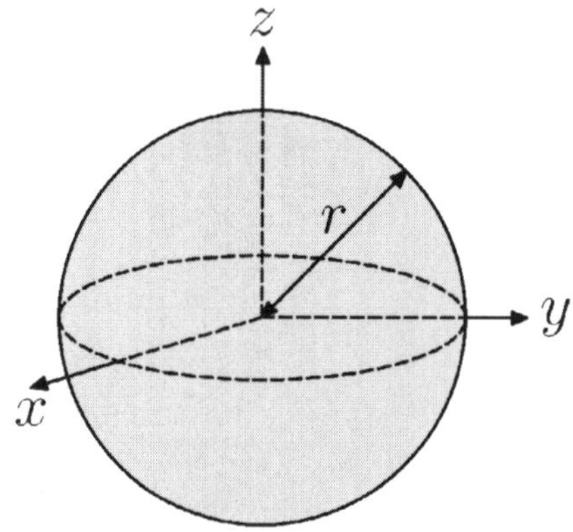

As one more example, consider the moment of inertia of a solid sphere of constant density about an axis through its centre of mass. This is determined by summing the moment of inertias of the thin discs that form the sphere. If the surface of the ball is defined by the equation

$$x^2 + y^2 + z^2 = R^2,$$

then the radius r of the disc at the cross-section z along the z-axis is

$$r(z)^2 = x^2 + y^2 = R^2 - z^2.$$

Therefore, the moment of inertia of the ball is the sum of the moment of inertias of the discs along the z-axis,

$$I_{C,\text{ball}} = \int_{-R}^{R} \frac{\pi\rho}{2} r(z)^4\,dz = \int_{-R}^{R} \frac{\pi\rho}{2}(R^2 - z^2)^2\,dz = \frac{\pi\rho}{2}(R^4 z - 2R^2 z^3/3 + z^5/5)\Big|_{-R}^{R}$$

$$= \pi\rho(1 - 2/3 + 1/5)R^5 = \frac{2}{5}mR^2,$$

where$m$ = $(4/3)\pi R^3\rho$ is the mass of the ball.

Moment of Inertia in Planar Movement of a Rigid Body

If a mechanical system is constrained to move parallel to a fixed plane, then the rotation of a body in the system occurs around an axis k perpendicular to this plane. In this case, the moment of inertia of the mass in this system is a scalar known as the *polar moment of inertia.* The definition of the polar moment of inertia can be obtained by considering momentum, kinetic energy and Newton's laws for the planar movement of a rigid system of particles. If a system of n particles, P_i, $i = 1,...,n$, are assembled into a rigid body, then the momentum of the system can be written in terms of positions relative to a reference point R, and absolute velocities v_i

$$\Delta r_i = r_i - R, \quad v_i = \omega \times (r_i - R) + V,$$

where ω is the angular velocity of the system and V is the velocity of R.

For planar movement the angular velocity vector is directed along the unit vector k which is perpendicular to the plane of movement. Introduce the unit vectors e_i from the reference point R to a point r_i, and the unit vector $t_i = k \times e_i$ so

$$\Delta r_i e_i = r_i - R, \quad v_i = \omega \Delta r_i t_i + V, \quad i = 1, \ldots, n.$$

This defines the relative position vector and the velocity vector for the rigid system of the particles moving in a plane.

Note on the cross product: When a body moves parallel to a ground plane, the trajectories of all the points in the body lie in planes parallel to this ground plane. This means that any rotation that the body undergoes must be around an axis perpendicular to this plane. Planar movement is often presented as projected onto this ground plane so that the axis of rotation appears as a point. In this case, the angular velocity and angular acceleration of the body are scalars and the fact that they are vectors along the rotation axis is ignored. This is usually preferred for introductions to the topic. But in the case of moment of inertia, the combination of mass and geometry benefits from the geometric properties of the cross product. For this reason, in this section on planar movement the angular velocity and accelerations of the body are vectors perpendicular to the ground plane, and the cross product operations are the same as used for the study of spatial rigid body movement.

Angular Momentum in Planar Movement

The angular momentum vector for the planar movement of a rigid system of particles is given by

$$\begin{aligned} L &= \sum_{i=1}^{n} m_i (r_i - R) \times v_i \\ &= \sum_{i=1}^{n} m_i \Delta r_i e_i \times (\omega \Delta r_i t_i + V) \\ &= \left(\sum_{i=1}^{n} m_i \Delta r_i^2\right) \omega \vec{k} + \left(\sum_{i=1}^{n} m_i \Delta r_i e_i\right) \times V. \end{aligned}$$

Use the centre of massC as the reference point so

$$\Delta r_i \mathrm{e}_i = \mathrm{r}_i - \mathrm{C}, \quad \sum_{i=1}^{n} m_i \Delta r_i \mathrm{e}_i = 0,$$

and define the moment of inertia relative to the centre of mass I_C as

$$I_C = \sum_{i=1}^{n} m_i \Delta r_i^2,$$

then the equation for angular momentum simplifies to

$$\mathrm{L} = I_C \omega \vec{k}.$$

The moment of inertia I_C about an axis perpendicular to the movement of the rigid system and through the centre of mass is known as the *polar moment of inertia.*

For a given amount of angular momentum, a decrease in the moment of inertia results in an increase in the angular velocity. Figure skaters can change their moment of inertia by pulling in their arms.

Thus, the angular velocity achieved by a skater with outstretched arms results in a greater angular velocity when the arms are pulled in, because of the reduced moment of inertia.

Kinetic Energy in Planar Movement

The kinetic energy of a rigid system of particles moving in the plane is given by

$$E_\mathrm{K} = \frac{1}{2} \sum_{i=1}^{n} m_i \mathrm{v}_i \cdot \mathrm{v}_i = \frac{1}{2} \sum_{i=1}^{n} m_i (\omega \Delta r_i \mathrm{t}_i + \mathrm{V}) \cdot (\omega \Delta r_i \mathrm{t}_i + \mathrm{V}).$$

This equation expands to yield three terms

$$E_\mathrm{K} = \frac{1}{2} \omega^2 \sum_{i=1}^{n} m_i \Delta r_i^2 (\mathrm{t}_i \cdot \mathrm{t}_i) + \omega \mathrm{V} \cdot (\sum_{i=1}^{n} m_i \Delta r_i \mathrm{t}_i) + \frac{1}{2} (\sum_{i=1}^{n} m_i) \mathrm{V} \cdot \mathrm{V}.$$

Let the reference point be the centre of mass C of the system so the second term becomes zero, and introduce the moment of inertia I_C so the kinetic energy is given by

$$E_\mathrm{K} = \frac{1}{2} I_C \omega^2 + \frac{1}{2} M \mathrm{V} \cdot \mathrm{V}.$$

The moment of inertia I_C is the *polar moment of inertia* of the body.

Figure: *A 1920's John Deere tractor with the spoked flywheel on the engine. The large moment of inertia of the flywheel smooths the operation of the tractor*

Newton's laws for a rigid system of N particles, P_i, $i = 1,...,N$, can be written in terms of a resultant force and torque at a reference point R, to yield

$$\mathrm{F} = \sum_{i=1}^{N} m_i \mathrm{A}_i, \quad \mathrm{T} = \sum_{i=1}^{N} (\mathrm{r}_i - \mathrm{R}) \times (m_i \mathrm{A}_i),$$

where r_i denotes the trajectory of each particle.

The kinematics of a rigid body yields the formula for the acceleration of the particle P_i in terms of the position R and acceleration A of the reference particle as well as the angular velocity vector ω and angular acceleration vector α of the rigid system of particles as,

$$\mathrm{A}_i = \alpha \times (\mathrm{r}_i - \mathrm{R}) + \omega \times \omega \times (\mathrm{r}_i - \mathrm{R}) + \mathrm{A}.$$

For systems that are constrained to planar movement, the angular velocity and angular acceleration vectors are directed along k perpendicular to the plane of movement, which simplifies this acceleration equation. In this case, the acceleration vectors can be simplified by introducing the unit vectors e_i from the reference point R to a point r_i and the unit vectors $t_i = k \times e_i$, so

$$\mathrm{A}_i = \alpha(\Delta r_i \mathrm{t}_i) - \omega^2 (\Delta r_i \mathrm{e}_i) + \mathrm{A}.$$

This yields the resultant torque on the system as

$$\tau = \sum_{i=1}^{N} (m_i \Delta r_i \mathrm{e}_i) \times (\alpha(\Delta r_i \mathrm{t}_i) - \omega^2 (\Delta r_i \mathrm{e}_i) + \mathrm{A}) = (\sum_{i=1}^{N} m_i \Delta r_i^2) \alpha \vec{k} + (\sum_{i=1}^{N} m_i \Delta r_i \mathrm{e}_i) \times \mathrm{A},$$

where ei × $e_i = 0$, and $e_i \times t_i = k$ is the unit vector perpendicular to the plane for all of the particles P_i .

Use the centre of massC as the reference point and define the moment of inertia relative to the centre of mass I_C, then the equation for the resultant torque simplifies to

$$\tau = I_C \alpha \vec{k}.$$

The parameter I_C is the *polar moment of inertia* of the moving body.

The Inertia Matrix for Spatial Movement of a Rigid Body

The scalar moments of inertia appear as elements in a matrix when a system of particles is assembled into a rigid body that moves in three-dimensional space. This inertia matrix appears in the calculation of the angular momentum, kinetic energy and resultant torque of the rigid system of particles.

An important application of the inertia matrix and Newton's laws of motion is the analysis of a spinning top. This is discussed in the article on gyroscopic precession. A more detailed presentation can be found in the article on Euler's equations of motion.

Let the system of particles P_i, $i = 1,...,n$ be located at the coordinates r_i with velocities v_i relative to a fixed reference frame. For a (possibly moving) reference point R, the relative positions are

$$\Delta r_i = r_i - R$$

and the (absolute) velocities are

$$v_i = \omega \times \Delta r_i + V_R$$

where ω is the angular velocity of the system, and V_R is the velocity of R.

Angular Momentum

If the reference point R in the assembly, or body, is chosen as the centre of mass C, then its angular momentum takes the form,

$$L = \sum_{i=1}^{n} m_i \Delta r_i \times v_i = \sum_{i=1}^{n} m_i \Delta r_i \times (\omega \times \Delta r_i),$$

where the terms containing V_R sum to zero by definition of the centre of mass.

In order to define the inertia matrix, introduce the skew-symmetric matrix $[B]$ constructed from a vector b that performs the cross product operation, such that

$$[B]y = b \times y.$$

This matrix $[B]$ has the components of $b = (b_x, b_y, b_z)$ as its elements, in the form

$$[B] = \begin{bmatrix} 0 & -b_z & b_y \\ b_z & 0 & -b_x \\ -b_y & b_x & 0 \end{bmatrix}.$$

Now construct the skew-symmetric matrix $[\Delta r_i] = [r_i - C]$ obtained from the relative position vector $\Delta r_i = r_i - C$, and use this skew-symmetric matrix to define,

$$L = (-\sum_{i=1}^{n} m_i [\Delta r_i]^2)\omega = [I_C]\omega,$$

where $[I_C]$ defined by

$$[I_C] = -\sum_{i=1}^{n} m_i [\Delta r_i]^2,$$

is the inertia matrix of the rigid system of particles measured relative to the centre of mass C.

Kinetic Energy

The kinetic energy of a rigid system of particles can be formulated in terms of the centre of mass and a matrix of mass moments of inertia of the system. Let the system of particles P_i, $i = 1,...,n$ be located at the coordinates r_i with velocities v_i, then the kinetic energy is

$$E_K = \frac{1}{2}\sum_{i=1}^{n} m_i v_i \cdot v_i = \frac{1}{2}\sum_{i=1}^{n} m_i (\omega \times \Delta r_i + V_C)\cdot(\omega \times \Delta r_i + V_C),$$

where $\Delta r_i = r_i - C$ is the position vector of a particle relative to the centre of mass.

This equation expands to yield three terms

$$E_K = \frac{1}{2}\sum_{i=1}^{n} m_i (\omega \times \Delta r_i)\cdot(\omega \times \Delta r_i)) + \sum_{i=1}^{n} m_i V_C \cdot (\omega \times \Delta r_i)) + \frac{1}{2}\sum_{i=1}^{n} m_i V_C \cdot V_C.$$

The second term in this equation is zero because C is the centre of mass. Introduce the skew-symmetric matrix $[\Delta r_i]$ so the kinetic energy becomes

$$E_K = \frac{1}{2}\sum_{i=1}^{n} m_i ([\Delta r_i]\omega)\cdot([\Delta r_i]\omega) + \frac{1}{2}(\sum_{i=1}^{n} m_i) V_C \cdot V_C.$$

$$E_K = \frac{1}{2}\sum_{i=1}^{n} m_i (\omega^T [\Delta r_i]^T [\Delta r_i]\omega) + \frac{1}{2}(\sum_{i=1}^{n} m_i) V_C \cdot V_C.$$

$$E_K = \frac{1}{2}\omega\cdot(-\sum_{i=1}^{n} m_i[\Delta r_i]^2)\omega + \frac{1}{2}(\sum_{i=1}^{n} m_i)V_C\cdot V_C.$$

Thus, the kinetic energy of the rigid system of particles is given by

$$E_K = \frac{1}{2}\omega\cdot[I_C]\omega + \frac{1}{2}MV_C^2.$$

where $[I_C]$ is the inertia matrix relative to the centre of mass and M is the total mass.

Resultant Torque

The inertia matrix appears in the application of Newton's second law to a rigid assembly of particles. The resultant torque on this system is,

$$\tau = \sum_{i=1}^{n} (r_i - R)\times(m_i a_i),$$

where a_i is the acceleration of the particle P_i. The kinematics of a rigid body yields the formula for the acceleration of the particle P_i in terms of the position R and acceleration A of the reference point, as well as the angular velocity vector ù and angular acceleration vector á of the rigid system as,

$$a_i = \alpha\times(r_i - R) + \omega\times\omega\times(r_i - R) + A_R.$$

Use the centre of mass C as the reference point, and introduce the skew-symmetric matrix $[\Delta r_i]=[r_i\text{-}C]$ to represent the cross product $(r_i - C)x$, in order to obtain

$$\tau = (-\sum_{i=1}^{n} m_i[\Delta r_i]^2)\alpha + \omega\times(-\sum_{i=1}^{n} m_i[\Delta r_i]^2)\omega$$

This calculation uses the identity

$$\Delta r_i \times(\omega\times(\omega\times\Delta r_i)) + \omega\times(\Delta r_i\times(\Delta r_i\times\omega)) = 0,$$

obtained from the Jacobi identity for the triple cross product.

Thus, the resultant torque on the rigid system of particles is given by

$$\tau = [I_C]\alpha + \omega\times[I_C]\omega,$$

where $[I_C]$ is the inertia matrix relative to the centre of mass.

Parallel Axis Theorem

The inertia matrix of a body depends on the choice of the reference point. There is a useful relationship between the inertia matrix relative to the centre of mass C and the inertia matrix relative to another point R. This relationship is called the parallel axis theorem.

Consider the inertia matrix $[I_R]$ obtained for a rigid system of particles measured relative to a reference point R, given by

$$[I_R] = -\sum_{i=1}^{n} m_i [r_i - R]^2.$$

Let C be the centre of mass of the rigid system, then

$$R = (R - C) + C = d + C,$$

where d is the vector from the centre of mass C to the reference point R. Use this equation to compute the inertia matrix,

$$[I_R] = -\sum_{i=1}^{n} m_i [r_i - C - d]^2.$$

Expand this equation to obtain

$$[I_R] = (-\sum_{i=1}^{n} m_i [r_i - C]^2) + (\sum_{i=1}^{n} m_i [r_i - C])[d] + [d](\sum_{i=1}^{n} m_i [r_i - C]) + (-\sum_{i=1}^{n} m_i)[d][d].$$

The first term is the inertia matrix $[I_C]$ relative to the centre of mass. The second and third terms are zero by definition of the centre of mass C. And the last term is the total mass of the system multiplied by the square of the skew-symmetric matrix [d] constructed from d.

The result is the parallel axis theorem,

$$[I_R] = [I_C] - M[d]^2,$$

where d is the vector from the centre of mass C to the reference point R.

Note on the minus sign: By using the skew symmetric matrix of position vectors relative to the reference point, the inertia matrix of each particle has the form $-m[r]^2$, which is similar to the mr^2 that appears in planar movement. However, to make this to work out correctly a minus sign is needed. This minus sign can be absorbed into the term $m[r]^T[r]$, if desired, by using the skew-symmetry property of [r].

The Inertia Matrix and the Scalar Moment of Inertia around an Arbitrary Axis

The relationship between the inertia matrix of a rigid body and the scalar moment of inertia of the same body about a specified axis is important and rarely presented in detail. The following calculation expands the derivation presented by Kane and Levinson.

Let a rigid assembly of rigid system of N particles, P_i, $i = 1,\ldots,N$, have coordinates r_i. Choose R as a reference point and compute the moment of inertia around an axis L defined by the unit vector S through the reference point R. The moment of inertia of the system around this line L=R+tS is computed by determining the perpendicular vector from this axis to the particle P_i given by

$$\Delta r_i^{\perp} = (r_i - R) - (S\cdot(r_i - R))S = [[I] - [SS^T]](\Delta r_i),$$

where [I] is the identity matrix and [SS^T] is the outer product matrix formed from the unit vector S along the line L.

In order to relate this scalar moment of inertia to the inertia matrix of the body, introduce the skew-symmetric matrix [S] such that [S]y=S x y, then we have the identity

$$-[S]^2 = [I] - [SS^T],$$

which relies on the fact that S is a unit vector.

The magnitude squared of the perpendicular vector is

$$|\Delta r_i^{\perp}|^2 = (-[S]^2(\Delta r_i))\cdot(-[S]^2(\Delta r_i)) = -S\cdot[\Delta r_i][\Delta r_i]S.$$

The simplification of this equation uses the identity

$$(S\times(S\times(\Delta r_i)))\cdot S\times(S\times(\Delta r_i)) = (S\times(S\times(\Delta r_i)))\times S\cdot(S\times(\Delta r_i)),$$

where the dot and the cross products have been interchanged. Expand the cross products to compute

$$-(\Delta r_i)\times S\cdot(S\times(\Delta r_i) = -S\cdot[\Delta r_i][\Delta r_i]S,$$

where [Δr_i] is the skew symmetric matrix obtained from the vector $\Delta r = r_i - R$.

Thus, the moment of inertia around the line L through R in the direction S is obtained from the calculation

$$I_L = \sum_{i=1}^{N} m_i |\Delta r_i^{\perp}|^2 = -\sum_{i=1}^{N} m_i S\cdot[\Delta r_i]^2 S,$$

or

$$I_L = S\cdot(-\sum_{i=1}^{N} m_i[\Delta r_i]^2)S = S\cdot[I_R]S = S^T[I_R]S,$$

where [I_R] is the moment of inertia matrix of the system relative to the reference point R.

This shows that the inertia matrix can be used to calculate the moment of inertia of a body around any specified rotation axis in the body.

The Inertia Tensor

The inertia matrix is often described as the inertia tensor, which consists of the same moments of inertia and products of inertia about the three coordinate axes. The inertia tensor is constructed from the nine component tensors, (the symbol $\otimes$ is the tensor product)

$$e_i \otimes e_j, \quad i, j = 1,2,3,$$

where e_i, i=1,2,3 are the three orthogonal unit vectors defining the inertial frame in which the body moves. Using this basis the inertia tensor is given by

$$I = \sum_{i=1}^{3}\sum_{j=1}^{3} I_{ij} e_i \otimes e_j.$$

This tensor is of degree two because the component tensors are each constructed from two basis vectors. In this form the inertia tensor is also called the *inertia binor*.

For a rigid system of particles P_k, $k = 1,\ldots,N$ each of mass m_k with position coordinates $r_k=(x_k, y_k, z_k)$, the inertia tensor is given by

$$I = \sum_{k=1}^{N} m_k((r_k \cdot r_k)E - r_k \otimes r_k),$$

where E is the identity tensor

$$E = e_1 \otimes e_1 + e_2 \otimes e_2 + e_3 \otimes e_3.$$

The inertia tensor for a continuous body is given by

$$I = \int_V \rho(r)\big((r \cdot r)E - r \otimes r\big)dV,$$

where r defines the coordinates of a point in the body and ρ (r) is the mass density at that point. The integral is taken over the volume V of the body. The inertia tensor is symmetric because $I_{ij} = I_{ji}$.

Alternatively it can also be written in terms of the hat operator as:

$$I = \int_V \rho(r)(\hat{r})^2 \, dV,$$

The inertia tensor can be used in the same way as the inertia matrix to compute the scalar moment of inertia about an arbitrary axis in the direction n,

$$I_n = n \cdot I \cdot n,$$

where the dot product is taken with the corresponding elements in the component tensors. A product of inertia term such as I_{12} is obtained by the computation

$$I_{12} = \mathrm{e}_1 \cdot \mathrm{I} \cdot \mathrm{e}_2,$$

and can be interpreted as the moment of inertia around the x-axis when the object rotates around the y-axis.

The components of tensors of degree two can be assembled into a matrix. For the inertia tensor this matrix is given by,

$$[I] = \begin{bmatrix} I_{11} & I_{12} & I_{13} \\ I_{21} & I_{22} & I_{23} \\ I_{31} & I_{32} & I_{33} \end{bmatrix} = \begin{bmatrix} I_{xx} & I_{xy} & I_{xz} \\ I_{xy} & I_{yy} & I_{yz} \\ I_{xz} & I_{yz} & I_{zz} \end{bmatrix}.$$

It is common in rigid body mechanics to use notation that explicitly identifies the x, y, and z axes, such as I_{xx} and I_{xy}, for the components of the inertia tensor.

Identities for a Skew-Symmetric Matrix

In order to compute moment of inertia of a mass around an axis, the perpendicular vector from the mass to the axis is needed. If the axis L is defined by the unit vector S through the reference point R, then the perpendicular vector from the line L to the point r is given by

$$\Delta \mathrm{r}_i^{\perp} = (\mathrm{r}_i - \mathrm{R}) - (\mathrm{S} \cdot (\mathrm{r}_i - \mathrm{R}))\mathrm{S} = [[I] - [\mathrm{SS}^T]](\Delta \mathrm{r}_i),$$

where [I] is the identity matrix and $[SS^T]$ is the outer product matrix formed from the unit vector S along the line L. Recall that skew-symmetric matrix [S] is constructed so that [S]y=S x y. The matrix $[I-SS^T]$ in this equation subtracts the component of $\Delta r = r - R$ that is parallel to S.

The previous sections show that in computing the moment of inertia matrix this operator yields a similar operator using the components of the vector Δr that is

$$[I \,|\, \Delta \mathrm{r}|^2 - \Delta \mathrm{r} \Delta \mathrm{r}^T].$$

It is helpful to keep the following identities in mind In order to compare the equations that define the inertia tensor and the inertia matrix.

Let [R] be the skew symmetric matrix associated with the position vector R=(x, y, z), then the product in the inertia matrix becomes

$$-[R]^2 = -\begin{bmatrix} 0 & -z & y \\ z & 0 & -x \\ -y & x & 0 \end{bmatrix}^2 = \begin{bmatrix} y^2+z^2 & -xy & -xz \\ -yx & x^2+z^2 & -yz \\ -zx & -zy & x^2+y^2 \end{bmatrix}.$$

This can be viewed as another way of computing the perpendicular distance from an axis to a point, because the matrix formed by the outer product [RR^T] yields the identify

$$-[R]^2 = |\mathrm{R}|^2[I] - [\mathrm{RR}^T] = \begin{bmatrix} x^2+y^2+z^2 & 0 & 0 \\ 0 & x^2+y^2+z^2 & 0 \\ 0 & 0 & x^2+y^2+z^2 \end{bmatrix} - \begin{bmatrix} x^2 & xy & xz \\ yx & y^2 & yz \\ zx & zy & z^2 \end{bmatrix},$$

where [I] is the 3x3 identity matrix.

Also notice, that

$$|\mathrm{R}|^2 = \mathrm{R}\cdot\mathrm{R} = \mathrm{tr}[\mathrm{RR}^T],$$

where*tr* denotes the sum of the diagonal elements of the outer product matrix, known as its trace.

The Inertia Matrix in Different Reference Frames

The use of the inertia matrix in Newton's second law assumes its components are computed relative to axes parallel to the inertial frame and not relative to a body-fixed reference frame. This means that as the body moves the components of the inertia matrix change with time. In contrast, the components of the inertia matrix measured in a body-fixed frame are constant.

Body Frame Inertia Matrix

Let the body frame inertia matrix relative to the centre of mass be denoted [I_C^B], and define the orientation of the body frame relative to the inertial frame by the rotation matrix [A], such that,

$$\mathrm{x} = [A]\mathrm{y},$$

where vectors y in the body fixed coordinate frame have coordinates x in the inertial frame. Then, the inertia matrix of the body measured in the inertial frame is given by

$$[I_C] = [A][I_C^B][A^T].$$

Notice that [A] changes as the body moves, while [I_C^B] remains constant.

Principal Axes

Measured in the body frame the inertia matrix is a constant real symmetric matrix. A real symmetric matrix has the eigen decomposition into the product of a rotation matrix [Q] and a diagonal matrix [Λ], given by

$$[I_C^B] = [Q][\Lambda][Q^T],$$

where

$$[\Lambda] = \begin{bmatrix} I_1 & 0 & 0 \\ 0 & I_2 & 0 \\ 0 & 0 & I_3 \end{bmatrix}.$$

The columns of the rotation matrix [Q] define the directions of the principal axes of the body, and the constants I_1, I_2 and I_3 are called the principal moments of inertia. This result was first shown by J. J. Sylvester (1852), and is a form of Sylvester's law of inertia.

For bodies with constant density an axis of rotational symmetry is a principal axis.

Inertia Ellipsoid

The moment of inertia matrix in body-frame coordinates is a quadratic form that defines a surface in the body called Poinsot's ellipsoid. Let $[\Lambda]$ be the inertia matrix relative to the centre of mass aligned with the principal axes, then the surface

$$\mathrm{x}^T[\Lambda]\mathrm{x} = 1,$$

or

$$I_1x^2 + I_2y^2 + I_3z^2 = 1,$$

defines an ellipsoid in the body frame. Write this equation in the form,

$$\frac{x^2}{(1/\sqrt{I_1})^2} + \frac{y^2}{(1/\sqrt{I_2})^2} + \frac{z^2}{(1/\sqrt{I_3})^2} = 1,$$

to see that the semi-principal diameters of this ellipsoid are given by

$$a = \frac{1}{\sqrt{I_1}}, \quad b = \frac{1}{\sqrt{I_2}}, \quad c = \frac{1}{\sqrt{I_3}}.$$

Let a point x on this ellipsoid be defined in terms of its magnitude and direction, x= | x | n, where n is a unit vector. Then the relationship presented above, between the inertia matrix and the scalar moment of inertia I_n around an axis in the direction n, yields

$$\mathrm{x}^T[\Lambda]\mathrm{x} = |\mathrm{x}|^2\, \mathrm{n}^T[\Lambda]\mathrm{n} = |\mathrm{x}|^2\, I_n = 1.$$

Thus, the magnitude of a point x in the direction n on the inertia ellipsoid is

$$|\mathrm{x}| = \frac{1}{\sqrt{I_n}}.$$

Deflection (Engineering)

In engineering, deflection is the degree to which a structural element is displaced under a load. It may refer to an angle or a distance.

The deflection distance of a member under a load is directly related to the slope of the deflected shape of the member under that load and can be calculated by integrating the function that mathematically describes the slope of the member under that load. Deflection can be calculated by standard formula (will only give the deflection of common beam configurations and load cases at discrete locations), or by methods such as virtual work, direct integration, Castigliano's method, Macaulay's method or the direct stiffness method, amongst others. The deflection of beam elements is usually calculated on the basis of the Euler–Bernoulli beam equation while that of a plate or shell element is calculated using plate or shell theory.

An example of the use of deflection in this context is in building construction. Architects and engineers select materials for various applications. The beams used for frame work are selected on the basis of deflection, amongst other factors.

Beam Deflection for Various Loads and Supports

End loaded cantilever beams:

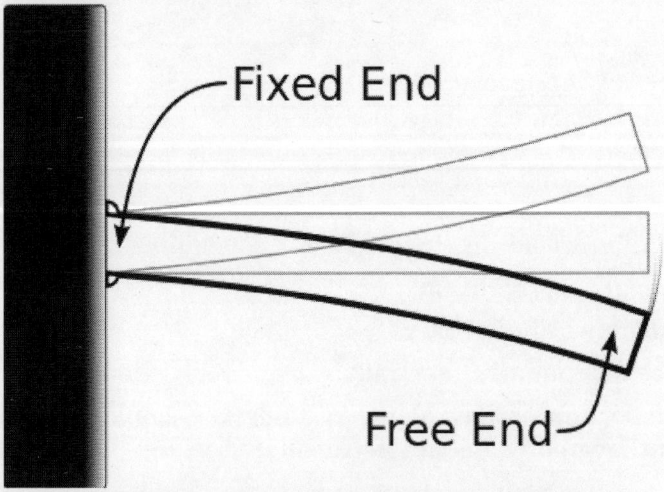

Figure: *Deflection of a cantilevered beam*

The elastic deflection δ and angle of deflection ϕ (in radians) in the example image, a (weightless) cantilever beam, with an end load, can be calculated (at the free end B) using:

$$\delta_B = \frac{FL^3}{3EI} \qquad \phi_B = \frac{FL^2}{2EI}$$

where

F = Force acting on the tip of the beam

L = Length of the beam (span)

E = Modulus of elasticity

I = Area moment of inertia

Note that if the span doubles, the deflection increases eightfold. The deflection at any point, x, along the span of an end loaded cantilevered beam can be calculated using:

$$\delta_x = \frac{Fx^2}{6EI}(3L - x)$$

$$\phi_x = \frac{Fx}{2EI}(2L - x)$$

Note that at $x = L$ (the end of the beam), the δ_x and ϕ_x equations are identical to the δ_B and ϕ_B equations above.

Uniformly Loaded Cantilever Beam

The deflection, at the free end B, of a cantilevered beam under a uniform load is given by:

$$\delta_B = \frac{qL^4}{8EI}$$

$$\phi_B = \frac{qL^3}{6EI}$$

where

q = Uniform load on the beam (force per unit length)

L = Length of the beam

E = Modulus of elasticity

I = Area moment of inertia

The deflection at any point, x, along the span of a uniformly loaded cantilevered beam can be calculated using:

$$\delta_x = \frac{qx^2}{24EI}(6L^2 - 4Lx + x^2)$$

$$\phi_x = \frac{qx}{6EI}(3L^2 - 3Lx + x^2)$$

Centre Loaded Beam

The elastic deflection (at the midpoint C) of a beam, loaded at its centre, supported by two simple supports is given by:

$$\delta_C = \frac{FL^3}{48EI}$$

where

F = Force acting on the centre of the beam

L = Length of the beam between the supports

E = Modulus of elasticity

I = Area moment of inertia

The deflection at any point, x, along the span of a centre loaded simply supported beam can be calculated using:

$$\delta_x = \frac{Fx}{48EI}(3L^2 - 4x^2)$$

for $\quad 0 \le x \le \frac{L}{2}$

<distance><leg>namereg<stadd>design<k>2.89>td tc=2.9>1+0.26<staticload>dc/dr<2.98<8.{2.56}+10.23

Intermediately Loaded Beam

The maximum elastic deflection on a beam supported by two simple supports, loaded at a distance a from the closest support, is given by:

$$\delta_{max} = \frac{Fa(L^2 - a^2)^{3/2}}{9\sqrt{3}LEI}$$

where

F = Force acting on the beam

L = Length of the beam between the supports

E = Modulus of elasticity

I = Area moment of inertia

a = Distance from the load to the closest support (i.e. $a \le L/2$)

This maximum deflection occurs at a distance x_1 from the closest support and is given by:

$$x_1 = \sqrt{\frac{L^2 - a^2}{3}}$$

Uniformly Loaded Beam

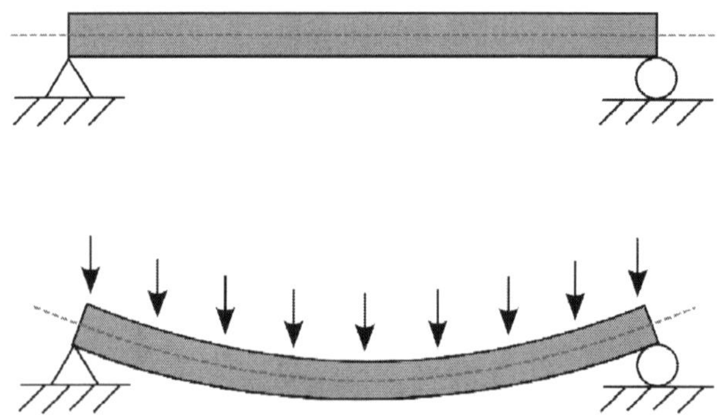

Figure: *Shows a statically determinate beam, deflecting under an evenly distributed load.*

The elastic deflection (at the midpoint C) on a beam supported by two simple supports, under a uniform load (as pictured) is given by:

$$\delta_C = \frac{5qL^4}{384EI}$$

where

q= Uniform load on the beam (force per unit length)

L = Length of the beam

E = Modulus of elasticity

I= Area moment of inertia

The deflection at any point, x, along the span of a uniformly loaded simply supported beam can be calculated using:

$$\delta_x = \frac{qx}{24EI}(L^3 - 2Lx^2 + x^3)$$

Structural Deflection

Building codes determine the maximum deflection, usually as a fraction of the span e.g. 1/400 or 1/600. Either the strength limit state (allowable stress) or the serviceability limit state (deflection considerations amongst others) may govern the minimum dimensions of the member required.

The deflection must be considered for the purpose of the structure. When designing a steel frame to hold a glazed panel, one allows only minimal deflection to prevent fracture of the glass.

The deflected shape of a beam can be represented by the moment diagram, integrated (twice, rotated and translated to enforce support conditions).

Bending Moment

A bending moment is a measure of the average internal stress induced in a structural element when an external force or moment is applied to the element causing the element to bend.

The internal stresses in a cross-section of a structural element can be resolved into a resultant force and a resultant couple. For equilibrium, the moment created by external forces (and external moments) must be balanced by the couple induced by the internal stresses. The resultant internal couple is called the bending moment while the resultant internal force is called the shear force (if it is transverse to the plane of element) or the normal force (if it is along the plane of the element).

The bending moment at a section through a structural element may be defined as "the sum of the moments about that section of all external forces acting to one side of that section". The forces and moments on either side of the section must be equal in order to counteract each other and maintain a state of equilibrium so the same bending moment will result from summing the moments, regardless of which side of the section is selected. If clockwise bending moments are taken as negative, then a negative bending moment within an element will cause "sagging", and a positive moment will cause "hogging". It is therefore clear that a point of zero bending moment within a beam is a point of contraflexure—that is the point of transition from hogging to sagging or vice versa.

Moments and torques are measured as a force multiplied by a distance so they have as unit newton-metres (N·m), or pound-foot or foot-pound (ft·lb). The concept of bending moment is very important in engineering (particularly in civil and mechanical engineering) and physics.

Background

Tensile and compressive stresses increase proportionally with bending moment, but are also dependent on the second moment of area of the cross-section of the structural element. Failure in bending will occur when the bending moment is sufficient to induce tensile stresses greater than the yield stress of the material throughout the entire cross-section. It is possible that failure of a structural element in shear may occur before failure in bending, however the mechanics of failure in shear and in bending are different. Moments are calculated by multiplying the external vector forces (loads or reactions) by the vector distance at which they are applied. When analysing an entire element, it is sensible to calculate moments at both ends of the element, at the beginning, centre and end of any uniformly distributed loads, and directly underneath any point loads. Of course any "pin-joints" within a structure allow free rotation, and so zero moment occurs at these points as there is no way of transmitting turning forces from one side to the other.

It is more common to use the convention that a clockwise bending moment to the left of the point under consideration is taken as positive. This then corresponds to the second derivative of a function which, when positive, indicates a curvature that is 'lower at the centre' i.e. sagging. When defining moments and curvatures in this way calculus can be more readily used to find slopes and deflections.

Critical values within the beam are most commonly annotated using a bending moment diagram, where negative moments are plotted to scale above a horizontal line and positive below. Bending moment varies linearly over unloaded sections, and parabolically over uniformly loaded sections. Engineering descriptions of the computation of bending moments can be confusing because of unexplained sign conventions and implicit assumptions. The descriptions below use vector mechanics to compute of moments of force and bending moments in an attempt to explain, from first principles, why particular sign conventions are chosen.

Computing the Moment of Force

An important part of determining bending moments in practical problems is the computation of moments of force. Let F be a force vector acting at a point A in a body. The moment of this force about a reference point (O) is defined as

$$M = r \times F$$

where M is the moment vector and r is the position vector from the reference point (O) to the point of application of the force (A). The $\times$symbol indicates the vector cross product. For many problems, it is more convenient to compute the moment of force about an axis that passes through the reference point O. If the unit vector along the axis is e, the moment of force about the axis is defined as

$$M = e \cdot M = e \cdot (r \times F)$$

where.indicates the vector dot product.

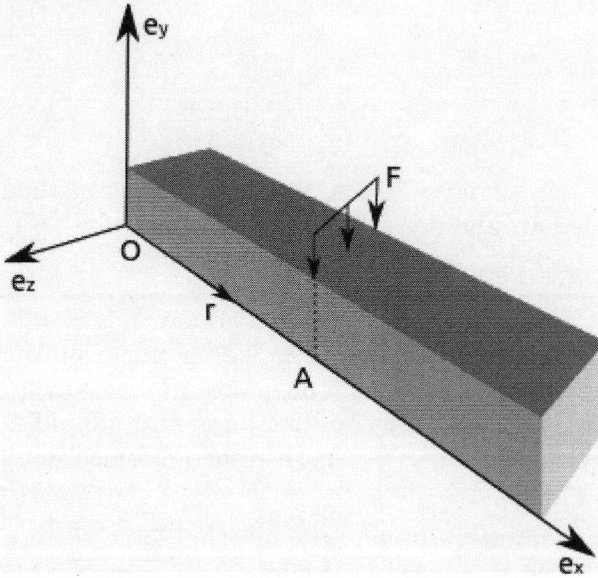

Figure: *Computing the moment of force in a beam.*

Example

The adjacent figure shows a beam that is acted upon by a force *F*. If the coordinate system is defined by the three unit vectors e_x, e_y, e_z, we have the following

$$F = 0e_x - Fe_y + 0e_z \quad \text{and} \quad r = xe_x + 0e_y + 0e_z.$$

Therefore,

$$M = r \times F = \begin{vmatrix} e_x & e_y & e_z \\ x & 0 & 0 \\ 0 & -F & 0 \end{vmatrix} = -Fxe_z.$$

The moment about the axis e_z is then

$$M_z = e_z \cdot M = -Fx.$$

Sign Conventions

The negative value suggests that a moment that tends to rotate a body clockwise around an axis should have a negative sign. However, the actual sign depends on the choice of the three axes e_x, e_y, e_z. For instance, if we choose another right handed coordinate system with $E_x = e_x, E_y = -e_z, E_z = e_y$, we have

$$F = 0E_x + 0E_y - FE_z \quad \text{and} \quad r = xE_x + 0E_y + 0E_z.$$

Then,

$$\mathrm{M} = \mathrm{r} \times \mathrm{F} = \begin{vmatrix} \mathrm{E}_x & \mathrm{E}_y & \mathrm{E}_z \\ x & 0 & 0 \\ 0 & 0 & -F \end{vmatrix} = Fx\mathrm{E}_y \quad \text{and} \quad M_y = \mathrm{E}_y \cdot \mathrm{M} = Fx.$$

For this new choice of axes, a positive moment tends to rotate body clockwise around an axis.

Computing the Bending Moment

In a rigid body or in an unconstrained deformable body, the application of a moment of force causes a pure rotation. But if a deformable body is constrained, it develops internal forces in response to the external force so that equilibrium is maintained. An example is shown in the figure below. These internal forces will cause local deformations in the body.

For equilibrium, the sum of the internal force vectors is equal to the applied external force and the sum of the moment vectors created by the internal forces is equal to the moment of the external force. The internal force and moment vectors are oriented in such a way that the total force (internal + external) and moment (external + internal) of the system is zero. The internal moment vector is called the bending moment.

Though bending moments have been used to determine the stress states in arbitrary shaped structures, the physical interpretation of the computed stresses is problematic. However, physical interpretations of bending moments in beams and plates have a straightforward interpretation as the stress resultants in a cross-section of the structural element. For example, in a beam in the figure, the bending moment vector due to stresses in the cross-section A perpendicular to the x-axis is given by

$$\mathrm{M}_x = \int_A \mathrm{r} \times (\sigma_{xx}\mathrm{e}_x + \sigma_{xy}\mathrm{e}_y + \sigma_{xz}\mathrm{e}_z)\,dA \quad \text{where} \quad \mathrm{r} = y\mathrm{e}_y + z\mathrm{e}_z.$$

Expanding this expression we have,

$$\mathrm{M}_x = \int_A \left(-y\sigma_{xx}\mathrm{e}_z + y\sigma_{xz}\mathrm{e}_x + z\sigma_{xx}\mathrm{e}_y - z\sigma_{xy}\mathrm{e}_x\right)dA =: M_{xx}\,\mathrm{e}_x + M_{xy}\,\mathrm{e}_y + M_{xz}\,\mathrm{e}_z.$$

We define the bending moment components as

$$\begin{bmatrix} M_{xx} \\ M_{xy} \\ M_{xz} \end{bmatrix} := \int_A \begin{bmatrix} y\sigma_{xz} - z\sigma_{xy} \\ z\sigma_{xx} \\ -y\sigma_{xx} \end{bmatrix} dA.$$

The internal moments are computed about an origin that is at the neutral axis of the beam or plate and the integration is through the thickness (h)

Example

In the beam shown in the adjacent figure, the external forces are the applied force at point A $(-F\mathrm{e}_y)$ and the reactions at the two support points O and B ($\mathrm{R}_O = R_O\mathrm{e}_y$ and $\mathrm{R}_B = R_B\mathrm{e}_y$). The reactions can be computed using balances of forces and moments about point A, i.e.,

$$R_O + R_B - F = 0 \quad \text{and} \quad -\mathrm{r}_A \times \mathrm{R}_O + \mathrm{r}_B \times \mathrm{R}_B = 0.$$

If L is the length of the beam, we have

$$\mathrm{r}_A = x_A \mathrm{e}_x \quad \text{and} \quad \mathrm{r}_B = (L - x_A)\mathrm{e}_x.$$

If we solve for the reactions we have

$$R_O = \left(1 - \frac{x_A}{L}\right)F \quad \text{and} \quad R_B = \frac{x_A}{L}F.$$

Looking at the free body diagram of the part of the beam to the left of point X, the total moment of the external forces about the point X is

$$\mathrm{M} = -(\mathrm{r}_X - \mathrm{r}_A)\times \mathrm{F} - \mathrm{r}_X \times \mathrm{R}_O = \left[(x_A - x)\mathrm{e}_x\right]\times\left(-F\mathrm{e}_y\right) - (x\mathrm{e}_x)\times\left(R_O\mathrm{e}_y\right).$$

If we compute the cross products, we have

$$\mathrm{M} = \begin{vmatrix} \mathrm{e}_x & \mathrm{e}_y & \mathrm{e}_z \\ x_A - x & 0 & 0 \\ 0 & -F & 0 \end{vmatrix} - \begin{vmatrix} \mathrm{e}_x & \mathrm{e}_y & \mathrm{e}_z \\ x & 0 & 0 \\ 0 & R_0 & 0 \end{vmatrix} = F(x - x_A)\mathrm{e}_z - R_0 x \mathrm{e}_z = -\frac{Fx_A}{L}(L - x)\mathrm{e}_z.$$

For this situation, the only non-zero component of the bending moment is

$$\mathrm{M}_{xz} = -\left[\int_z \int_{-h/2}^{h/2} y\sigma_{xx}\,dy\,dz\right]\mathrm{e}_z.$$

For the sum of the moments at X about the axis e_z to be zero, we require

$$\mathrm{M} + \mathrm{M}_{xz} = 0 \quad \text{or} \quad -\frac{Fx_A}{L}(L - x) + M_{xz} = 0 \quad \text{or} \quad M_{xz} = \frac{Fx_A}{L}(L - x).$$

At $x = x_A$, we have $M_{xz} = Fx_A(L - x_A)/L$.

Sign Convention

In the above discussion, it is implicitly assumed that the bending moment is positive when the top of the beam is compressed. That can be seen if we consider a linear distribution of stress in the beam and find the resulting bending moment. Let the top of the beam be in compression with a stress $-\sigma_0$ and let the bottom of the beam have a stress σ_0. Then the stress distribution in the beam is $\sigma_{xx}(y) = -y\sigma_0$. The bending moment due to these stresses is

$$M_{xz} = -\left[\int_z \int_{-h/2}^{h/2} y(-y\sigma_0)dydz\right] = \sigma_0 I$$

where I is the area moment of inertia of the cross-section of the beam. Therefore the bending moment is positive when the top of the beam is in compression.

Many authors follow a different convention in which the stress resultant M_{xz} is defined as

$$\mathrm{M}_{xz} = \left[\int_z \int_{-h/2}^{h/2} y\sigma_{xx}\, dydz\right]\mathrm{e}_z.$$

In that case, positive bending moments imply that the top of the beam is in tension. Of course, the definition of top depends on the coordinate system being used. In the examples above, the top is the location with the largest y-coordinate.

Girder

A girder is a support beam used in construction. Girders often have an I-beamcross section for strength, but may also have a box shape, Z shape or other forms. Girder is the term used to denote the main horizontal support of a structure which supports smaller beams. A girder is commonly used many times in the building of bridges, and planes.

In traditional timber framing a girder is called a girt.

The Warren type girder combines strength with economy of materials and can therefore be relatively light. Patented in 1848 by its designers James Warren and Willoughby Theobald Monzani, its structure consists of longitudinal members joined only by angled cross-members, forming alternately inverted equilateral triangle-shaped spaces along its length, ensuring that no individual strut, beam, or tie is subject to bending or torsional straining forces, but only to tension or compression. It is an improvement over the Neville truss which uses a spacing configuration of isosceles triangles.

Plate Girder Bridge

A plate girder bridge is a bridge supported by two or more plate girders.

Overview

In a plate girder bridge, the plate girders are typically I-beams made up from separate structural steel plates (rather than rolled as a single cross-section), which are welded or, in older bridges, bolted or riveted together to form the vertical web and horizontal flanges of the beam. In some cases, the plate girders may be formed in a Z-shape rather than I-shape. The first tubular wrought iron plate girder bridge was built in 1846-47 by James Millholland for the Baltimore and Ohio Railroad.

Plate girder bridges are suitable for short to medium spans and may support railroads, highways, or other traffic. Plate girders are usually prefabricated and the length limit is frequently set by the mode of transportation used to move the girder from the bridge shop to the bridge site.

Generally, the depth of the girder is no less than 1/15 the span, and for a given load bearing capacity, a depth of around 1/12 the span minimizes the weight of the girder. Stresses on the flanges near the centre of the span are greater than near the end of the span, so the top and bottom flange plates are frequently reinforced in the middle portion of the span. Vertical stiffeners prevent the web plate from buckling under shear stresses. These are typically uniformly spaced along the girder with additional stiffeners over the supports and wherever the bridge supports concentrated loads.

Deck-Type Plate Girder Bridge

In the deck-type bridge, a wood, steel or reinforced concrete bridge deck is supported on top of two or more plate girders, and may act compositely with them. In the case of railroad bridges, the railroad ties themselves may form the bridge deck, or the deck may support ballast on which the track is laid. Additional beams may span across between the main girders, for example in the form of bridge known as ladder-deck construction. Also, further elements may be attached to provide cross-bracing and prevent the girders from buckling.

Half-Through Plate Girder Bridge

In the half-through bridge (also called a pony truss), the bridge deck is supported between two plate girders, often on top of the bottom flange. The overall bridge then has a 'U'-shape in cross-section. As cross-bracing cannot normally be added, vertical stiffeners on the girders are normally used to prevent buckling (technically described as 'U-frame behaviour'). This form of bridge is most often used on railroads as the construction depth (distance between the underside of the vehicle, and the underside of the bridge) is much less. This allows obstacles to be cleared with less change in height.

Multi-Span Plate Girder Bridge

Multispan plate-girder bridges may be an economical way to span gaps longer than can be spanned by a single girder. Piers serve as intermediate abutments between the end Abutments of bridge. Separate plate girder bridges span between each pair of abutments in order to allow for expansion joints between the spans. Concrete is commonly used for low piers, while steel trestle work may be used for high bridges.

Box Girder

A box or tubular girder is a girder that forms an enclosed tube with multiple walls, rather than an I or H-beam. Originally constructed of riveted wrought iron, they are now found in rolled or welded steel, aluminium extrusions or pre-stressed concrete.

Compared to an I-beam, the advantage of a box girder is that it better resists torsion. Having multiple vertical webs, it can also carry more load than an I beam of equal height (although it will use more material than a taller I beam of equivalent capacity).

The distinction in naming between a box girder and a tubular girder is imprecise. Generally the term *box* girder is used, especially if it is rectangular in section. Where the girder carries its "content" *inside* the box, such as the Britannia Bridge, it is termed a *tubular* girder. *Tubular* girder is also used if the girder is round or oval in cross-section, such as the Royal Albert Bridge.

Where a large box girder contains more than two walls, i.e. with multiple boxes, it is referred to as a *cellular* girder.

Development of the Box Girder

The theoretical basis of the box girder was largely the work of the engineer Sir William Fairbairn, with the aid of the mathematician Eaton Hodgkinson, around 1830. They sought an optimal design for the most efficient design of beam in the new material of riveted wrought iron plates.

Cellular Construction

Most girders are statically loaded such that one web is in compression, the other in tension. Fairbairn's original cranes used a cellular construction for the compression face for their jib, so as to resist buckling. This jib was curved, tapered and formed of riveted wrought iron plates. Three cells were formed inside the concave (lower) face of this girder, again of riveted plates.

Where a tubular girder is used as a bridge span (i.e. loaded in the centre rather than at one end, like a crane) the compressive force is in the top web of the girder and so the cells are placed at the top. Dynamic forces (moving loads, wind) may also require both faces to be cellular, as may be seen in the Britannia Bridge section.

In some ways this isn't a "cellular girder" as such (compared to a spaceframe or geodesic construction) as the cells don't share loads from the entire girder, but merely act to stiffen one plate in isolation. Design of such complex integrated structures requires mathematical modelling techniques in advance of Fairbairn's day.

Box Girders in Bridges

Fairbairn's theoretical girder appeared at just the right time for the increasing demand for long railway bridges. Robert Stephenson engaged both him and Hodgkinson as consultants to assist with his Britannia and Conwy bridges, both of which contained the railway track within a large tubular girder. Shortly afterwards Brunel also chose to use a pair of small diameter round girders as part of a larger truss at Chepstow. The Coronado Bay Bridge has the tallest box girder.

Safety Concerns over Box Girder Bridges

In the early 1970s, a number of box girder bridges collapsed during construction: the Cleddau Bridge in Wales, West Gate Bridge in Australia and the Koblenz Bridge in Germany. These led to extensive studies of the safety of box girders and serious concern over their continued use. This was also an early use of computer modelling and a spur to the development of finite element analysis in civil engineering.

I-beam

An I-beam, also known as H-beam, W-beam (for "wide flange"), Universal Beam (UB), Rolled Steel Joist (RSJ), or double-T (especially in Polish, Bulgarian, Spanish, Italian and German), is a beam with an I- or H-shaped cross-section. The horizontal elements of the "I" are flanges, while the vertical element is termed the "web". I-beams are usually made of structural steel and are used in construction and civil engineering.

The web resists shear forces, while the flanges resist most of the bending moment experienced by the beam. Beam theory shows that the I-shaped section is a very efficient form for carrying both bending and shear loads in the plane of the web. On the other hand, the cross-section has a reduced capacity in the transverse direction, and is also inefficient in carrying torsion, for which hollow structural sections are often preferred.

Figure: *This I-beam is used to support the first floor of a house.*

History

The method of producing an I-beam, as rolled from a single piece of steel, was patented by Alphonse Halbou of the company *Forges de la Providence* in 1849.

Bethlehem Steel was a leading supplier of rolled structural steel of various cross-sections in American bridge and skyscraper work of the mid-twentieth century. Today, rolled cross-sections have been partially displaced in such work by fabricated cross-sections.

Overview

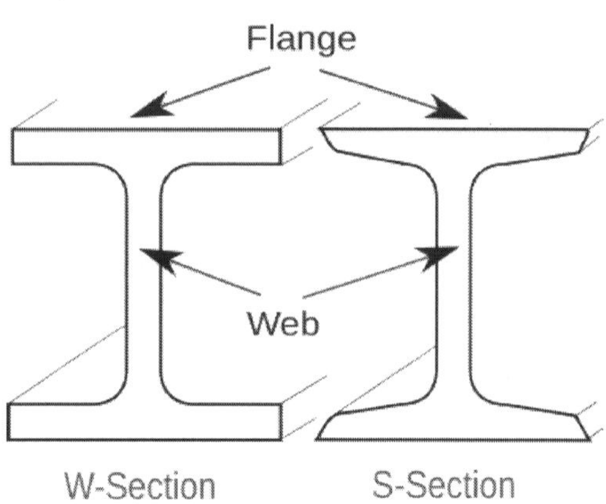

Figure: *Typical cross-sections of I-beams.*

There are two standard I-beam forms:

- Rolled I-beam, formed by hot rolling, cold rolling or extrusion (depending on material).
- Plate girder, formed by welding (or occasionally bolting or riveting) plates.

I-beams are commonly made of structural steel but may also be formed from aluminium or other materials. A common type of I-beam is the *rolled steel joist* (RSJ)—sometimes incorrectly rendered as *reinforced steel joist*. British and European standards also specify Universal Beams (UBs) and Universal Columns (UCs). These sections have parallel flanges, as opposed to the varying thickness of RSJ flanges which are seldom now rolled in the UK. Parallel flanges are easier to connect to and do away with the need for tapering washers. UCs have equal or near-equal width and depth and are more suited to being oriented vertically to carry axial load such as columns in multi-storey construction, while UBs are significantly deeper than they are wide are more suited to carrying bending load such as beam elements in floors.

I-joists — I-beams engineered from wood with fiberboard and/or laminated veneer lumber — are also becoming increasingly popular in construction, especially residential, as they are both lighter and less prone to warping than solid wooden joists. However there has been some concern as to their rapid loss of strength in a fire if unprotected.

Design

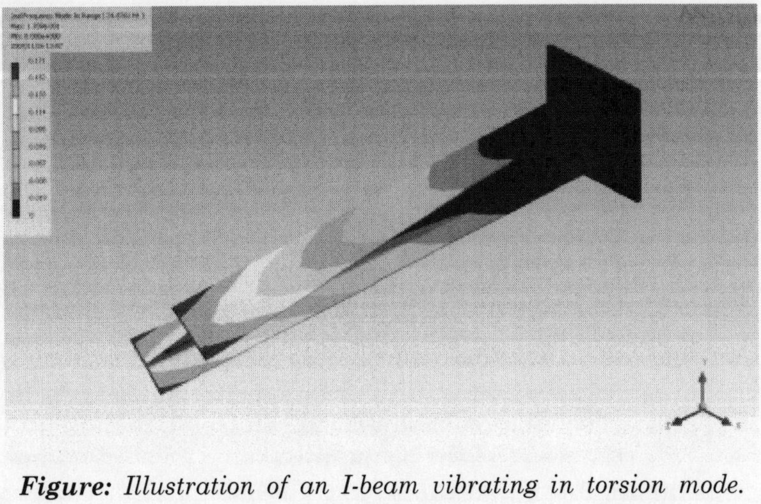

Figure: *Illustration of an I-beam vibrating in torsion mode.*

I-beams are widely used in the construction industry and are available in a variety of standard sizes. Tables are available to allow easy selection of a suitable steel I-beam size for a given applied load. I-beams may be used both as beams and as columns.

I-beams may be used both on their own, or acting compositely with another material, typically concrete. Design may be governed by any of the following criteria:

- deflection: the stiffness of the I-beam will be chosen to minimize deformation
- vibration: the stiffness and mass are chosen to prevent unacceptable vibrations, particularly in settings sensitive to vibrations, such as offices and libraries
- bending failure by yielding: where the stress in the cross section exceeds the yield stress
- bending failure by lateral torsional buckling: where a flange in compression tends to buckle sideways or the entire cross-section buckles torsionally
- bending failure by local buckling: where the flange or web is so slender as to buckle locally
- local yield: caused by concentrated loads, such as at the beam's point of support
- shear failure: where the web fails. Slender webs will fail by buckling, rippling in a phenomenon termed tension field action, but shear failure is also resisted by the stiffness of the flanges
- buckling or yielding of components: for example, of stiffeners used to provide stability to the I-beam's web.

Design for Bending

A beam under bending sees high stresses along the axial fibers that are farthest from the neutral axis. To prevent failure, most of the material in the beam must be located in these regions. Comparatively little material is needed in the area close to the neutral axis. This observation is the basis of the I-beam cross-section; the neutral axis runs along the centre of the web which can be relatively thin and most of the material can be concentrated in the flanges.

The ideal beam is the one with the least cross-sectional area (and hence requiring the least material) needed to achieve a given section modulus. Since the section modulus depends on the value of the moment of inertia, an efficient beam must have most of its material located as far from the neutral axis as possible. The farther a given amount of material is from the neutral axis, the larger is the section modulus and hence a larger bending moment can be resisted.

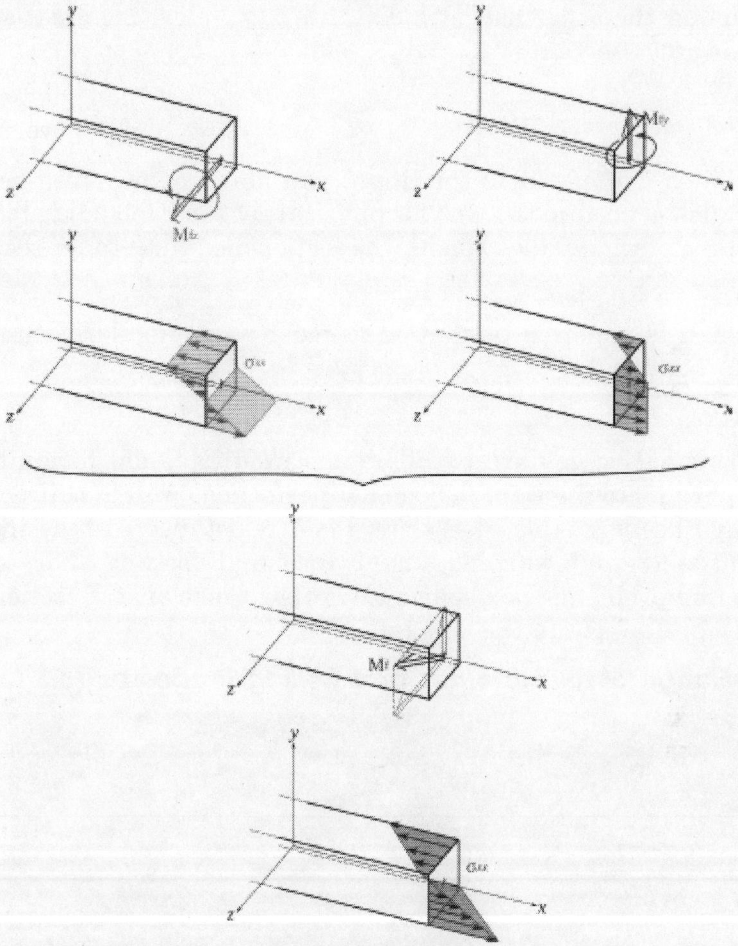

Figure: *The largest stresses (σ_{xx}) in a beam under bending are in the locations farthest from the neutral axis.*

When designing a symmetric I-beam to resist stresses due to bending the usual starting point is the required section modulus. If the allowable stress is σ_{max} and the maximum expected bending moment is M_{max}, then the required section modulus is given by

$$S = \frac{M_{max}}{\sigma_{max}} = \frac{I}{c}$$

where I is the moment of inertia of the beam cross-section and c is the distance of the top of the beam from the neutral axis.

For a beam of cross-sectional area a and height h, the ideal cross-section would have half the area at a distance $h/2$ above the cross-section and the other half at a distance $h/2$ below the cross-section For this cross-section

$$I = \frac{ah^2}{4}\ ;\ S = 0.5ah$$

However, these ideal conditions can never be achieved because material is needed in the web for physical reasons, including to resist buckling. For wide-flange beams, the section modulus is approximately

$$S \approx 0.35ah$$

which is superior to that achieved by rectangular beams and circular beams.

Issues

Though I-beams are excellent for unidirectional bending in a plane parallel to the web, they do not perform as well in bidirectional bending. These beams also show little resistance to twisting and undergo sectional warping under torsional loading. For torsion dominated problems, box beams and other types of stiff sections are used in preference to the I-beam.

Wide-Flange Steel Materials and Rolling Processes (U.S.)

Figure: *Rusty riveted steel I-beam*

In the United States, the most commonly mentioned I-beam is the wide-flange (W) shape.

These beams have flanges in which the planes are nearly parallel. Other I-beams include American Standard (designated S) shapes, in which flange surfaces are not parallel, and H-piles (designated HP), which are typically used as pile foundations.

Wide-flange shapes are available in grade ASTM A992, which has generally replaced the older ASTM grades A572 and A36. Ranges of yield strength:

- A36: 36,000 psi (250 MPa)
- A572: 42,000–60,000 psi (290–410 MPa), but 50,000 psi (340 MPa) is the most common
- A588: Similar to A572
- A992: 50,000–65,000 psi (340–450 MPa)

Like most steel products, I-beams often contain some recycled content.

The American Institute of Steel Construction (AISC) publishes the Steel Construction Manual for designing structures of various shapes.

It documents the common approaches, Allowable Strength Design (ASD) and Load and Resistance Factor Design (LRFD), (starting with 13th ed.) to create such designs.

Standards

The following standards define the shape and tolerances of I-beam steel sections:

Euronorms

- EN 10024, Hot rolled taper flange I sections - Tolerances on shape and dimensions.
- EN 10034, Structural steel I and H sections - Tolerances on shape and dimensions.
- EN 10162, Cold rolled steel sections - Technical delivery conditions - Dimensional and cross-sectional tolerances

Other

- DIN 1025-5
- ASTM A6, American Standard Beams

- BS 4-1
- IS 808 - Dimensions hot rolled steel beam, column, channel and angle sections

Designation and Terminology

- In the United States, steel I-beams are commonly specified using the depth and weight of the beam. For example, a "W10x22" beam is approximately 10 inches (25 cm) in depth (nominal height of the I-beam from the outer face of one flange to the outer face of the other flange) and weighs approximately 22 lb/ft (33 kg/m). It should be noted that wide flange section often vary from their nominal depth. In the case of the W14 series, they may be as deep as 22.84 inches (58.0 cm).
- In Canada, steel I-beams are now commonly specified using the depth and weight of the beam in metric terms. For example, a "W250x33" beam is approximately 250 millimetres (9.8 in) in depth (height of the I-beam from the outer face of one flange to the outer face of the other flange) and weighs approximately 33 kg/m (22 lb/ft). I-beams are still available in U.S. sizes from many Canadian manufacturers.
- In India I-beams are designated as ISMB, ISJB, ISLB, ISWB. ISMB: Indian Standard Medium Weight Beam, ISJB: Indian Standard Junior Beams, ISLB: Indian Standard Light Weight Beams, and ISWB: Indian Standard Wide Flange Beams. Beams are designated as per respective abbreviated reference followed by the depth of section, such as for example "ISMB 450", where 450 is the depth of section in millimetres (mm). The dimensions of these beams are classified as per IS:808 (as per BIS).
- In the United Kingdom, these steel sections are commonly specified with a code consisting of the major dimension (usually the depth)-x-the minor dimension-x-the mass per metre-ending with the section type, all measurements being metric. Therefore a 152x152x23UC would be a column section (UC = universal column) of approximately 152 millimetres (6.0 in) depth 152 mm width and weighing 23 kilograms per metre (46 lb/yd) of length.
- In Australia, these steel sections are commonly referred to as Universal Beams (UB) or Columns (UC). The designation for each is given as the height of the beam, the type (beam or column) and then the unit metre rate (e.g., a 460UB67 is a 460 millimetres (18.1 in) deep universal beam that weighs 67 kilograms per metre (135 lb/yd)).

Indian Standard Beams ISMB

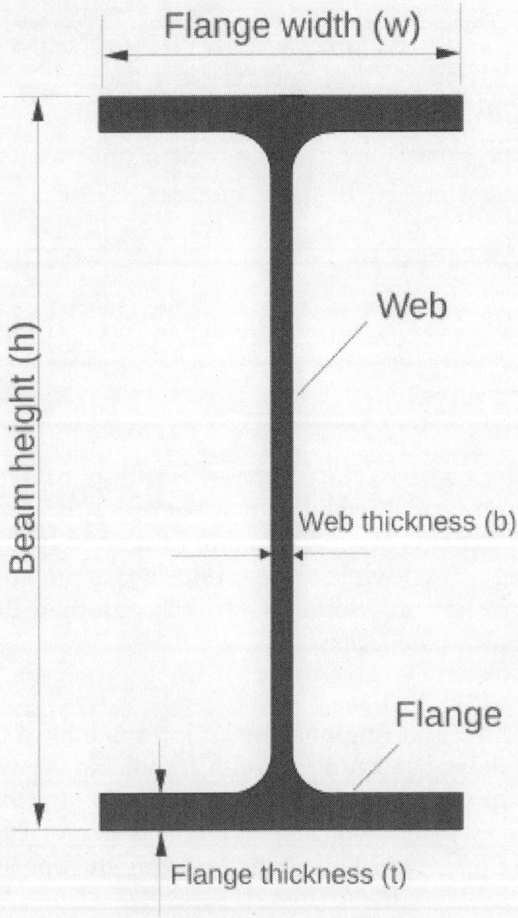

Figure: *Wide-flange I-beam.*

Type	*Beam height (mm)*	*Flange width (mm)*	*Web thickness (mm)*	*Flange thickness (mm)*	*Weight (kg/m)*	*Cross-section area (cm^2)*	*Moment of inertia in torsion (J) (cm^4)*
ISMB 80	80	46	3.8	5.2	6.0	7.64	0.70
ISMB 100	100	55	4.1	5.7	8.1	10.3	1.10
ISMB 120	120	70	4.4	6.3	10.4	13.2	1.71
ISMB 140	140	73	4.7	6.9	12.9	16.4	2.54
ISMB 750 × 137	753	263	11.5	17	137	175	137.1
ISMB 750 × 147	753	265	13.2	17	147	188	161.5
ISMB 750 × 173	762	267	14.4	21.6	173	221	273.6
ISMB 750 × 196	770	268	15.6	25.4	196	251	408.9

European Wide Flange Beams HEA and HEB

Type	*Beam height (mm)*	*Flange width (mm)*	*Web thickness (mm)*	*Flange thickness (mm)*	*Weight (kg/m)*	*Cross-section area (cm^2)*	*Moment of inertia in torsion (J) (cm^4)*
HE 100 A	96	100	5	8	16.7	21.2	5.24
HE 120 A	114	120	5	8	19.9	25.3	5.99
HE 140 A	133	140	5.5	8.5	24,7	31.4	8.13
HE 160 A	152	160	6	9	30.4	38.8	12.19
HE 1000 × 415	1020	304	26	46	415	528.7	2714
HE 1000 × 438	1026	305	26.9	49	437	557.2	3200
HE 1000 × 494	1036	309	31	54	494	629.1	4433
HE 1000 × 584	1056	314	35.6	64	584	743.7	7223

Cellular Beams

Cellular beams are the modern version of the traditional "castellated" beam which results in a beam approximately 40–60% deeper than its parent section. The exact finished depth, cell diameter and cell spacing are flexible. A cellular beam is up to 1.5 times stronger than its parent section and is therefore utilised to create efficient large span constructions.

Joist

In architecture and engineering, a joist is one of the horizontal supporting members that run between foundations, walls, or beams to support a ceiling or floor. They may be made of wood, engineered wood, steel, or concrete. Typically, a beam is bigger than a joist and joists are often supported by beams laid out in repetitive patterns.

Overview

Joists need to be designed large enough to support the anticipated load over a long period of time and in some countries such as the United States to meet standards for stiffness. The wider the span between the supporting structures, the deeper the joist will need to be if it is not to deflect under load. Lateral support called dwang, blocking, or strutting also increases its strength. There are approved formulas for calculating the depth required and reducing the depth as needed; however, a rule of thumb for calculating the depth of a wooden floor joist for a residential property is half the span in feet plus two inches; for example, the joist depth required for a 14 foot span is 9 inches. Many steel joist manufacturers supply load tables in order to allow designers to select the proper joist sizes for their projects.

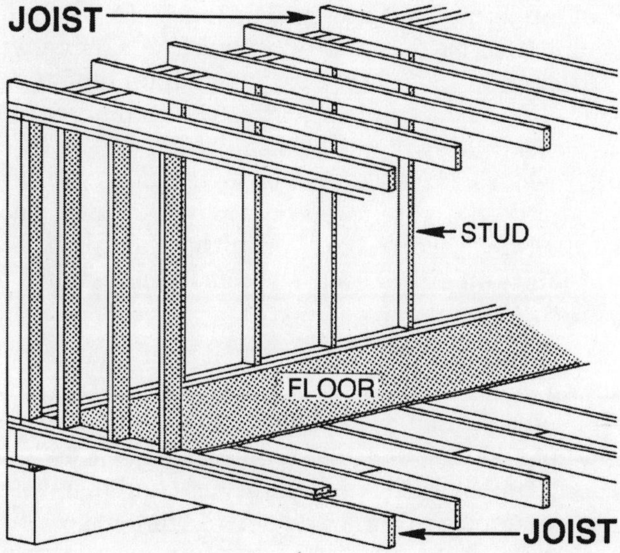

Engineered wood products such as I-joists gain strength from the depth of the floor or the height of each joist, as well as the larger bottom and top chords, as compared to standard dimensional lumber joists. A common saying regarding structural design is that "deeper is cheaper", referring to the more cost-effective design of a given structure by using deeper but more expensive joists, because fewer joists needed and longer spans are achieved, which more than makes up for the added cost of deeper joists.

Bandsill is another term for joist used by construction workers and home inspectors in the southeastern United States.

In traditional timber framing there may be a single set of joists which carry both a floor and ceiling called a single floor (single joist floor, single framed floor) or two sets of joists, one carrying the floor and another carrying the ceiling called a double floor (double framed floor). The term binding joist is sometimes used to describe beams at floor level running perpendicular to the ridge of a gable roof and joined to the intermediate posts. Joists which land on a binding joist are called bridging joists. A large beam in the ceiling of a room carrying joists is a summer beam. A ceiling joist may be installed flush with the bottom of the beam or sometimes below the beam. Joists left exposed and visible from below are called "naked flooring" or "articulated" (a modern U.S. term) and were typically planed smooth (wrought) and sometimes chamfered or beaded. Joists may join to their supporting beams in many ways: joists resting on top of the supporting beams are said to be "lodged"; dropped in using a butt cog joint (a type of lap joint), half-dovetail butt cog, or a half-dovetail lap joint. Joists may also be tenoned in during the raising with a soffit tenon or a tusk tenon (possibly with a housing). Joists can also be joined by being slipped into mortises after the beams are in place such as a chase mortise (pulley mortise), L-mortise, or "short joist". Also, in some Dutch-American work, ground level joists are be placed on a foundation and then a sill placed on top of the joists such as what Jack Sobon called an "inverted sill" or with a "plank sill". Joists can have different joints on either ends such as being tenoned on one end and lodged on the other end.

A reduction in the under-side of cogged joist-ends may be square, sloped or curved. Typically joists do not tie the beams together, but sometimes they are pinned or designed to hold under tension. Joists on the ground floor were sometimes a pole (pole joist, half round joist, log joist. A round timber with one flat surface) and in barns long joists were sometimes supported on a sleeper (a timber not joined to but supporting other beams). Joists left out of an area form an opening called a "well" as in a stairwell or chimney-well. The joists forming the well are the heading joist (header) and trimming joist (trimmer). Trimmers take the name of the feature such as hearth trimmer, stair trimer, etc. Shortened joists are said to be crippled. The term rim joist is rare before the 1940s in America, it forms the edge of a floor. The outermost joist in half timber construction may be of a more durable species than the interior joists. In a barn, loose poles above the drive floor are called a scaffold. Between the joists, the area called a joist-bay, and above the ceiling in some old houses is material called pugging which was used to deaden sound, insulate, and resist the spread of fire.

5

Compression and Beam-Column Members

Structural Material

1. Ultimate strength
2. Yield strength-corresponds to yield point.
3. Rupture
4. Strain hardening region
5. Necking region.

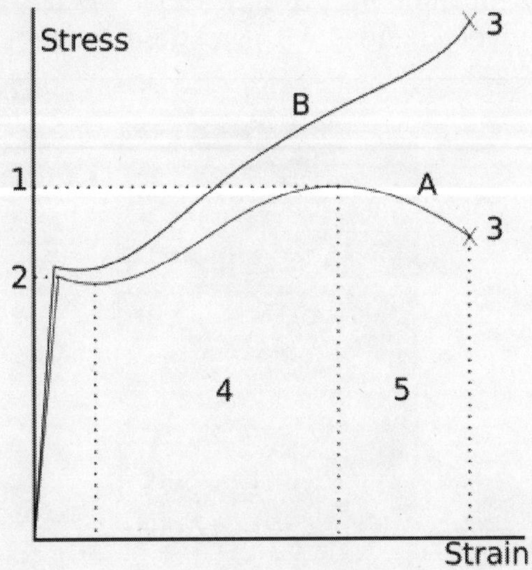

Figure: *Stress-strain curve for low-carbon steel. Hooke's law is only valid for the portion of the curve between the origin and the yield point.*

A: Apparent stress (F/A_0)

B: Actual stress (F/A)

Structural engineering depends on the knowledge of materials and their properties, in order to understand how different materials support and resist loads.

Common structural materials are:

Iron

Wrought Iron: Wrought iron is the simplest form of iron, and is almost pure iron (typically less than 0.15% carbon). It usually contains some slag. Its uses are almost entirely obsolete, and it is no longer commercially produced.

Wrought iron is very poor in fires. It is ductile, malleable and tough. It does not corrode as easily as steel.

Cast Iron

Cast iron is a brittle form of iron which is weaker in tension than in compression. It has a relatively low melting point, good fluidity, castability, excellent machinability and wear resistance. Though almost entirely replaced by steel in building structures, cast irons have become an engineering material with a wide range of applications, including pipes, machine and car parts.

Cast iron retains high strength in fires, despite its low melting point. It is usually around 95% iron, with between 2.1-4% carbon and between 1-3% silicon. It does not corrode as easily as steel.

Steel

Figure: *The 630 foot (192 m) high, stainless-clad (type 304) Gateway Arch in Saint Louis, Missouri*

Steel is an iron alloy with between 0.2 and 1.7% carbon. Steel is used extremely widely in all types of structures, due to its relatively low cost, high strength to weight ratio and speed of construction.

Steel is a ductile material, which will behave elastically until it reaches yield (point 2 on the stress-strain curve), when it becomes plastic and will fail in a ductile manner (large strains, or extensions, before fracture at point 3 on the curve). Steel is equally strong in tension and compression.

Steel is weak in fires, and must be protected in most buildings. Despite its high strength to weight ratio, steel buildings have as much thermal mass as similar concrete buildings.

The elastic modulus of steel is approximately 205 GPa

Steel is very prone to corrosion (rust).

Stainless Steel

Stainless steel is an iron-carbon alloy with a minimum of 10.5% chromium content. There are different types of stainless steel, containing different proportions of iron, carbon, molybdenum, nickel. It has similar structural properties to steel, although its strength varies significantly.

It is rarely used for primary structure, and more for architectural finishes and building cladding.

It is highly resistant to corrosion and staining.

Concrete

Figure: *The interior of the Sagrada Familia, constructed of reinforced concrete to a design by Gaudi*

Figure: *A "cage" of reinforcing steel*

Concrete is used extremely widely in building and civil engineering structures, due to its low cost, flexibility, durability, and high strength. It also has high resistance to fire.

Concrete is a non-linear, non-elastic and brittle material. It is strong in compression and very weak in tension. It behaves non-linearly at all times. Because it has essentially zero strength in tension, it is almost always used as reinforced concrete, a composite material. It is a mixture of sand, aggregate, cement and water. It is placed in a mould, or form, as a liquid, and then it sets (goes off), due to a chemical reaction between the water and cement. The hardening of the concrete is called curing. The reaction is exothermic (gives off heat).

Concrete increases in strength continually from the day it is cast. Assuming it is not cast under water or in constantly 100% relative humidity, it shrinks over time as it dries out, and it deforms over time due to a phenomenon called creep. Its strength depends highly on how it is mixed, poured, cast, compacted, cured (kept wet while setting), and whether or not any admixtures were used in the mix. It can be cast into any shape that a form can be made for. Its colour, quality, and finish depend upon the complexity of the structure, the material used for the form, and the skill of the worker.

The elastic modulus of concrete can vary widely and depends on the concrete mix, age, and quality, as well as on the type and duration of loading applied to it. It is usually taken as approximately 25 GPa for long-term loads once it has attained its full strength (usually considered to be at 28 days after casting). It is taken as approximately

38 GPa for very short-term loading, such as footfalls. Concrete has very favourable properties in fire - it is not adversely affected by fire until it reaches very high temperatures. It also has very high mass, so it is good for providing sound insulation and heat retention (leading to lower energy requirements for the heating of concrete buildings). This is offset by the fact that producing and transporting concrete is very energy intensive.

Reinforced Concrete

Reinforced concrete is concrete in which steel reinforcement bars ("rebars"), plates or fibers have been incorporated to strengthen a material that would otherwise be brittle. In industrialised countries, nearly all concrete used in construction is reinforced concrete. Due to its weakness in tension capacity, concrete will fail suddenly and in brittle manner under flexural (bending) or tensile force unless adequately reinforced with steel.

Prestressed Concrete

Prestressed concrete is a method for overcoming the concrete's natural weakness in tension. It can be used to produce beams, floors or bridges with a longer span than is practical with ordinary reinforced concrete. Prestressing tendons (generally of high tensilesteelcable or rods) are used to provide a clamping load which produces a compressive stress that offsets the tensile stress that the concrete compression member would otherwise experience due to a bending load.

Aluminium

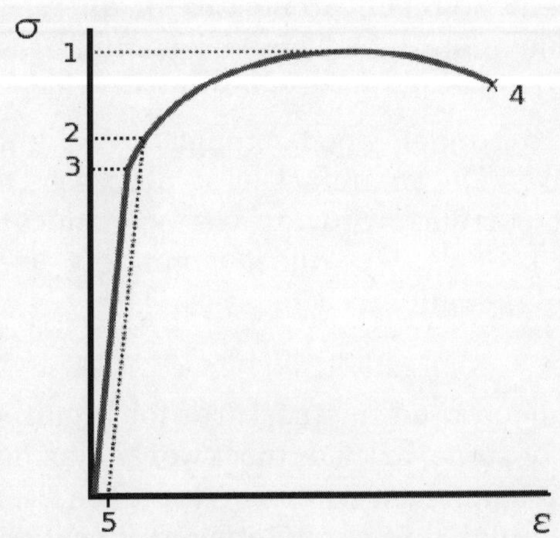

Figure: *Stress vs. Strain curve typical of aluminum*

1. Ultimate strength
2. Yield strength
3. Proportional Limit Stress
4. Rupture
5. Offset strain (typically 0.002).

Aluminium is a soft, lightweight, malleable metal. The yield strength of pure aluminium is 7–11 MPa, while aluminium alloys have yield strengths ranging from 200 MPa to 600 MPa. Aluminium has about one-third the density and stiffness of steel. It is ductile, and easily machined, cast, and extruded.

Corrosion resistance is excellent due to a thin surface layer of aluminium oxide that forms when the metal is exposed to air, effectively preventing further oxidation. The strongest aluminium alloys are less corrosion resistant due to galvanic reactions with alloyed copper.

Aluminium is used in some building structures (mainly in facades) and very widely in aircraft engineering because of its good strength to weight ratio. It is a relatively expensive material.

In aircraft it is gradually being replaced by carbon composite materials.

Composites

Composite materials are used increasingly in vehicles and aircraft structures, and to some extent in other structures. They are increasingly used in bridges, especially for conservation of old structures such as Coalport cast iron bridge built in 1818. Composites are often anisotropic (they have different material properties in different directions) as they can be laminar materials. They most often behave non-linearly and will fail in a brittle manner when overloaded.

They provide extremely good strength to weight ratios, but are also very expensive. The manufacturing processes, which are often extrusion, do not currently provide the economical flexibility that concrete or steel provide. The most commonly used in structural applications are glass-reinforced plastics.

Masonry

Masonry has been used in structures for hundreds of years, and can take the form of stone, brick or blockwork. Masonry is very strong in compression but cannot carry tension (because the mortar between bricks or blocks is unable to carry tension). Because it cannot carry structural tension, it also cannot carry bending, so masonry walls become unstable at relatively small heights. High masonry structures require stabilisation against lateral loads from buttresses (as with the flying buttresses seen in many European medieval churches) or from windposts.

Historically masonry was constructed with no mortar or with lime mortar. In modern times cement based mortars are used. The mortar glues the blocks together, and also smooths out the interface between the blocks, avoiding localised point loads that might have led to cracking.

Since the widespread use of concrete, stone is rarely used as a primary structural material, often only appearing as a cladding, because of its cost and the high skills needed to produce it. Brick and concrete blockwork have taken its place.

Masonry, like concrete, has good sound insulation properties and high thermal mass, but is generally less energy intensive to produce. It is just as energy intensive as concrete to transport.

Timber

Figure: *The reconstructed Globe Theatre, London, by BuroHappold*

Timber is the oldest of structural materials, and though mainly supplanted by steel, masonry and concrete, it is still used in a significant number of buildings. The properties of timber are non-linear and very variable, depending on the quality, treatment of wood, and type of wood supplied. The design of wooden structures is based strongly on empirical evidence.

Wood is strong in tension and compression, but can be weak in bending due to its fibrous structure. Wood is relatively good in fire as it chars, which provides the wood in the centre of the element with some protection and allows the structure to retain some strength for a reasonable length of time.

Compression Member

A compression member are structural elements that are subjected only to axial compressive forces; that is, the loads are applied longitudinal axis through the centroid of the member cross section, and the load over the cross sectional area gives the stress on the compressed member.

Description

For a compression member, such as a column, the principal stress comes mainly from axial forces, that is forces that fall along one line, usually the centreline.

The loading capacity of a short column is determined by strength limit of the material. The strength of a column of intermediate size is limited by its degree of inelasticity. A long column is constrained by the elastic limit (that is by amount of buckling).

Behaviour of Structural Members in Compression

Steel columns are often slender because of the high strength of the material.

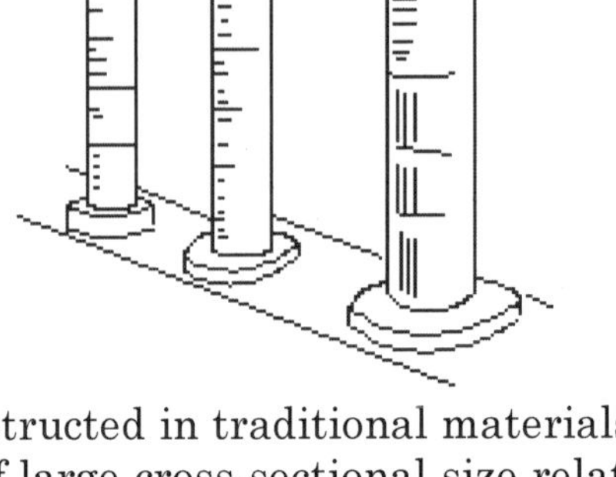

Columns constructed in traditional materials such as stone and brick tend to be of large cross-sectional size relative to their length. A column in structural steel may also be stocky but the much greater strength of steel compared with stone enables the safe design of much more slender columns.

The load capacity of a stocky column is related to its material strength and area of cross-section. When loaded, a stocky column will shorten elastically until a point is reached when any further increase in load causes a disproportionately high reduction in length. At this stage the column is described as having reached the yield point. On further loading the column will reach its collapse load. For the column to behave in this way it must be very short in relation to the size of its cross-section. Under this condition, the failure of the column is due to the failure of steel as a material in compression and the axial load capacity of the column to carry load is determined as follows:

- Axial load capacity of the column = yield stress of steel in compression x cross-sectional area of column.

However, in practice such columns are rarely found since the high strength of steel requires a relatively small cross-sectional area for the column. A typical steel column is therefore likely to be slender. This contrasts with columns constructed in relatively weak materials which require a large cross-section and are therefore generally stocky.

In practice, the failure load of steel columns is associated with buckling; this is related to the column slenderness.

In columns of practical proportions failure occurs well before the crushing strength of steel is reached. As the compressive stress is gradually increased, a value is reached at which the column, instead of just axially shortening in length, buckles and deforms perpendicular to its axis. This value of load is called the buckling load. When a column has reached its buckling load it has effectively failed as a structural element and is incapable of sustaining the load. The aim of column design, therefore, is to predict the load at which the column may collapse and to ensure that there is always an adequate factor of safety compared with the applied load.

Column end Fixity and Effective Length

The restraint at the ends of a column has a significant influence on the effective length and therefore the buckling strength.

Because of the significance of the effective length, the way a column buckles under load will depend on how the column is constructed. If one end of a slender column is fixed at its base and the other end is totally free of any restraint, the column will be extremely prone to lateral buckling under load. In contrast, a column which is fixed at both ends will be much more stable. End conditions can be idealised as free, pinned or fixed.

Idcaliscd cnd conditions for a column are described as free, pinned or fixed. Each of these conditions is described below.

Pinned ends allow free rotation.

When a column end is connected in such a way that the joint allows the column to rotate freely but restrains it from translation (horizontal movement), the connection could be described as being pinned. Such a joint could be produced in practice in several ways. For example the figure below shows a typical arrangement where the column is free to rotate in the xx direction, but it is secured in position at point A.

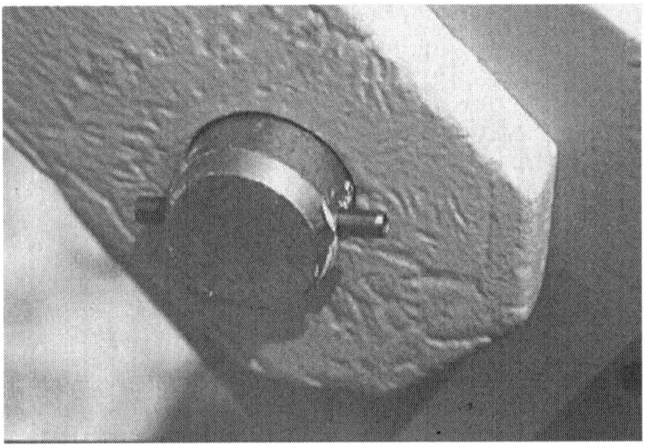

Figure: *Fixed ends do not allow any rotation.*

A fixed end does not allow the column end to move in any manner, ie. to translate or to rotate. Such a joint could be bolted or welded.

Figure: *Free ends allow both lateral movement and rotation.*

Such a condition can apply to only one end of a column and the other end must be fixed to ensure stability of the column. A free end is totally unrestrained and both translation and rotation may take place.

The end conditions influence the shape of the buckled form; the effective length corresponds to that part of the column which deforms as a single curve in the shape of a pin ended buckle.

The type of end condition will dictate how much restraint to lateral buckling is offered to the column. The pin ended column can buckle more freely than a fixed ended column and is therefore less strong.

The effective length describes the part of the column which is free to buckle laterally. For pinned ends, the effective length L_E and the actual length L are the same. For partial or nominally full rotational end restraint, the effective length L_E = 0.85 L and 0.7 L respectively. These figures are established by laboratory experiment. For a cantilever column the effective length can be deduced by comparing with a column which has both the ends pinned. When the deflection configuration is completed, it becomes clear that the effective length L_E is 2L. Table 24 of BS 5950 lists a number of end conditions for columns and the related factors for the effective lengths.

Combined Bending and Direct Compression

Columns are often subject to some bending in addition to compression. Two factors contribute towards the bending in a practical column; these are eccentricity of real beam connection details, and the effect of wind loading.

Bending can develop in columns due to eccentricity of beam connections; this is accounted for by assuming that the beam reaction is applied at a prescribed distance from the face of the column. The way the floor beams are connected to a column causes inevitable eccentric application of loads on the column. Beam end reactions are generally applied on the supporting brackets or cleats and consequently they are at a distance e from the centre of the column.

This eccentric load would cause the column to bend towards the applied load resulting in tension on one face and compression on the other along with a uniform compression all along the section due to the direct effect of the compressive load alone. Whether both faces will be in compression, albeit unequal amounts of compression, or the outside face will be in tension would depend on the value of the beam reaction and the value of eccentricity.

Figure: *The effect of connection eccentricity is negligible for internal columns supporting an approximately symmetrical arrangement of beams.*

In the figure below, the column would be subjected to bending in directions opposed to each other. If W_1 and W_2 were equal and they were an equal distance away from the column centre, ie.e_1 and e_2 were equal, then the bending produced by the two beams would be equal and in the opposite directions and would cancel each other out.

If the column is an internal column with beams running in xx and yy axes the column would experience a combined effect of some or all of the following:

- Direct compression due to vertical loads imposed by the end reactions of beams 1, 2, 3 and 4.
- Compression and tension due to bending about the xx axis.
- Compression and tension due to bending about the yy axis.

In principle a simplified interaction relationship is used to assess the strength of a column in such cases, but this is beyond the scope of the material covered here.

In buildings which depend on rigid frame action for lateral stability, bending will develop in the columns due to wind loads.

All building structures need to resist horizontal loads due to wind. If the structure cannot readily transfer the wind loads to braced parts of the building such as stair towers, lift wells and shear walls, it is generally necessary to rely on rigid frame action. The effect of this is to cause bending in both the beams and columns. In such cases the bending moment distribution must be determined by analysis, and the columns designed for combined axial load and bending.

Tension Members

Tension Member

Tension members are structural elements that are subjected to axial tensile forces. They are usually used in different types of structures. Examples of tension members are: bracing for buildings and bridges, truss members, and cables in suspended roof systems.

In an axially loaded tension member, the stress is given by:

$F = P/A$

where P is the magnitude 10 (Hati) of the load and A is the cross-sectional area. The stress given by this equation is exact, knowing that the cross section is not adjacent to the point of application of the load nor having holes for bolts or other discontinuity. For example,

if you are given a 8 x 11.5 plate is used as a tension member (section a-a) and it is connected to a gusset plate with two 7/8-inch-diameter bolts (section b-b). So, The area at section a - a (gross area of the member) is 8 x ½ = 4 in2

However, the area at section b - b (net area) is (8 – 2 x 7/8) x ½ = 3.12 in2

knowing that the higher stress is located at section b - b due to its smaller area.

To understand more about tension member, it would be useful to look at the stress-strain behaviour of steel:

Where, F_y is the yield stress, F_{su} is the ultimate stress, ε_y is the yield strain and ε_{su} is the ultimate strain.

Designing a Tension Member

In order to design tension members, it is important to analyse how the member would fail under both yielding (excessive deformation) and fracture which considered the limit states. The limit state that produces the smallest design strength is considered the controlling limit state; which also prevent the steel structure from failure.

Using AISC (American Institute of Steel Construction), we could obtain the recommended load and resistance factor design approaches.

The ultimate load on a structure can be calculated from one of the following combination:

1.4 D

1.2 D + 1.6 L + 0.5 (L_r or S)

1.2 D + 1.6 (L_r or S) + (0.5 L or 0.8 W)

1.2 D + 1.6 W + 0.5 L + 0.5 (L_r or S)

0.9 D + 1.6 W

L= 14

- D... is the dead load or the weight of the structure itself
- L... is the live load which vary for different structures
- S... is the snow load
- W... is the wind load

the central problem of designing a member is to find a cross section for which the required strength doesn't exceed the available strength:

$P_u < ¢ P_n$ where P_u is the sum of the factored loads.

to prevent yielding

$0.90\ F_y\ A_g > P_u$

to avoid fracture,

$0.75\ F_u A_e > P_u$

therefore, to design a tension member, we have to consider the loads applied to this member, the design forces acting on this member (M_u, P_u, and V_u), and the point where this member would fail.

Tension (Physics)

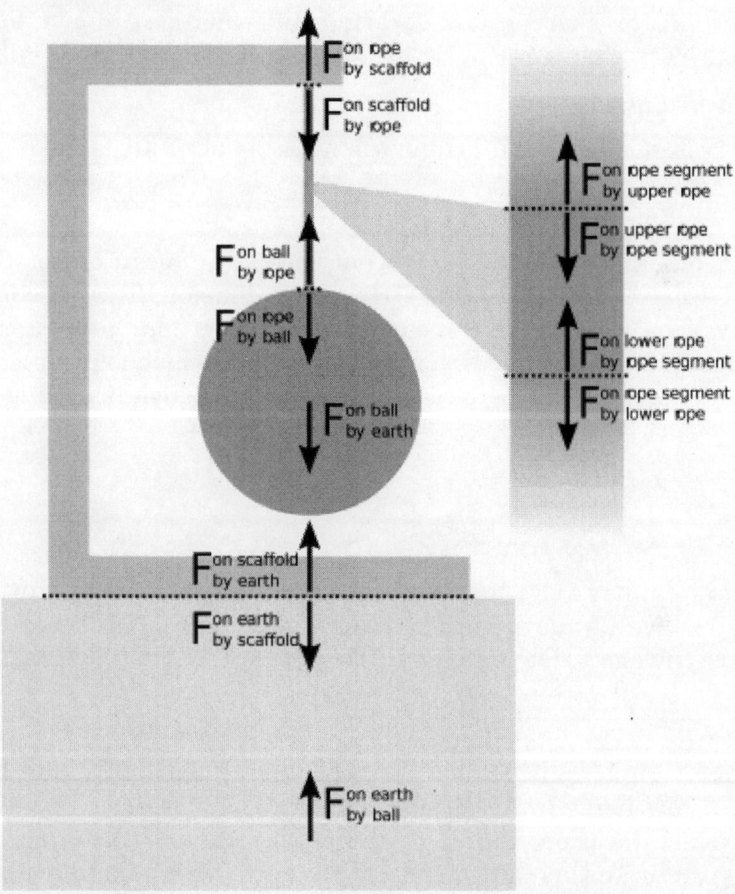

Figure: *The forces involved in supporting a ball by a rope. Tension is the force of the rope on the scaffold, the force of the rope on the ball, and the balanced forces acting on and produced by segments of the rope.*

In physics, tension is the pulling force exerted by a string, cable, chain, or similar solid object on another object. It results from the net electrostatic attraction between the particles in a solid when it is deformed so that the particles are further apart from each other than when at equilibrium, where this force is balanced by repulsion due to electron shells; as such, it is the pull exerted by a solid trying to restore its original, more compressed shape. Tension is the opposite of compression. Slackening is the reduction of tension. As tension is the magnitude of a force, it is measured in newtons (or sometimes pounds-force) and is always measured parallel to the string on which it applies. There are two basic possibilities for systems of objects held by strings: either acceleration is zero and the system is therefore in equilibrium, or there is acceleration and therefore a net force is present. Note that a string is assumed to have negligible mass.

System in Equilibrium

A system is in equilibrium when the sum of all forces is zero.

$$\sum \vec{F} = 0$$

For example, consider a system consisting of an object that is being lowered vertically by a string with tension, T, at a constant velocity. The system has a constant velocity and is therefore in equilibrium because the tension in the string (which is pulling up on the object) is equal to the force of gravity, mg, which is pulling down on the object.

$$\sum \vec{F} = \vec{T} + m\vec{g} = 0$$

System Under Net Force

A system has a net force when an unbalanced force is exerted on it, in other words the sum of all forces is not zero. Acceleration and net force always exist together.

$$\sum \vec{F} \neq 0$$

For example, consider the same system as above but suppose the object is now being lowered with an increasing velocity downwards (positive acceleration) therefore there exists a net force somewhere in the system. In this case, negative acceleration would indicate that $|mg|>|T|$.

$$\sum \vec{F} = T - mg \neq 0$$

In another example, suppose that two bodies A and B having masses m_1 and m_2 respectively are connected with each other by an inextensible string over a frictionless pulley. There are two forces acting on the body A: its weight ($w_1 = m_1 g$) pulling down, and the tension T in the string pulling up. If body A has greater mass than body B, $m_1 > m_2$. Therefore, the net force F_1 on body A is $w_1 - T$, so $m_1 a = m_1 g - T$.

Strings in Modern Physics

String-like objects in relativistic theories, such as the strings used in some models of interactions between quarks, or those used in the modern string theory, also possess tension. These strings are analyzed in terms of their world sheet, and the energy is then typically proportional to the length of the string. As a result, the tension in such strings is independent of the amount of stretching.

In an extensible string, Hooke's law applies.

Types of Structural Steel Tension Members

A tension member is designed as a structural member subjected to tensile force in a direction parallel to its longitudinal axis. A tension member is also called a tie member or simply a tie.

Types of Tension Members

The types of structure and method of end connections determine the type of a tension member. Tension members used may be broadly grouped into four groups.

1. Wires and cables,
2. Rods and bars
3. Single structural shapes and plates
4. Built-up members

Wires and Cables

The wire types are used for hoists, derricks, rigging slings, guy wires and hangers for suspension bridges.

Rods and Bars

The square and round bars are shown in figures are quite often used for small tension members. The round bars with threaded ends are used with pin-connections at the ends instead of threads.

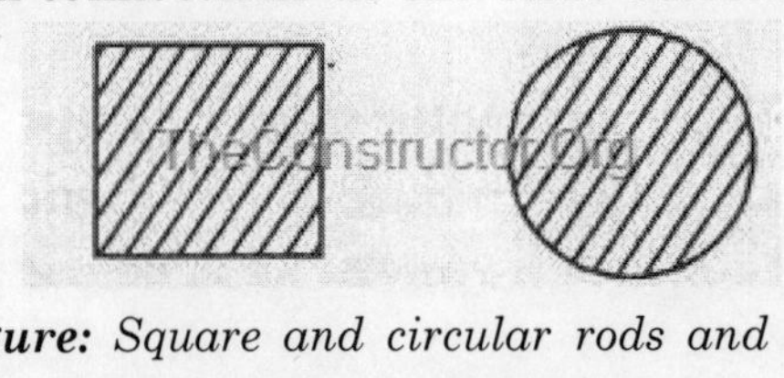

Figure: *Square and circular rods and bars*

The ends of rectangular bars or plates are enlarged by forging and bored to form eye bars. The eye bars are used with pin connections. The rods and bars have the disadvantage of inadequate stiffness resulting in noticeable sag under the self weight.

Single Structural Shapes and Plates

The single structural shapes, i.e. angle sections and tee-sections as shown in figures are used as tension members. The angle sections are considerably more rigid than the wire ropes, rods and bars. When the length of tension member is too ling, then the single angle section also becomes flexible.

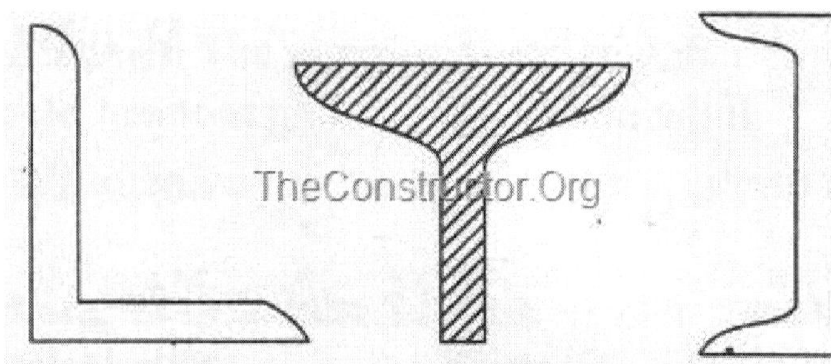

Figure: *Single structural shapes for tension member*

The single angle sections have the disadvantage of eccentricity in both planes in a riveted connection.

The channel section has eccentricity in one axis only. Single channel sections have high rigidity in the direction of web and low rigidity in the direction of flange.

Occasionally, I-sections are sued as tension members. The I-sections have more rigidity, and single I-sections are more economical than built up sections.

Built-up Sections

Two or more than two members are used to form built up members. When the single rolled steel section can not furnish the required area, then built-up sections are used.

The double angle sections of unequal legs shown in the figure are extensively used as tension members in the roof trusses. The angle sections are placed back to back on two sides of a gusset plate. When both the angle sections are attached on the same side of the gusset, then built-up section has eccentricity in one plane and is subjected to tension and bending simultaneously. The two angle sections may be arranged in the star shape (i.e. the angles are placed diagonally opposite to each other with leg on outer sides). The star shape angle sections may be connected by batten plates. The batten plates are alternatively placed in two perpendicular directions.

The star arrangement provides a symmetrical and concentric connection.

Two angle sections as shown in the figure (a) are used in the two-plane trusses where two parallel gussets are used at each connection. Two angle sections as shown in figure (b) have the advantage that the distance between them could be adjusted to suit connecting members at their ends.

Four angle sections as shown in figure (c) are also used in the two-plane trusses. The angles are connected to two parallel gussets. For angle sections connected by plates as shown in figure (d) are used as tension members in bridge girders.

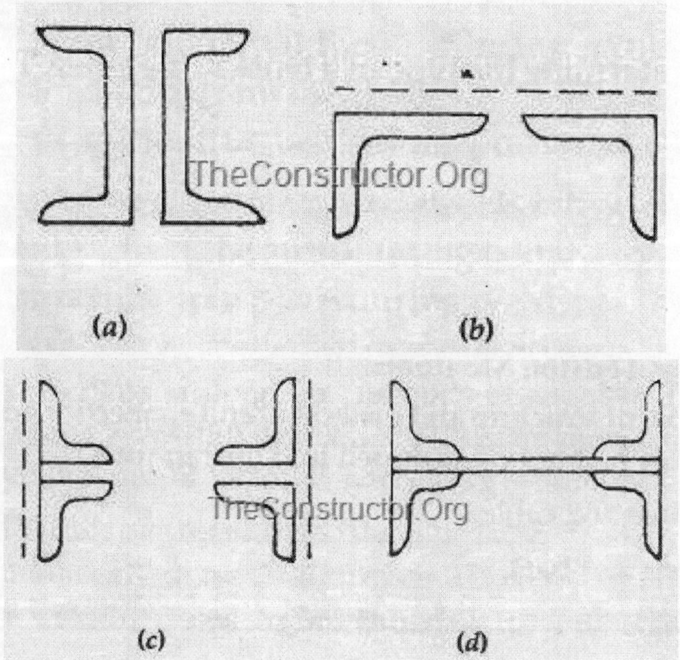

Figure: *Built up steel sections*

A built-up section may be made of two channels placed back to back with a gusset in between them. Such sections are used for medium loads in a single plane-truss. In two-plane trusses, two channels are arranged at a distance with their flange turned inward. It simplifies the transverse connections and also minimizes lacing.

The flanges of two channels are kept outwards, as in the case of chord members or long span girders, in order to have greater lateral rigidity.

The heavy built-up tension members in the bridge girder trusses are made of angles and plates. Such members can resist compression in reversal of stress takes place.

Behaviour of Tension Members Under Loads

Behaviour of tension members under the action of loads for tension ties and is discussed below:

Behaviour of Tension Tie:

The tie is the simplest type of member in tension. When a tensile load is applied on one side of a tie fixed at one end as shown in figure below, the applied load causes reaction at one end and the extension of members will take place.

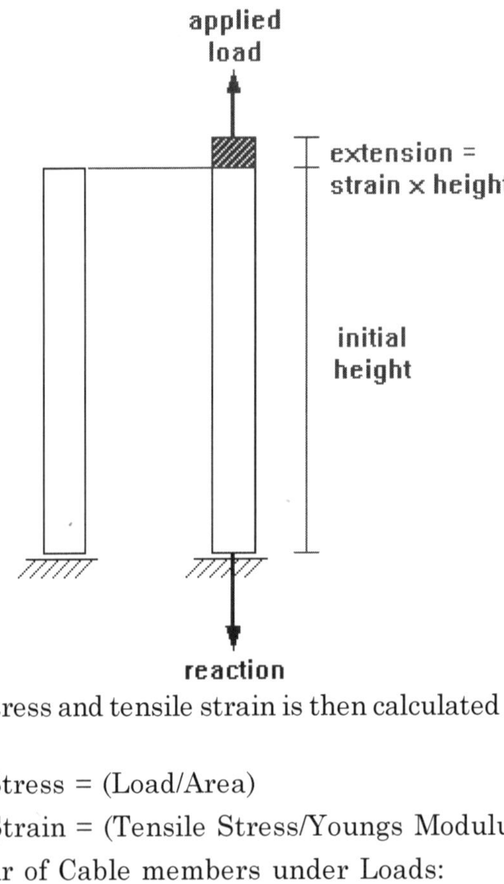

Tensile stress and tensile strain is then calculated for the members as below:

Tensile Stress = (Load/Area)

Tensile Strain = (Tensile Stress/Youngs Modulus)

Behaviour of Cable members under Loads:

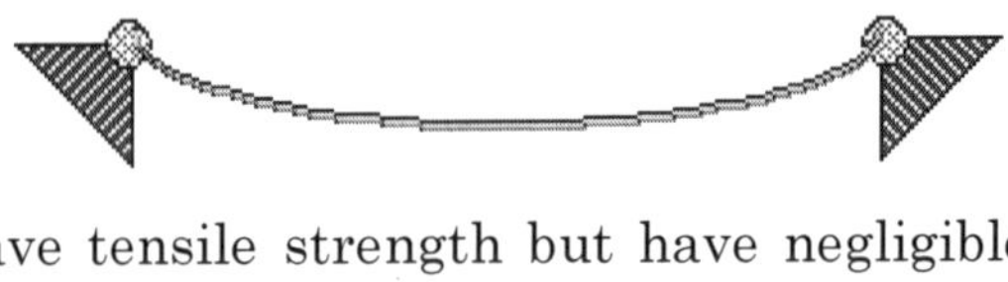

Cables have tensile strength but have negligible resistance to bending moments and shear. Cables hangs in catenary due to self-weight as shown in figure above.

Cables when subjected to a point load (P) the working line would change shape as shown since it can only act in pure tension.

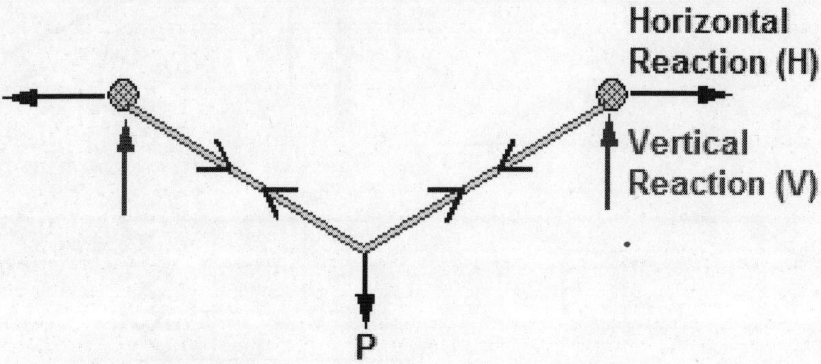

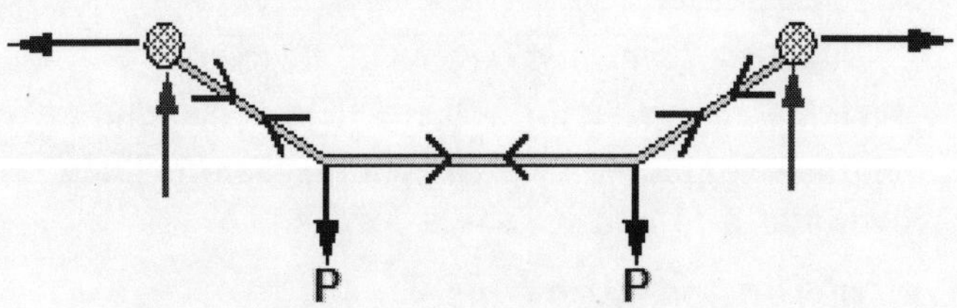

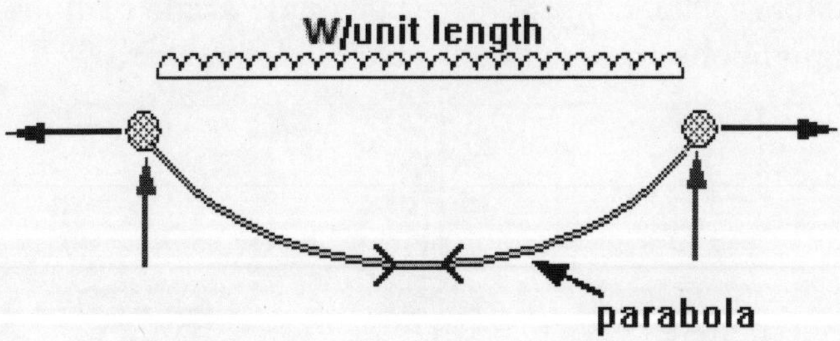

The figures with two point loads, uniformly distributed load are shown here. It can be seen that tensile structures of this type deflect to a profile which is same shape as the equivalent bending moment diagram.

Tension Members-Cross-Section & Behaviour

Tension members are considered to be carrying only axial force. Tension members are subjected to uniform stress because their entire cross-section is used for carrying axial forces. Ties, suspenders of cable stayed and suspension bridges, building suspenders hung from central core, sag rods of roof purlins, members of trusses etc. are examples of tension members.

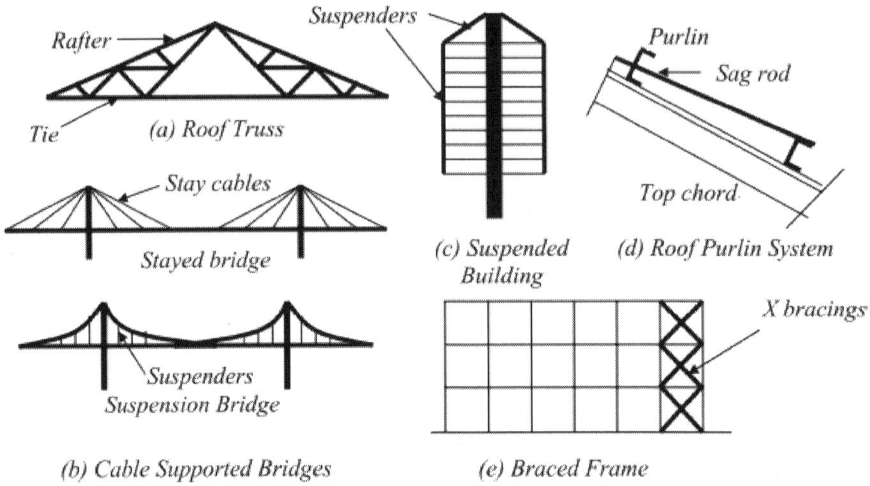

Figure: *Tension members in structures*

Members used as bracings for lateral load as shown in above figure(e) experiences tension as well as compression. They are design to resist both tensile and compressive forces.

Cross-Section of Tension Members

Angle section and channel section as shown below are used as tension members. They can be used as single angle or double angle section or single channel or double channel sections.

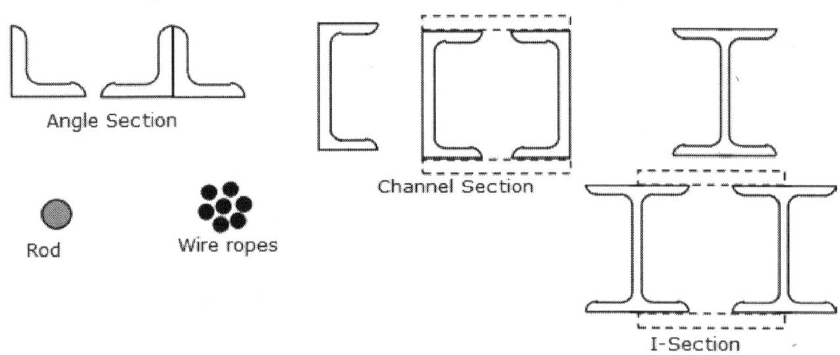

Circular rods, wire ropes are used tension members. Wire ropes are used as suspenders in cable suspended bridges.

Tension members in bridges are used as built-up sections of channel section or I-sections.

Stress-strain behaviour of tension members are same as basic stress-strain behaviour of steel because tension members are subjected to uniform stresses.

Behaviour of Concrete Under Tension

Concrete is not normally designed to resist direct tension. However, tensile stresses develops in concrete members as a result of flexure, shrinkage and temperature changes. Principal tensile stresses may also result from multi-axial states of stress.

Normally cracking in concrete occurs when tensile strength exceeds its limiting value. Pure shear in concrete causes tension on diagonal planes, thus the value of direct tensile strength of concrete is useful in estimating the shear strength of beams with unreinforced webs, etc. Also, the flexural tensile strength of concrete is required for estimation of the moment at first crack required for the computation of deflections and crack widths in flexural members.

Concrete is very weak in tension, the direct tensile strength is only about 7 to 15 percent of its compressive strength. It is difficult to perform a direct tension test on a concrete specimen, as it requires a purely axial tensile force to be applied, free of any misalignment and secondary stress in the specimen at the grips of the testing machine. Hence, indirect tension tests are conducted to determine direct tensile strength of concrete, usually by the flexure test or the cylinder splitting test.

Stress-Strain Curve of Concrete in Tension

Concrete has a low failure strain in uniaxial tension. It is found to be in the range of 0.0001 to 0.0002. The stress-strain curve in tension is generally approximated as a straight line from the origin to the failure point. The modulus of elasticity in tension is taken to be the same as that in compression. As the tensile strength of concrete is very low, and often ignored in design, the tensile stress-strain relation is of little practical value.

Reinforcement Detailing of RCC Members

Steel trusses are being used for both buildings and bridges. But the design principles are different for different uses. Many books are course oriented and not with a practical principles. Now an attempt has been made to gather information on the design principles from various references ON THE LAYOUT AND OTHER DESIGN PRINCIPLES.

Steel roof trusses are used for mainly for the Industrial buildings where free space requirement are essential for more working areas. The span of truss varies from 10'-0" to 300'-0" depending on the type of requirement and the available spaces.

The following steps should be considered when designing a truss:

1. Select the general layout of truss members and truss spacing.
2. Estimate external loads to be applied including self weight of truss, purlins and roof covering together with wind loads.
3. Determine critical (worst combination) loading. It is usual to consider Dead loads alone and then Dead and Imposed loads combined.
4. Analyze the frame work to find forces in all members.
5. Select material and section to produce in each member a stress value which does not exceed the permissible value. Particular care must be taken with compression members (struts) or members normally in tension but subject to stress reversal due to wind uplift.

6

Connection

Simple Connections

Types of Simple Connections

Simple connections are nominally pinned connections that transmit end shear only and have negligible resistance to rotation and therefore do not transfer significant moments at the ultimate limit state. This definition underlies the design of multi-storey braced frames in the UK designed as 'simple construction', in which the beams are designed as simply-supported and the columns are designed for axial load and the small moments induced by the end reactions from the beams. Stability is provided to the frame by the presence of bracing or by the inclusion of a concrete core.

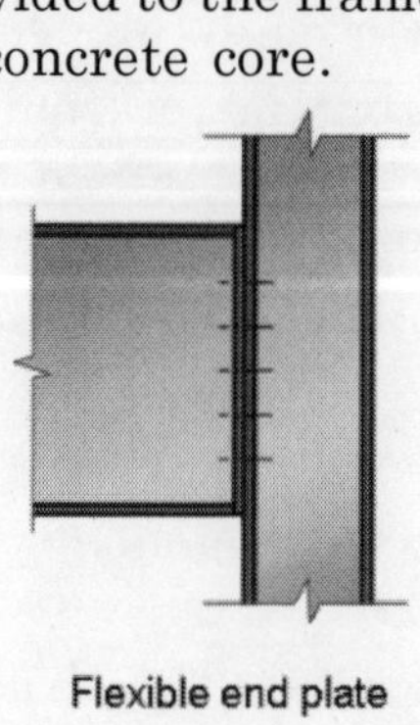

Flexible end plate

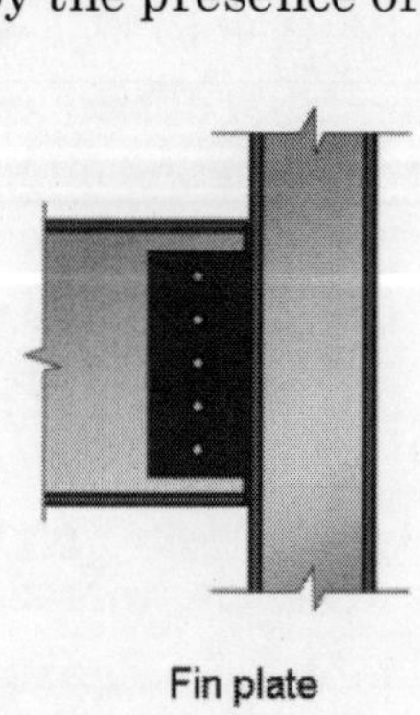

Fin plate

Figure: *Simple connections*

Two principle forms of simple connection (as shown on the right) are used in the UK, these being:

- Flexible end-plates and
- Fin plates.

Commonly encountered simple connections include:

- Beam-to-beam and beam-to-column connections using:
    - o Partial depth end plates
    - o Full depth end plates
    - o Fin plates
- Column splices (bolted cover plates and end plates)
- Column bases
- Bracing connections (Gusset plates).

Simple connections can also be needed for skewed joints, beams eccentric to columns and connection to column webs. These are classed as special connections and are treated separately.

Design Procedures

The design of simple connections is based on BS EN 1993-1-8 and its accompanying National Annex. The capacities of the fasteners and fittings are based on the rules given in clause 3.6. The spacing of the fasteners comply with clause 3.5 and follow the recommendations presented in the 'Green Book' (SCI P358).

ECCS publication No. 126 also provides useful guidance on the design of simple connections to Eurocode 3.

Joint Considerations

Joint Classification

According to BS EN 1993-1-8, nominally pinned joints:

- Should be capable of transmitting the internal forces, without developing significant moments which might adversely affect the members or the structure as a whole and
- Be capable of accepting the resulting rotations under the design loads

In addition, the joint must:

- provide the directional restraint to members which has been assumed in the member design
- have sufficient robustness to satisfy the structural integrity requirements (tying resistance).

BS EN 1993-1-8 requires that all connections must be classified; by stiffness, which is appropriate for elastic global analysis, or by strength, which is appropriate for rigid plastic global analysis, or by both stiffness and strength, which is appropriate for elastic-plastic global analysis.

Classification by Stiffness: The initial rotational stiffness of the connection, calculated in accordance with BS EN 1993-1-8 , 6.3.1 is compared with the classification boundaries given in BS EN 1993-1-8, 5.2.2.

Alternatively, joints may be classified based on experimental evidence, experience of previous satisfactory performance in similar cases or by calculations based on test evidence.

Classification by strength:

The following two requirements must be satisfied in order to classify a connection as nominally pinned, based on its strength:

- The design moment resistance of the connection does not exceed 25% of the design moment resistance required for a full-strength joint
- The joint should be capable of accepting the rotations resulting from the design loads.

The UK National Annex to BS EN 1993-1-8 states that connections designed in accordance with the 'Green Book' (SCI P358) may be classified as nominally pinned joints.

All the standard connections given in the 'Green Book' (SCI P358) may be classified as nominally pinned based on the strength requirements together with extensive experience of details used in practice. Care should be taken before amending the standard details as the resulting connection may fall outside the provisions of the UK National Annex. In particular:

- The rotation capacity of the standard fin plate details have been demonstrated by test; modified details may not be ductile
- The thickness of the full depth end plates have been limited to ensure the moment resistance is less than 25% of a full strength joint, and can thus be classified as nominally pinned.

Structural Integrity

The UK Building Regulations require that all buildings should be designed to avoid disproportionate collapse. Commonly, this is achieved by designing the joints in a steel frame (the beam-to-column connections and the column splices) for tying forces. Guidance on the design values of tying forces is given in BS EN 1991-1-7 Annex A, and its UK National Annex.

The requirements relate to the building Class, with a design value of horizontal tying force generally not less than 75 kN, and usually significantly higher. Full depth end plate details have been developed to provide an increased tying resistance compared to partial depth end plate details. Further details on structural robustness are presented in SCI P391 .

Selection of Connection Types

The selection of beam end connections can often be quite involved. The relative merits of the three connection types (partial depth end plates, full depth end plates and fin plates) are summarised in the table below.

Relative merits of beam end connection types			
	Partial depth end plate	***Full depth end plate***	***Fin plate***
Design			
Shear resistance - percentage of beam resistance	Up to 75%	100%	Up to 50% *Up to 75% with two vertical lines of bolts*
Tying resistance	Fair	Good	Good
Special considerations			
Skewed Joints	Fair	Fair	Good
Beams eccentric to columns	Fair	Fair	Good
Connection to column webs	Good	Good	Fair *To facilitate erection, flange stripping may be required. Stiffening may be required for long fin plates*
Fabrication and treatment			
Fabrication	Good	Good	Good *Stiffening may be required for long fin plates*
Surface treatment	Good	Good	Good
Erection			
Ease of erection	Fair *Care needed for two-sided connections*	Fair *Care needed for two-sided connections*	Good
Site adjustment	Fair	Fair	Fair
Temporary stability	Fair	Good	Fair

Composite Floors

It is recognised that interaction with a composite floor will affect the behaviour of a simple connection. Common practice is to design such connections without utilising the benefits of the continuity of reinforcement through the concrete slab. However, SCI P213 enables reinforcement continuity to be allowed for in providing relatively simple full depth end plate connections with substantial moment resistance.

Costs

Simple connections are invariably cheaper to fabricate than moment-resisting connections, because they provide a significant degree of simplicity and standardisation. Giving specific guidance on costs is difficult, as a Steelwork Contractor's workmanship rates can vary considerably and are dependant upon the level of investment in plant and machinery.

However, the main objective is to minimise work content. The material cost for fittings and bolts is small compared with workmanship costs. In a typical fabrication workshop the cost of fabrication of connections may be 30% to 50% of the total fabrication cost.

Sustainability

Standardised connections are efficient in their production. Steelwork Contractors equip their workshops with specialist machinery that increases the speed of fabrication, allowing them to produce fittings and prepare the members much more quickly than they would if the connection configuration were different each time.

The standardised details mean the steelwork is straightforward to erect, which provides a safer working environment for the steel erectors. Due to the nature of most bolted joints, the connections are demountable at the end of the service life of the structure. The steelwork can be dismantled, reused or recycled, therefore reducing the environmental impact of the construction.

Standardised Connections

The Benefits of Standardisation

In a typical braced multi-storey frame, the connections may account for less than 5% of the frame weight, and 30% or more of the total cost. Efficient connections will therefore have the lowest detailing, fabrication and erection labour content.

Recommended components		
Component	***Preferred Option***	***Notes***
Fittings	Material of grade S275	Recommended sizes of end plates and fin plates – see table below
Bolts	M20 8.8 Bolts, fully threaded	Some heavily loaded connections may need larger diameter bolts Foundation bolts may be M20, M24, M30, 8.8 or 4.6
Holes	Generally 22 mm diameter, punched or drilled	26 mm diameter for M24 bolts 6 mm oversize for foundation bolts
Welds	Fillet welds generally 6 mm or 8 mm leg length	Larger welds may be needed for some column bases

Recommended sizes of end plates and fin plates			
Fittings		***Location***	
Size (mm)	***Thickness (mm)***	***End plate***	***Fin plate***
100	10		•
120	10		•
150	10	•	•
160	10		•
180	10	•	•
200	12	•	

Beam-to-Beam and Beam-to-Column Connections

The design procedures given below are suitable for either hand calculation or for the preparation of computer software. Designing connections by hand can be a laborious process and so a full set of resistance tables has been included in the 'Green Book' (SCI P358). Verifying the strength of a nominally pinned joint involves three stages:

1. Ensuring that the joint is detailed such that it develops only nominal moments which do not adversely affect the members or the joint itself. The joint should be detailed so that it behaves in a ductile manner.
2. Identifying the load path through the joint i.e. from the beam to the supporting member.
3. Checking the resistance of each component.

For normal design there are ten design procedure checks for all the parts of a beam to beam or beam to column joint for vertical shear. A further six checks are necessary to verify the tying resistance of the joint. Beam to column connections must be able to resist lateral tying forces unless these forces are resisted by other means within the structure, such as the floor slabs.

The table below summarises the design procedure checks required for partial depth end plates, full depth end plates and fin plates. The design procedures are described fully in the 'Green Book' (SCI P358).

Design procedure for beam connections - Summary table			
Design procedure checks	***Partial-depth end plate***	***Full-depth end plate***	***Fin plate***
1 Recommended detailing practice			
2 Supported beam	Welds	Welds	Bolt Group
3 Supported beam	N/A	N/A	Fin plate
4 Supported beam	Web in shear		
5 Supported beam	Resistance at a notch	N/A	Resistance at a notch
6 Supported beam	Local stability of notched beam	N/A	Local stability of notched beam
7 Unrestrained supported beam	Overall stability of notched beam	N/A	Overall stability of notched beam
8 Connection	Bolt group	Bolt group	Welds
9 Connection	End plate in shear	N/A	N/A
10 Supporting beam/column	Shear and bearing		
11 Tying resistance	Plate and bolts		
12 Tying resistance	Supported beam web		
13 Tying resistance	Welds		
14 Tying resistance	Supporting column web (UKC or UKB)		
15 Tying resistance	Supporting column wall (RHS or SHS)		
16 Tying resistance	N/A	N/A	Supporting column wall (CHS)

Notes: Checks on the bending, shear, local and lateral buckling resistance of a notched beam section are included in this table as it is usually at the detailing stage that the requirement for notches is established, following which, a check must be made on the reduced section

Flexible end Plate Connections

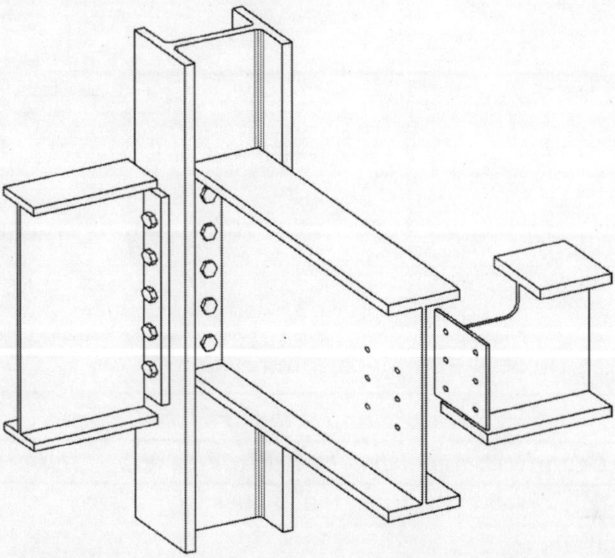

Figure: *End plate beam to column and beam to beam connections*

Typical flexible end plate connections are shown in the figure right. The end plate, which may be partial depth or full depth, is welded to the supported beam in the workshop. The beam is then bolted to the supporting beam or column on site.

This type of connection is relatively inexpensive but has the disadvantage that there is no room for site adjustment. Overall beam lengths need to be fabricated within tight limits, although packs can be used to compensate for fabrication tolerances and erection tolerances.

End plates are probably the most popular of the simple beam connections currently in use in the UK. They can be used with skewed beams and can tolerate moderate offsets in beam to column joints.

Flowdrill or Hollo-Bolts are used for connections to hollow section columns.

Detailing requirements and design checks for partial depth and full depth end-plates connections, which are applicable to beam-to-beam connections as well as beam-to-column connections, are comprehensively covered in the 'Green Book' (SCI P358). These include procedures, worked examples, detailing and design resistance tables.

A Partial depth end plate designer tool is also available.

Standard flexible end-plate details (full depth and partial depth end plates) are shown in the figure below, together with recommended dimensions and fittings.

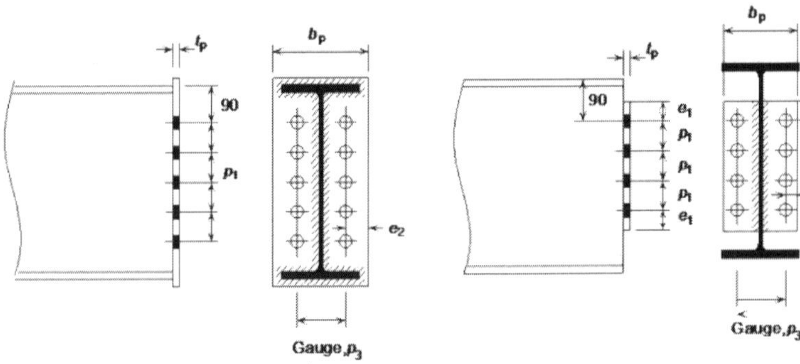

Standard flexible end-plate connections

Ordinary and Flowdrill Bolts		
Supported Beam	***Recommended End Plate Size $b_p \times t_p$***	***Bolt Gauge p_3***
Up to 533 UKB	150 × 10	90
533 UKB and above	200 × 12	140

Bolts:	M20 in 22 mm diameter holes
End plate:	S275 steel, minimum length $0.6h_{b1}$ where h_{b1} is the depth of the supported beam
Vertical pitch:	p_1=70 mm
End distance:	e_1=40 mm
Edge distance:	e_2=30 mm

Hollo-Bolts		
Supported Beam	***Recommended End Plate Size $b_p \times t_p$***	***Bolt gauge p_3***
Up to 533 UKB	180 × 10	90
533 UKB and above	200 × 12	110

End plate:	S275 steel, minimum length $0.6h_{b1}$ where h_{b1} is the depth of the supported beam
Vertical pitch:	p_1=80 mm
End distance:	e_1=45 mm
Edge distance:	e_2=45 mm

Fin Plates

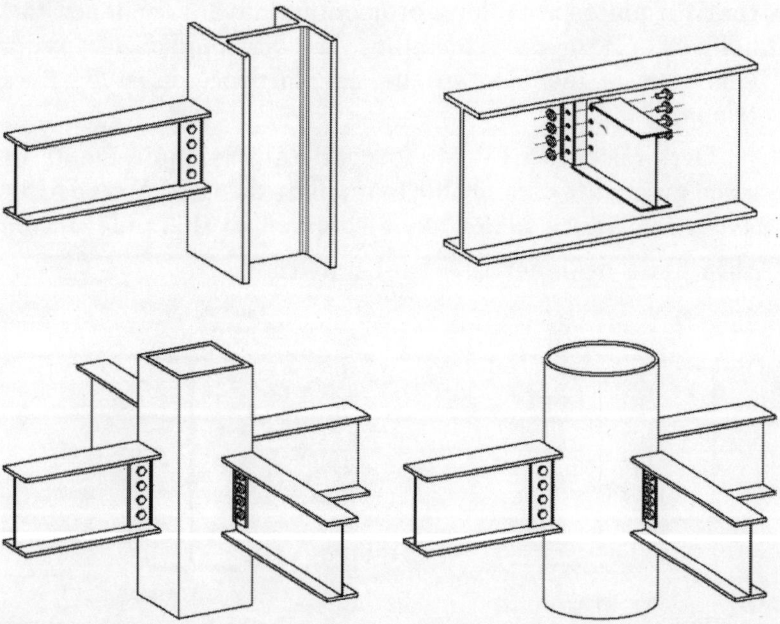

Figure: *Fin plate beam to column and beam to beam connections*

Fin plate connections are economical to fabricate and simple to erect. These connections are popular, as they can be the quickest connections to erect and overcome the problem of shared bolts in two-sided connections.

A fin plate connection consists of a length of plate welded in the workshop to the supporting member, to which the supported beam web is bolted on site, as shown in the figure below. There is clearance between the ends of the supported beams and the supporting column, thus ensuring an easy fit.

In the design of a fin plate connection it is important to identify the appropriate line of action for the shear. There are two possibilities: either the shear acts at the face of the column or it acts along the centre of the bolt group connecting the fin plate to the beam web.

For this reason all critical sections should be checked for a minimum moment taken as the product of the vertical shear and the distance between the face of the column (or beam web) and the centre of the bolt group. The critical sections are then checked for the resulting moment combined with the vertical shear.

Fin plate connections derive their in-plane rotational capacity from the bolt deformation in shear, from the distortion of the bolt holes in bearing and from the out-of-plane bending of the fin plate. Note that fin plates with long projections have a tendency to twist and fail by lateral torsional buckling. An additional check to consider this behaviour is included in the design procedures for fin plate connections.

The 'Green Book' (SCI P358) covers detailing requirements, design checks and procedures applicable to fin plate design. Worked examples and design resistance tables are also given in this publication.

A Fin plate designer tool is also available.

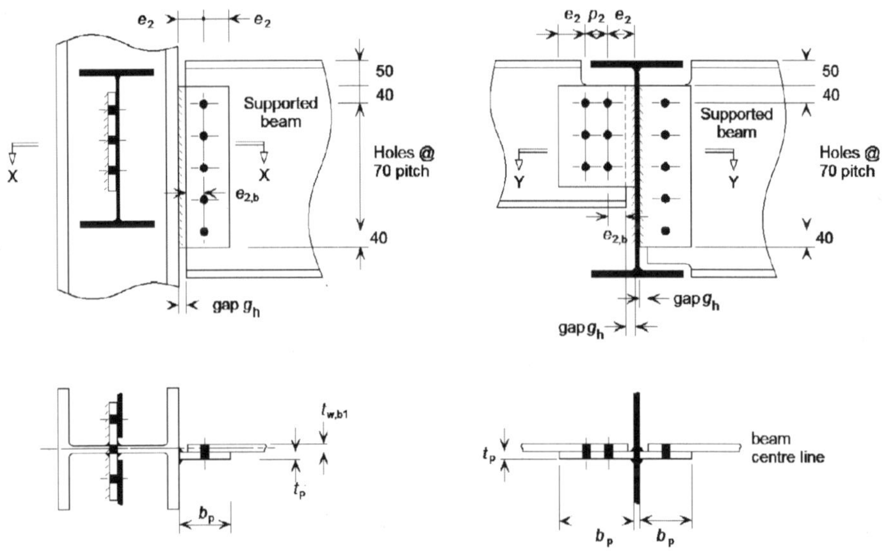

Section XX

Section YY

Standard fin plate connections details				
Supported beam nominal depth (mm)	***Vertical bolt lines n_2***	***Recommended fin plate size (mm)***	***Horizontal bolt spacing, e_2/e_2 or $e_2/p_2/e_2$ (mm)***	***Gap, g_h (mm)***
?610	1	100 × 10	50/50	10
>610*	1	120 × 10	60/60	20
?610	2	160 × 10	50/60/50	10
>610*	2	180 × 10	60/60/60	20

Bolts: M20 8.8 in 22 mm diameter holes

Plate: S275 steel, minimum length $0.6h_{b1}$ where h_{b1} is the depth of the supported beam

Weld: Two 8 mm fillets for 10 mm thick plates

* For beams over 610 mm nominal depth the span to depth ratio of beam should not exceed 20 and the vertical distance between extreme bolts should not exceed 530 mm

Column Splices

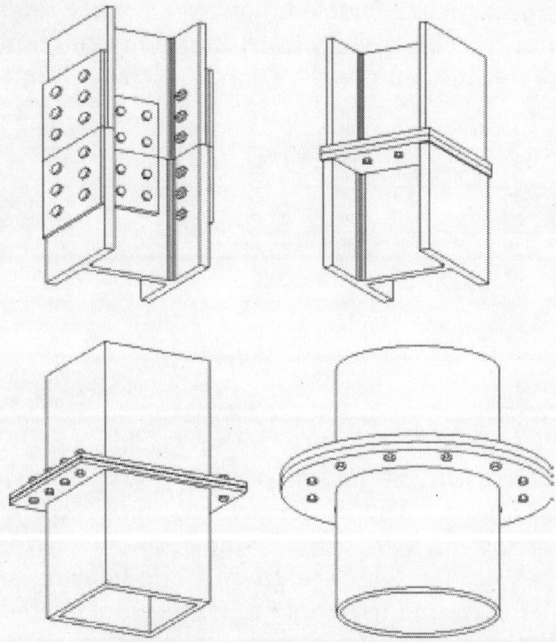

Figure: *Splice connections*

Column splices in multi-storey construction are required to provide continuity of both strength and stiffness about both axes of the columns. Typical bolted column splices used for rolled I section and hollow section members are shown in the figure on the right.

Splices are typically provided every two or three storeys and are usually located approximately 600mm above floor level. This results in convenient lengths for fabrication, transport and erection, and gives easy access from the adjacent floor for bolting up on site. The provision of splices at each storey level is seldom economical since the saving in column material is generally far outweighed by the material, fabrication and erection costs of making the splice.

Bolted Cover Plate Splices for I Sections

There are two categories for this type of connection:

- bearing type
- non-bearing type.

In the bearing type splice the loads are transferred in direct bearing from the upper shaft either directly or through a division plate. The 'bearing type' splice is the simpler connection, usually having fewer bolts than the non-bearing splice, and is therefore the one most commonly used in practice. When no net tension is present, a standard connection can be used, however BS EN 1993-1-8 imposes the requirement for the splice plates and bolts to transmit at least 25% of the maximum compressive force in the column.

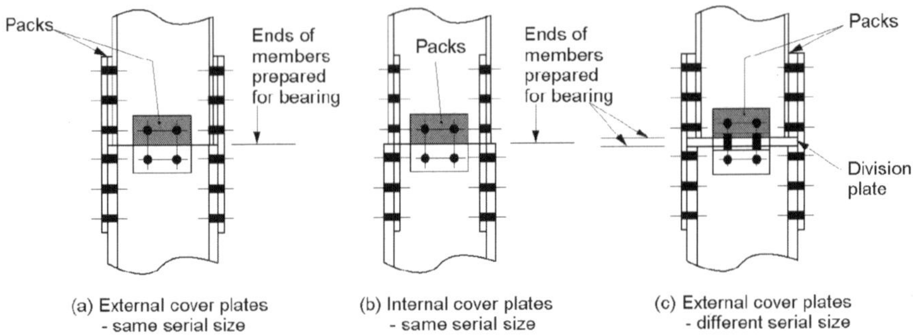

Figure: *Bearing column splices for rolled I sections*

Splices categorised as non-bearing type transfer loads via the bolts and splice plates. Any direct bearing between the members is ignored, the connection sometimes being detailed with a physical gap between the two shafts. The design of a non-bearing splice is more involved, as all forces and moments must be transmitted through the bolts and splice plates.

As splices are generally provided just above floor levels the moment due to strut action is considered insignificant. The moments induced in splices placed at other positions, however, should be taken into account.

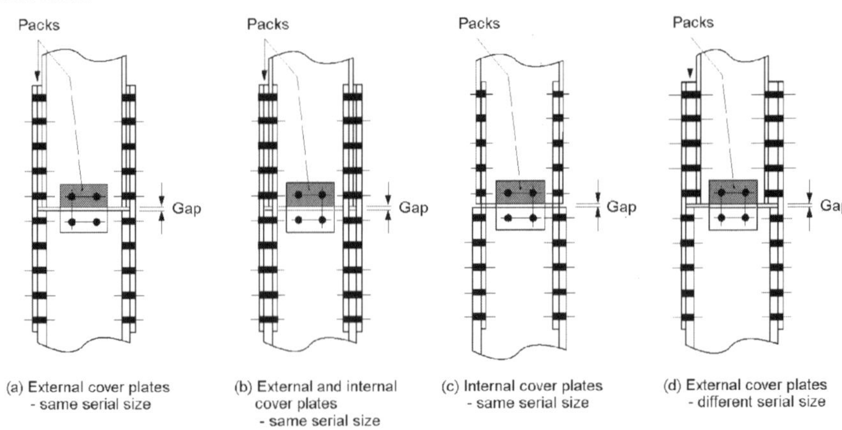

Figure: *Non bearing column splices for rolled I sections*

Column splices should hold the connected members in line, and wherever practical, the members should be arranged so that the centroidal axis of the splice material coincides with the centroidal axis of the column sections above and below the splice. If the column sections are offset (for example to maintain a constant external line) the moment due to the eccentricity should be accounted for in the joint design.

Design checks required for bolted cover plate column splices as well as helpful procedures, worked examples, detailing requirements and design resistance tables are available in chapter 6 of the 'Green Book' (SCI P358).

Bolted 'Cap and Base' or 'End Plate' Splices for Tubular and Rolled I Sections

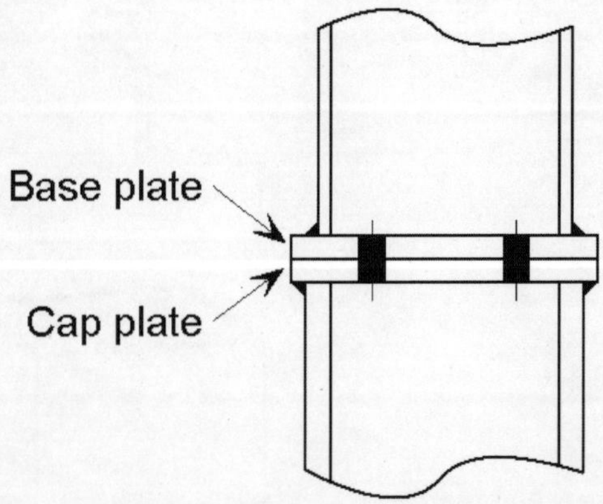

Figure: *'Cap and base' or 'End plate' splice*

This type of splice, consisting of plates which are welded to the ends of the lower and upper columns and then simply bolted together on site, is commonly used in tubular construction but may also be used for open sections.

The most simple form of the connection is as shown in the figure right and is satisfactory as long as the ends of each shaft are prepared in the same way as for a bearing type splice. The possibility of load reversal should be considered, in addition to stability during erection and tying requirements.

Although commonly used, it is difficult to demonstrate that cap and base splices meet the requirements of BS EN 1993-1-8 clause 6.2.7.1(13) and (14). If these types of splices are used, common practice is to ensure that the plates are thick, and that bolts are located close to the flanges to increase the stiffness of the connection. Extended plates, with bolts outside the profile of the section may be used. If cap and base plate splices are located away from a point of restraint, special consideration should be given to ensuring adequate stiffness so that the member design is not invalidated. 'Cap and base' or 'end plate' column splices are covered in chapter 6 of the 'Green Book' (SCI P358). Detailing requirements, design procedures, worked examples and design resistance tables are given.

Column Bases

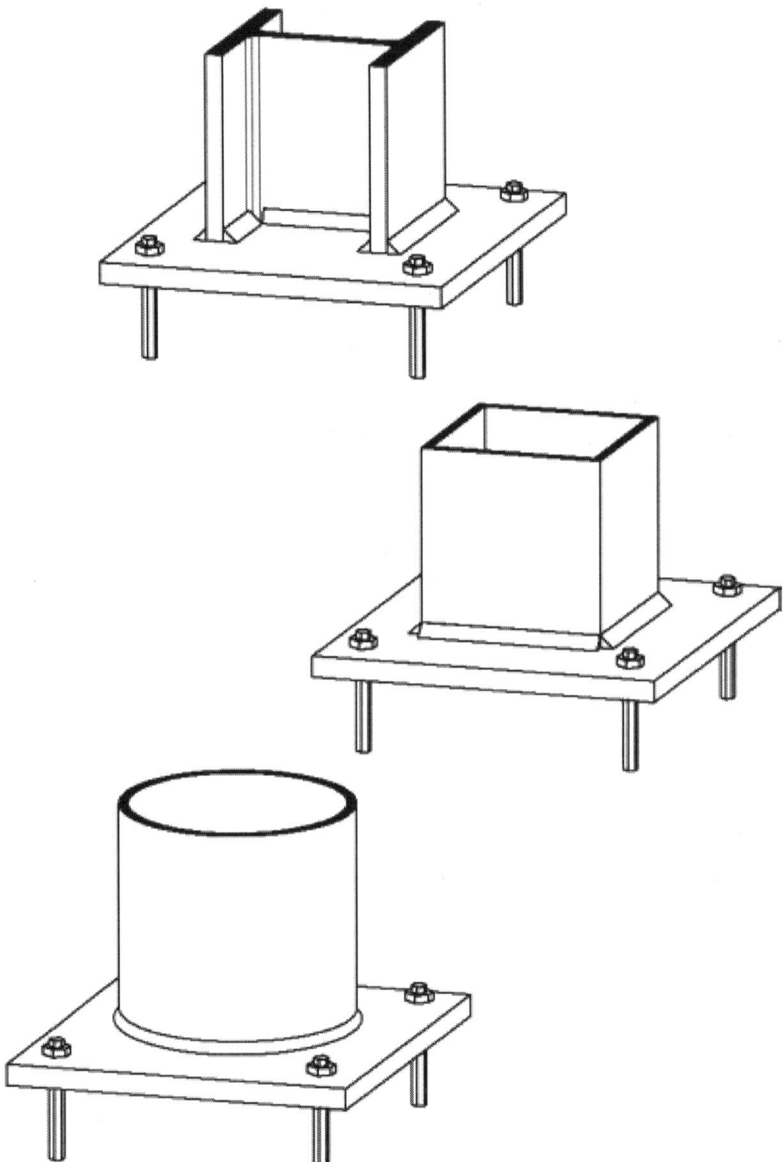

Figure: *Typical column bases*

Typical column bases, as shown in the figure on the right, consist of a single plate fillet welded to the end of the column and attached to the foundation with four holding down bolts. The bolts are cast into the concrete base in location tubes or cones and are fitted with anchor plates to prevent pull out. High strength grout is poured into the space below the plate.

Such column bases are often only subject to axial compression and shear. However, uplift may be a design case for column bases in braced bays.

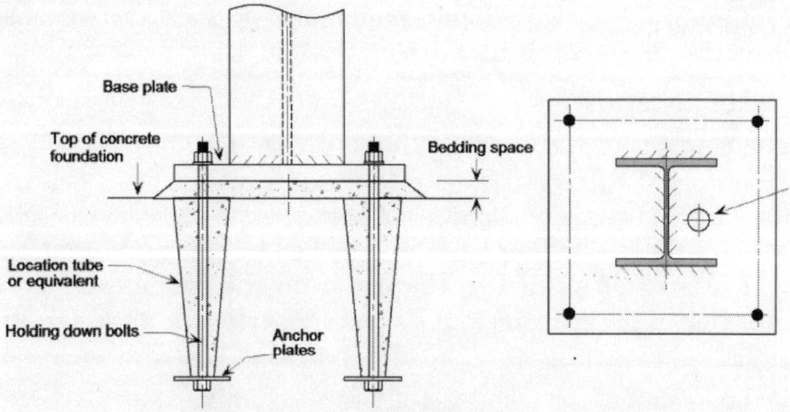

Figure: *Column base holding down bolts*

A simple rectangular or square base plate is almost universally used for columns in simple construction. The base plate should be of sufficient size, stiffness and strength to transmit the axial compressive force from the column to the foundation through the bedding material, without exceeding the local bearing resistance of the foundation.

A Base plate design tool is available.

Column bases are generally designed to transfer the force from the column to the base plate in direct bearing. Holding down systems are designed to stabilise the column during construction, and resist any uplift in braced bays. In some cases it is assumed that horizontal shear is also carried by the holding down bolts.

Horizontal Shear Transfer

The way in which horizontal shear forces are transferred to the foundation is not well researched. Some designers check the resistance of the holding down bolts, and ensure that they are adequately grouted. This practice has been successfully followed for portal frame bases, which carry a significant shear.

Braced bays may have relatively high shear forces. Designers may opt to provide a shear stub welded to the underside of the base plate, though the recess may complicate the casting of the foundation, and special attention must be paid to the grouting operation. Design methods that cover this type of detail are given in the 'Green Book' (SCI P398).

Shear between the column end and the base plate can be transmitted by friction or by nominal welds between the column and the base plate. Welds may be provided to the web only, or around parts of the profile - it is generally found that the weld resistance is more than adequate for small shear forces.

Bracing Connections

Bracing members include flats, angles, channels, I sections, and hollow sections. Bracing arrangements may involve the bracing members working in tension alone, or in both tension and compression. In most cases, the bracing member is attached by bolting to a gusset plate, which is itself welded to the beam, to the column, or more commonly welded to the beam and its end connection as shown in the figure right.

Bracing systems are usually analysed assuming that all forces intersect on member centrelines. However, realising this assumption in the connection details may result in a connection with a very large gusset plate, especially if the bracing is shallow or steep.

It is often more convenient to arrange the member intersections to make a more compact joint and check locally for the effects of eccentricities which are introduced.

Bracing connections are generally made with non-preloaded bolts in clearance holes. In theory at least, this allows some movement in the connection, but in practice this is ignored in orthodox construction. In some cases it may be that movement on reversal is unacceptable - preloaded connections should be used in these circumstances.

The general design process is:

- Identify the load path through the connection
- Arrange the connection to ensure that the design intent of the members is realised, e.g. the beam connections remain nominally pinned
- Include the effects of any significant eccentricity
- Check the components in the connection.

Special Connections

Steelwork connections for simple construction, illustrated above, will generally produce the most economic steel frame. A departure from these connections will inevitably result in an increase in overall cost. The increase in detail drawing, fabrication and erection costs can be more than 100% if non standard connections form the majority of the connections used.

The need for special connections can often be avoided by judicious selection of member sizes. A minimum weight structure is unlikely to be the most cost effective. It is therefore good economic practice to ensure that steelwork can be placed with centrelines on established grids. The top flanges of beams should, where possible, be at a constant level, but this is less critical to cost than eccentric connections.

When designing special connections, it may be possible to use a modified version of one of the standardised connections given in the Green Book, subject to additional design checks. Design principles and component sizing rules given in the Green Book should be incorporated in connection design as much as possible.

Moment Resisting Connections

The use of standard connections for beam-to-column and beam-to-beam connections is considered and an overview of the design procedures is presented, based on Eurocode 3. Both bolted and welded connections are considered. Column splices and column bases are also presented.

Types of Moment Resisting Connections

Moment resisting connections are used in multi-storey un-braced buildings and in single-storey portal frame buildings. Connections in multi-storey frames are most likely to be full depth end plate connections and extended end plate connections. Where a deeper connection is required to provide a larger lever-arm, a haunched connection can be used. However, as extra fabrication will result, this situation should be avoided if possible.

For portal frame structures, haunched moment resisting connections at the eaves and apex of a frame are almost always used, as in addition to providing increased connection resistances, the haunch increases the resistance of the rafter.

The most commonly used moment resisting connections are those which are bolted end plate beam-to-column connections; these are shown in the figure below.

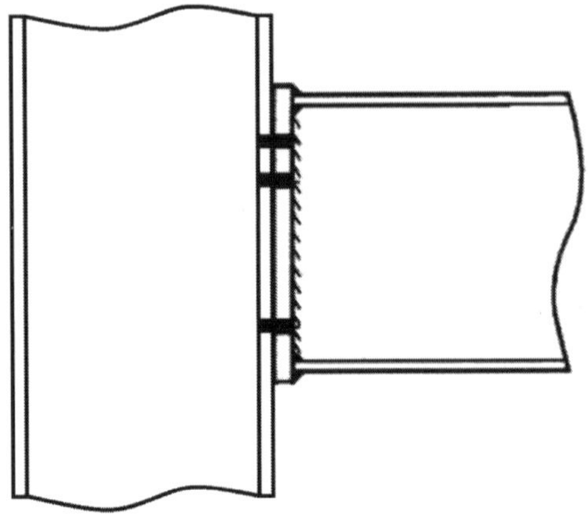

Figure: *Full depth end plate*

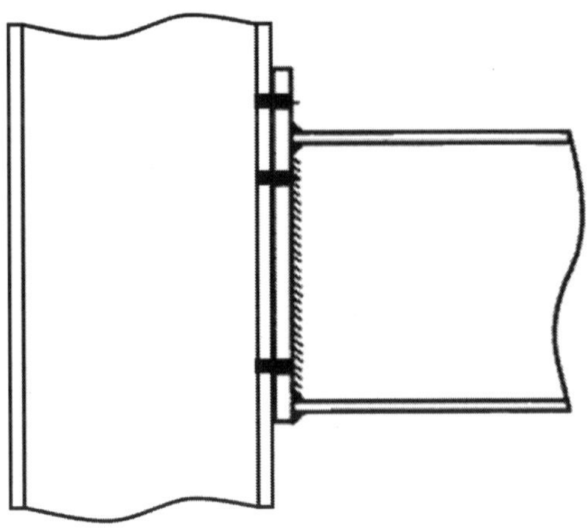

Figure: *Extended end plate*

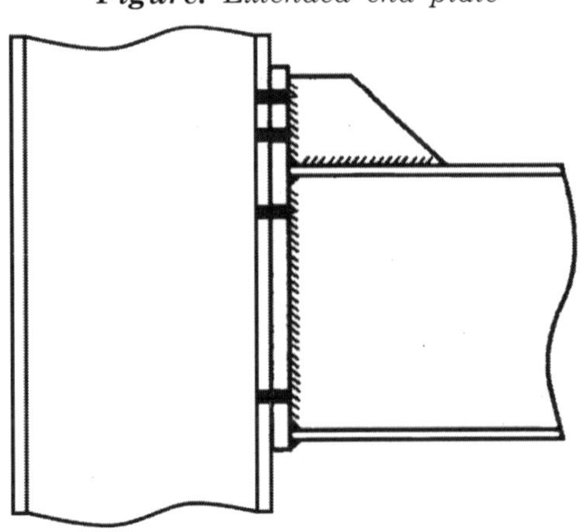

Figure: *Stiffened extended end plate*

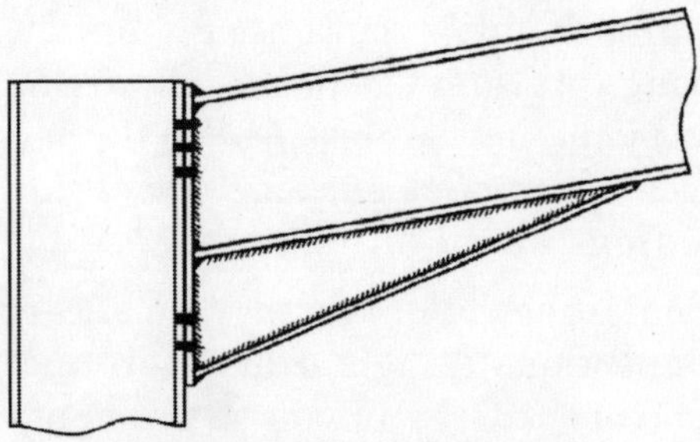

Figure: *Haunched beam*

Instead of bolted beam-to-column connections, welded connections can be used. These connections can provide full moment continuity but are expensive to produce, especially on site. Welded beam-to-column connections can be specified where construction in a factory environment is required. Welded connections are also used for the construction of buildings in seismic areas.

Other types of moment resisting connections include:

- Splices in columns and beams, including apex connections in portal frames, and
- Column bases.

One aspect that is not covered in this article is welded joints between hollow sections. However, guidance on the design of welded joints for Celsius®355 and Hybox®355 hollow sections is available from Tata Steel.

Joint Classification

Design of joints in steel structures in the UK is covered by BS EN 1993-1-8 and its National Annex.

BS EN 1993-1-8 requires that joints are classified by stiffness (as rigid, semi-rigid or nominally pinned) or by strength (as full strength, partial strength or nominally pinned). The stiffness classification is relevant for elastic analysis of frames, the strength classification is for frames analysed plastically. The Standard defines joint models as simple, semi-continuous or continuous, depending on stiffness and strength. Moment-resisting joints will usually be rigid and either full or partial strength and thus the joints are either continuous or semi-continuous.

In most situations, the design intent would be that moment-resisting joints are rigid, and modelled as such in the frame analysis. If the joints were in fact semi-rigid, the behaviour of the joint would need to be taken into account in the frame analysis but the UK NA discourages this approach until experience is gained with the numerical method of calculating rotational stiffness. Clause 5.2.2.1(2) of BS EN 1993-1-8 notes that a joint may be classified on the basis of experimental evidence, experience of previous satisfactory performance in similar cases or by calculations based on test evidence.

The UK National Annex offers further clarification, and in NA.2.6 comments that connections designed in accordance with SCI P207. (the BS 5950 version of the Green Book on moment connections) may be classified in accordance with the recommendations in that publication. SCI P207 has been updated to take account of BS EN 1993-1-8 and has been reissued as SCI P398 .

Rigid Joint Classification

Well-proportioned connections that follow the recommendations for standardisation given in SCI P398 and designed for strength alone can generally be assumed to be rigid for joints in single-storey portal frames . For multi-storey unbraced frames, rotational stiffness is fundamental to the determination of frame stability. The designer must therefore either evaluate connection stiffness (in accordance with BS EN 1993-1-8) and account for this in the frame design and assessment of frame stability or, if rigid joints have been assumed in the frame analysis, ensure that the connection design matches this assumption. For an end plate connection, it may be assumed that the connection is rigid if both the following requirements are satisfied:

- Adopting relatively thick end plates and potentially a stiffened column flange
- The column web panel shear force does not exceed 80% of the design shear resistance. If this is not possible, a stronger column should be used, or suitable strengthening should be provided.

Where a rigid joint cannot be assumed, the joint should be assumed to be 'semi-rigid' and the flexibility of the joints allowed for in the assessment of frame stability.

Costs

Moment-resisting joints are invariably more expensive to fabricate than simple (shear only) connections . Although the material cost of the components in the connection (the plates, the bolts etc) may not be significant, moment-resisting joints generally involve more welding than other connections. Welding is an expensive operation and also involves inspection after completing the welds.

Local strengthening adds further expense: increasing the resistance of the main members should always be considered as a cost-effective alternative. Local strengthening often makes the connections to the minor axis more difficult to achieve, adding further cost.

Haunches involve a large amount of welding and are therefore expensive. When used to increase the resistance of the member, such as in a portal frame rafter, their use is justified, but haunches can be an expensive option if provided only to make a bolted connection feasible.

Standard Connections

Although there are no standard moment-resisting connections, the principles of standardisation remain important for structural efficiency, cost-effective construction and safety. The following guidance is generally recommended, at least for initial design purposes:

- M20 or M24 property class 8.8 bolts, fully threaded
- Bolts at 90 or 100 mm cross-centres ('gauge')
- Bolts at 90 mm vertical centres ('pitch')
- S275 or S355 fittings (end plates, splice plates and stiffeners)
- 20 mm end plates with M20 bolts; 25 mm end plates with M24 bolts.

Bolted Beam-to-Column Connections

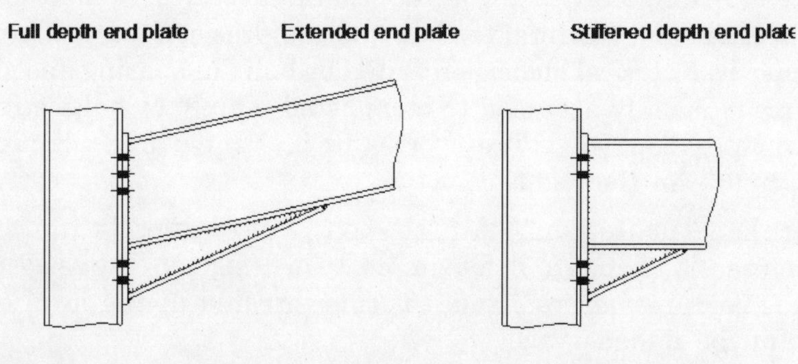

Figure: *Typical bolted endplate beam-to-column connections*

Bolted end plate joints between I-section or H-section beams and columns as shown in the figure below are designed using the approach described in BS EN 1993-1-8. Bolted end plate splices and apex connections, which use similar design procedures, are covered in the section on splices.

Design Basis

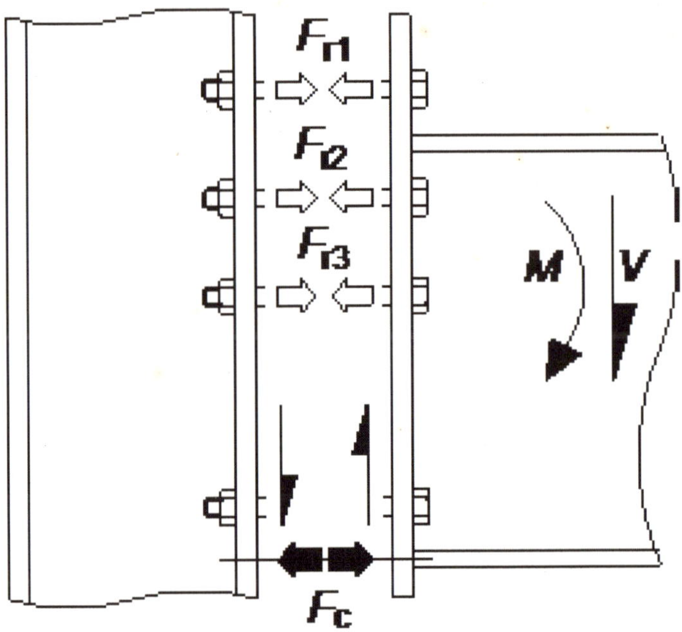

Figure: *Forces in an end-plate connection*

The resistance of a bolted end plate connection is provided by a combination of tension forces in the bolts adjacent to one flange and compression forces in bearing at the other flange. Unless there is axial force in the beam, the total tension and compression forces are equal and opposite. Vertical shear is resisted by bolts in bearing and shear; the force is usually assumed to be resisted mainly by bolts adjacent to the compression flange. These forces are illustrated diagrammatically in the figure on the right.

At the ultimate limit state, the centre of rotation is at, or near, the compression flange and, for simplicity in design, it may be assumed that the compression resistance is concentrated at the level of the centre of the flange.

The bolt row furthest from the compression flange will tend to attract the greatest tension force and design practice in the past has been to assume a 'triangular' distribution of forces, pro rata to the distance from the bottom flange. However, where either the column flange or the end plate is sufficiently flexible (as defined by NA.2.7 of the UK NA) that ductile behaviour is achieved, the full resistances of the lower rows may be used (this is sometimes referred to as a 'plastic distribution of bolt row forces').

Distribution of forces in the bolts

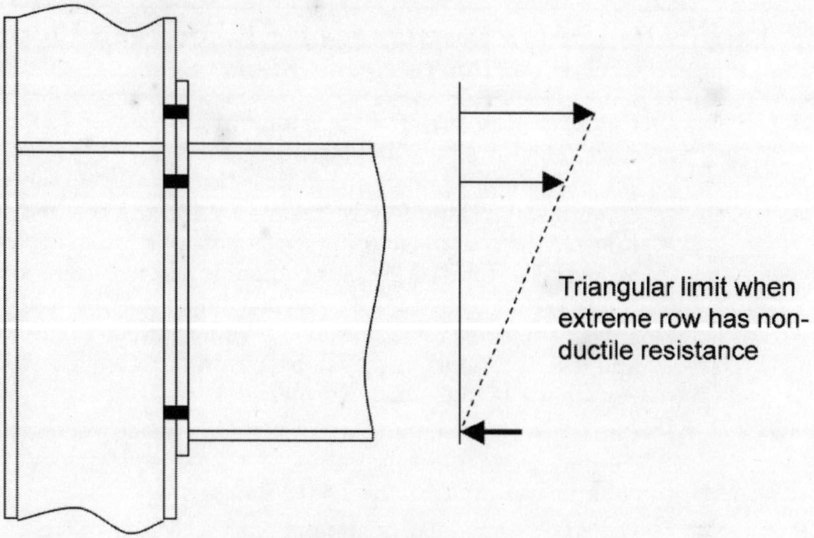

Figure: *Triangular distribution*

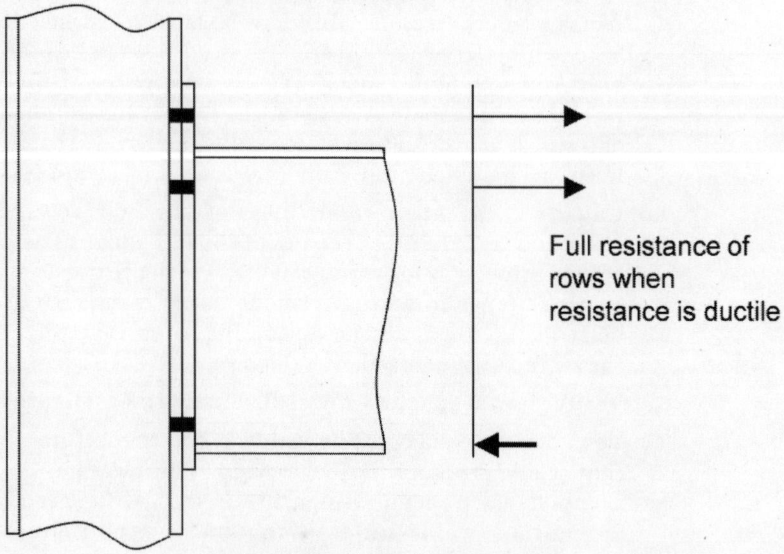

Figure: *'Plastic' distribution*

Design Method

The full design method for an endplate connection is necessarily an iterative procedure: a configuration of bolts and, if necessary, stiffeners are selected; the resistance of that configuration is evaluated; the configuration is then modified for greater resistance or greater economy, as appropriate; the revised configuration is re-evaluated, until a satisfactory solution is achieved.

Verification of the Resistance of a Welded end Plate Connection in Seven Steps	
STEP 1	Calculate the effective tension resistances of the bolt rows. This involves calculating the resistance of the bolts, the end plate, the column flange, the beam web and the column web. The effective resistance for any row may be that for the row in isolation, or as part of a group of rows, or may be limited by a 'triangular' distribution from compression flange level. The conclusion of this stage is a set of tension resistances, one value for each bolt row, and the summation of all bolt rows to give the total resistance of the tension zone.
STEP 2	Calculate the resistances of the compression zone of the column web, taking account of the shear force in the column web and of the beam flange.
STEP 3	Calculate the shear resistance of the column web.
STEP 4	If the total tension resistance exceeds the compression resistance, (Step 2) or column web shear resistance (Step 3), calculate reduced effective tension resistances for the bolt rows, where necessary to ensure equilibrium. Calculate the moment resistance. This is the summation of the products of bolt row force multiplied by its respective lever arms, calculated from the centre of compression.
STEP 5	Calculate the shear resistance of the bolt rows. The resistance is taken as the sum of the full shear resistance of the bottom row (or rows) of bolts (which are not assumed to resist tension) and 28% of the shear resistance of the bolts in the tension zone (assuming, conservatively, that they are fully utilised in tension).
STEP 6	Verify the adequacy of any stiffeners in the configuration.
STEP 7	Verify the adequacy of the welds in the connection. (Note that welds sizes are not critical in the preceding Steps). Components in compression in direct bearing need only a nominal weld, unless moment reversal must be considered.

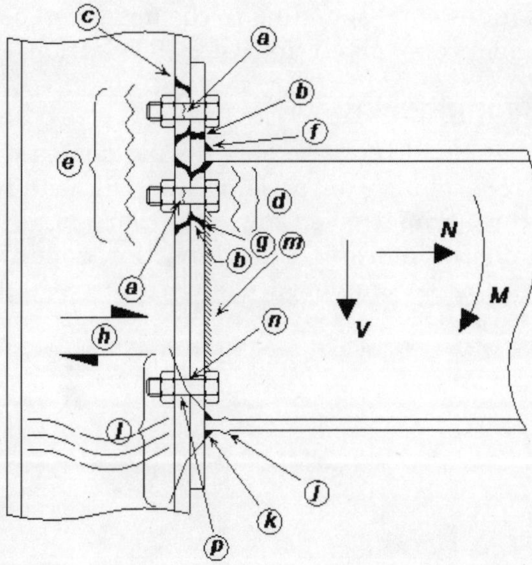

Figure: *Components to be evaluated in design procedure*

The verification of the resistance of a welded end plate connection considering each of the components that make up the connection is illustrated in the figure right and accompanying table below.

Components to be evaluated in design procedure			
Zone	***Ref***	***Component***	***Procedure***
Tension	a	Bolt tension	Step 1a
	b	End plate bending	Step 1a
	c	Column flange bending	Step 1a
	d	Beam web tension	Step 1b
	e	Column web tension	Step 1b
	f	Flange to end plate weld	Step 7
	g	Web to end plate weld	Step 7
Horizontal shear	h	Column web panel shear	Step 3
Compression	j	Beam flange compression	Step 2
	k	Beam flange weld	Step 7
	I	Column web	Step 2
Vertical shear	m	Web to end plate weld	Step 7
	n	Bolt shear	Step 5
	p	Bolt bearing (plate or flange)	Step 5

The calculations corresponding to the design steps set out above are described comprehensively in SCI P398 Section 2.5.

Methods of Strengthening

Careful selection of the members during design will often avoid the need for strengthening of the joint, and will lead to a more cost-efficient structure. Sometimes, however, there is no alternative to strengthening one or more of the connection zones. The range of stiffeners which can be employed is shown in the figures below.

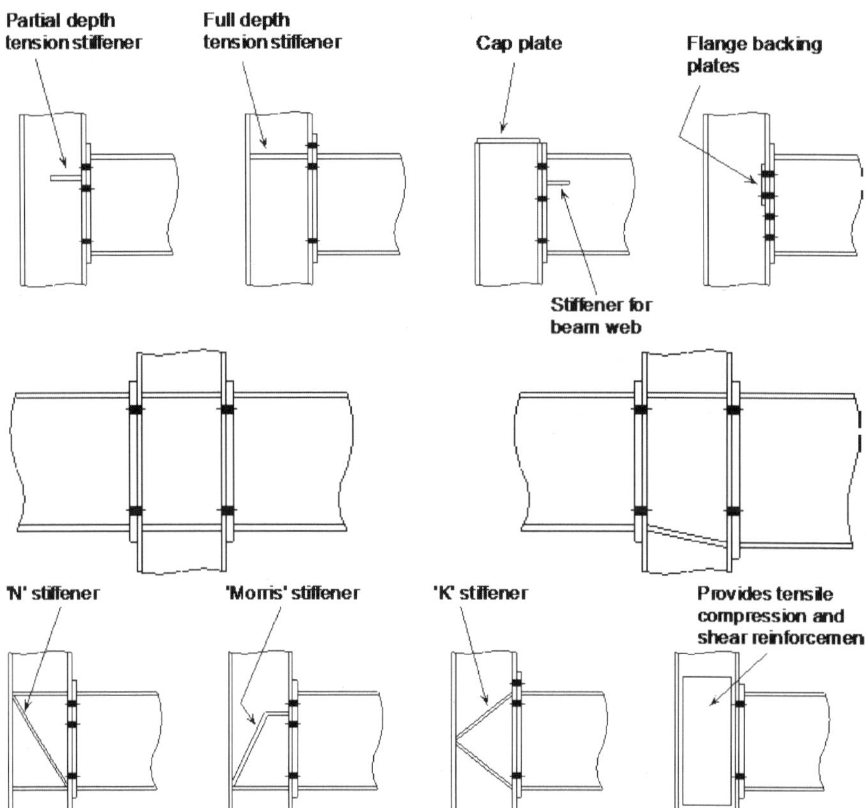

Figure: *Methods of strengthening*

The type of strengthening must be chosen such that it does not clash with other components at the connection. This is often a problem with conventional stiffeners when secondary beams connect into the column web.

There are usually several ways of strengthening each zone and many of them can contribute to overcoming a deficiency in more than one area, as shown in the table below.

Methods of strengthening columns					
Type of column stiffener		***Deficiency***			
		Web in tension	***Flange in bearing***	***Web in compression***	***Web in shear***
Horizontal stiffeners:	Full depth	•	•	•	
	Partial depth	•	•	•	
Supplementary web plates		•		•	•
Diagonal stiffeners (N & K)		•	•		•
Morris stiffeners		•	•		•
Flange backing plates			•		

Welded Beam-to-Column Connections

The intention with shop welded construction is to ensure that the main beam-to-column connections are made in a factory environment and can be full strength rigid joints. To achieve this, while still keeping the piece sizes small enough for transportation, short stubs of the beam section are welded to the columns. The connection of the stub to the rest of the beam is normally made with a bolted cover plate splice. Note that the bolted splices should be preloaded bolt assemblies.

A typical arrangement for a multi-storey building is shown in the figure below.

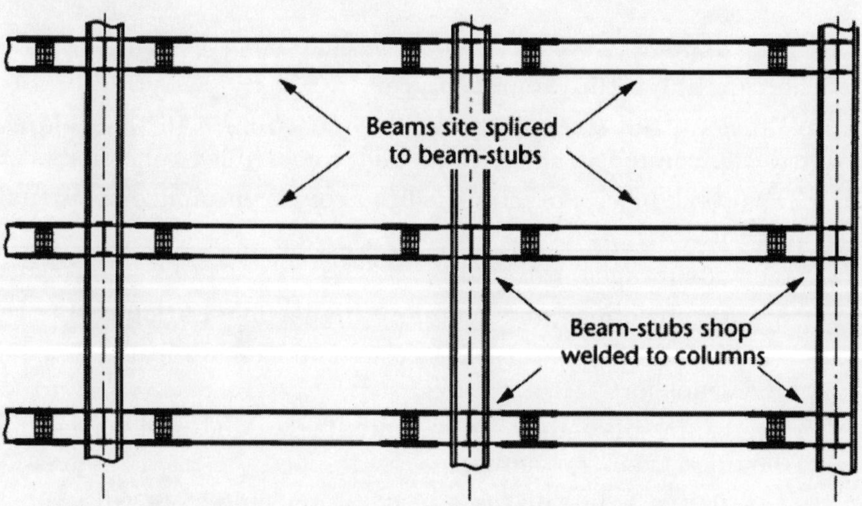

Figure: *Shop welded beam to column connections*

Shop Welded Connections

A typical factory welded connection, as shown in the figure on the right, consists of a short section beam stub factory welded on to the column flanges, and a tapered stub welded into the column inner profile on the other axis. The stub sections are prepared for bolting or welding with cover plates to the central portion of beam.

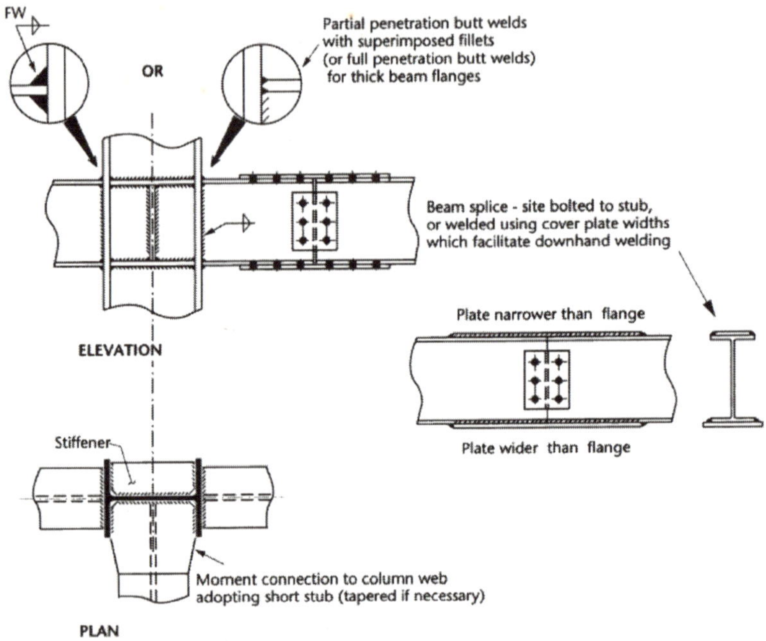

Figure: *Factory welded beam stub connection*

The benefits of this approach are:

- Efficient, full strength moment connections - all the welding to the column is carried out under controlled conditions
- The work piece can be turned to avoid or minimize positional welding.

The disadvantages are:

- More connections and therefore higher fabrication costs
- The 'Column Tree' stubs make the component difficult to handle and transport
- The beam splices have to be bolted or welded in the air some distance from the column
- The flange splice plates and bolts may interfere with some types of flooring such as pre-cast units or metal decking.

Practical Considerations

Continuous fillet welds are the usual choice for most small and medium sized beams with flanges up to 17 mm thick. However, many steelwork contractors prefer to switch to partial penetration butt welds with superimposed fillets, or full penetration butt welds, rather than use fillet welds larger than 12 mm.

To help provide good access for welding during fabrication, the column shafts can be mounted in special manipulators and rotated to facilitate welding in a down hand position to each stub.

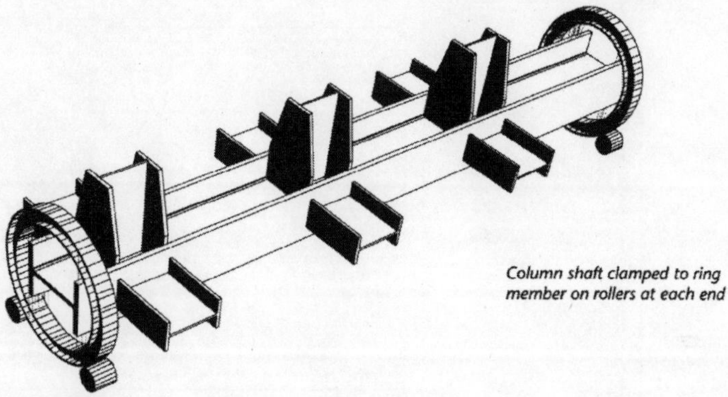

Figure: *Column manipulator for welding beam stubs to columns*

Design Method

In statically determinate frames, a partial strength connection, adequate to resist the design moment is satisfactory. If the frame is statically indeterminate, the connections must have sufficient ductility to accommodate any inaccuracy in the design moment arising, for example, from frame imperfections or settlement of supports. To achieve this, the welds in the connection must be made full strength.

Verification of the Resistance of a Welded Beam-to-Column Connection Outlined in Five Steps

STEP 1	Calculate the design forces in the tension and compression flanges of the beam. The presence of the web may be neglected when determining these forces.
STEP 2	Calculate the resistances in the tension zone and verify their adequacy. If, for an unstiffened column, the resistances are inadequate, determine the resistance for a stiffened column and verify its adequacy.
STEP 3	Calculate the resistances in the compression zone and verify their adequacy. If, for an unstiffened column, the resistances are inadequate, determine the resistance for a stiffened column and verify its adequacy.
STEP 4	Verify the adequacy of the column web panel in shear. If the unstiffened panel is inadequate, it may be stiffened, as for an end-plate connection .
STEP 5	Verify the adequacy of the welds to the flanges and web.

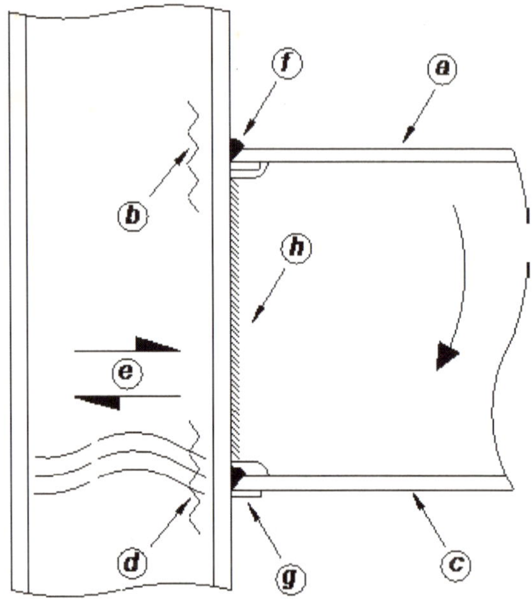

Figure: *Components to be evaluated in design procedure*

The verification of the resistance of a welded beam to column connection considering each of the components that make up the connection is illustrated in the figure right and accompanying table below.

Components to be evaluated in design procedure			
Zone	***Ref***	***Component***	***Procedure***
Tension	a	Beam flange	Step 2
	b	Column web	Step 2
Compression	c	Beam flange	Step 3
	d	Column web	Step 3
Horizontal shear	e	Column web panel shear	Step 4
Welds	f, g	Flange welds	Step 5
	h	Web weld	Step 5

The calculations corresponding to design steps set out above are described comprehensively in SCI P398 Section 3.4.

Splices

The design of beam and column splices between H or I sections that are subjected to bending moment, axial force and transverse shear force include the following types of joint:

- Bolted cover plate splices
- Bolted end plate splices
- Welded splices.

The design of bolted column splices that are subject to predominant compressive forces is covered in the simple connections article and in greater detail in SCI P358 .

Bolted Cover Plate Splices

Connection Details

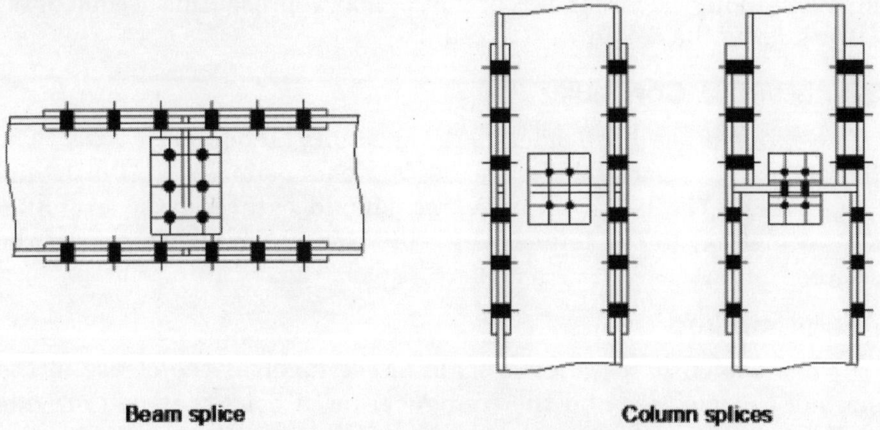

Figure: *Typical bolted cover plate splices*

Typical bolted cover plate splice arrangements are shown in the figure right.

In a beam splice there is a small gap between the two beam ends. For small beam sections, single cover plates may be adequate for the flanges and web. For symmetric cross sections, a symmetric arrangement of cover plates is normally used, irrespective of the relative magnitudes of the design forces in the flanges.

Column splices can be either of bearing or non-bearing type. Design guidance for bearing type column splices is given in SCI P358. Non-bearing column splices may be arranged and designed as for beam splices.

Design Basis

A beam splice (or a non-bearing column splice) resists the coexisting design moment, axial force and shear in the beam by a combination of tension and compression forces in the flange cover plates and shear, bending and axial force in the web cover plates.

To achieve a rigid joint classification, the connections must be designed as slip resistant connections. It is usually only necessary to provide slip resistance at SLS (Category B according to BS EN 1993 1 8, 3.4.1) although if a rigid connection is required at ULS, slip resistance at ULS must be provided (Category C connection). In elastically analysed structures, bolted cover plate splices are not required to provide the full strength of the beam section, only to provide sufficient resistance against the design moments and forces at the splice location. Note, however that when splices are located in a member away from a position of lateral restraint, a design bending moment about the minor axis of the section, representing second order effects, must be taken into account.

Stiffness and Continuity

Splices must have adequate continuity about both axes. The flange plates should therefore be, at least, similar in width and thickness to the beam flanges, and should extend for a minimum distance equal to the flange width or 225mm, on either side of the splice.

Design Method

The design process for a beam splice involves the choice of the sizes of cover plates and the configuration of bolts that will provide sufficient design resistance of the joint. The process has a number of distinct stages, which are outlined below.

The Design Process for a Beam Splice Outlined in Five Steps

STEP 1	Calculate design tension and compression forces in the two flanges, due to the bending moment and axial force (if any) at the splice location. These forces can be determined on the basis of an elastic stress distribution in the beam section or, conservatively, ignoring the contribution of the web. Calculate the shear forces, axial forces and bending moment in the web cover plates. The bending moment in the cover plates is that portion of the moment on the whole section that is carried by the web (irrespective of any conservative redistribution to the flanges) plus the moment due to the eccentricity of the bolt group resisting shear from the centreline of the splice. Calculate the forces in the individual bolts
STEP 2	Determine the bolt resistances and verify their adequacy, in the flanges and in the web.
STEP 3	Verify the adequacy of the tension flange at the splice and the cover plates.
STEP 4	Verify the adequacy of the compression flange at the splice and the cover plates.
STEP 5	Ensure that there is a minimum resistance for continuity of the beam.

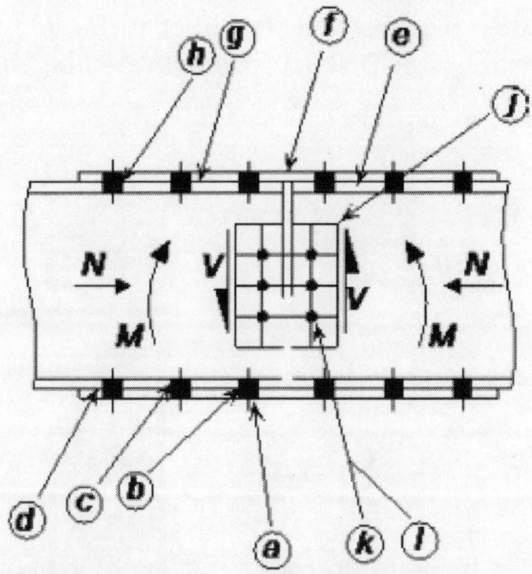

Figure: *Splice components to be verified*

The above steps involve the determination of resistance values of 11 distinct components of a bolted splice, as illustrated in the figure right and accompanying table below.

Splice components to be verified			
Zone	***Ref***	***Component***	***step***
Tension	a	Flange cover plate(s)	3
	b	Bolt shear	2
	c	Bolt bearing	2
	d	Flange	3
Compression	e	Flange	4
	f	Flange cover plate(s)	4
	g	Bolt shear	2
	h	Bolt bearing	2
Shear	j	Web cover plate(s)	1
	k	Bolt shear	1
	l	Bolt bearing	1

The calculations corresponding to design steps set out above are described in detail in SCI P398 Section 4.2.

Bolted End-Plate Splices

Connection Details

Bolted end plate connections, as splices or as apex connections in portal frames, are effectively the beam side of the beam to column connections, mirrored to form a pair. This form of connection has the advantage over the cover plate type in that preloaded bolts (and the consequent required preparation of contact surfaces) are not required. However they are less stiff than cover plate splice details.

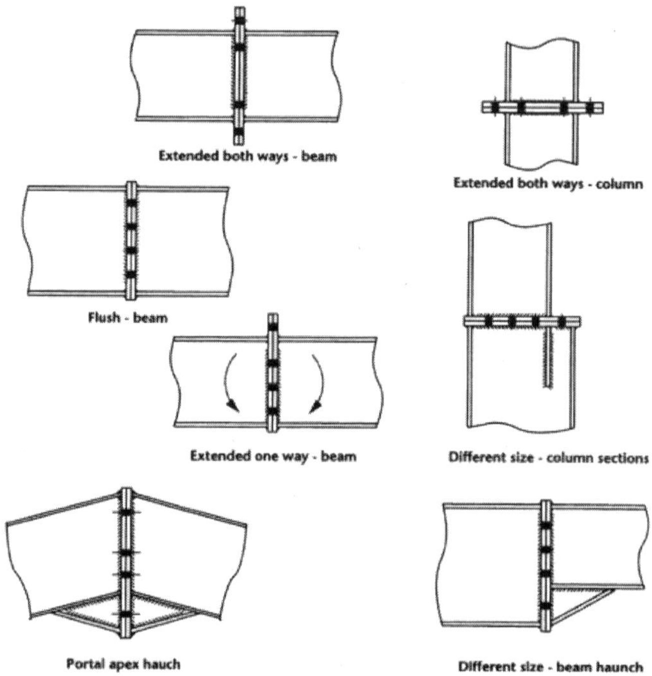

Figure: *Typical bolted end plate splices*

The 'portal apex haunch' splice is regularly used in single storey portal frames and is commonly assumed to be 'Rigid' for the purposes of elastic global analysis.

Design Method

The design method is essentially that described for beam-to-column connections , omitting the evaluation of column resistances. The relevant steps and the corresponding calculations are described in SCI P398 Section 4.3.

Beam-through-Beam Moment Connections

Connection Details

Beam-through-beam joints are usually made using end plate connections with non-preloaded bolts; typical details are shown in the figure below. Non-preloaded bolts may be used when there are only end plates but when a cover plate is used as well, preloaded bolts should be used, to prevent slip at ULS.

Typical Beam through Beam Splices

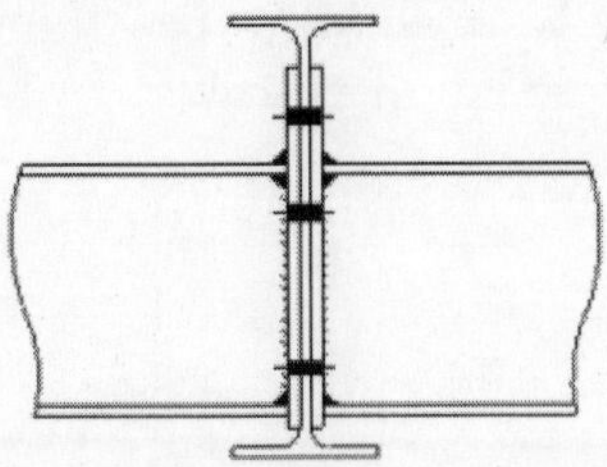

Extended end plates to beam we

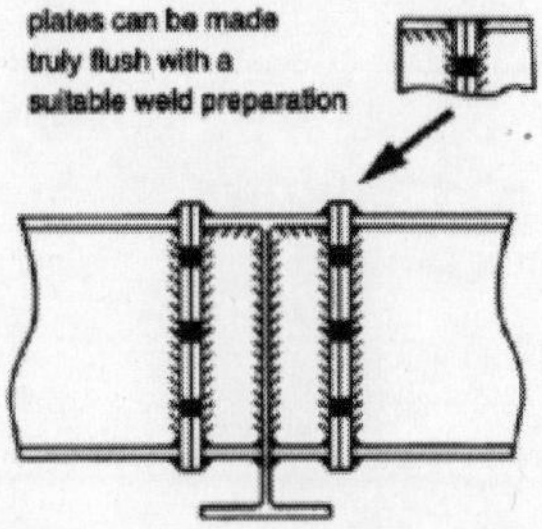

Flush end plates to beam web stiffeners

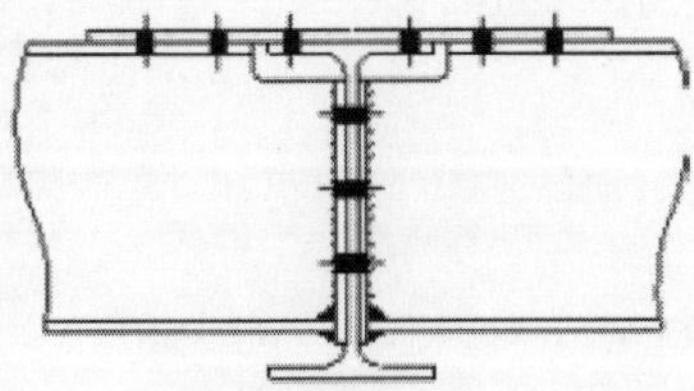

End plate and cover plate

Design Method

Where there is no cover plate, the design method for end plate splices may be used. Where a cover plate is used, it should be designed as for a cover plate splice; it may be assumed conservatively that the end plate bolts carry only vertical shear.

The connection between the cover plate and the supporting beam is usually only nominal, as the moment transferred in torsion to the supporting beam is normally very modest.

The relevant steps and the corresponding calculations are described in SCI P398 Section 4.4.

Welded Splices

Connection Details

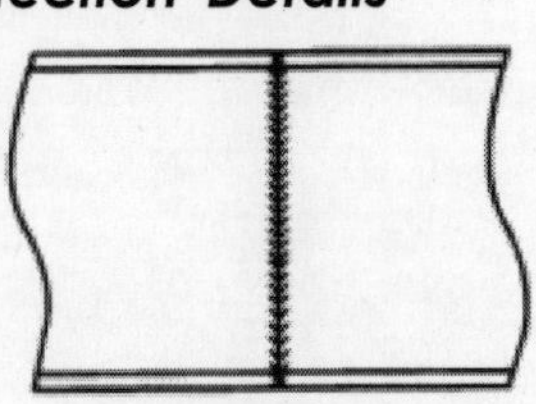

Butt welded beam

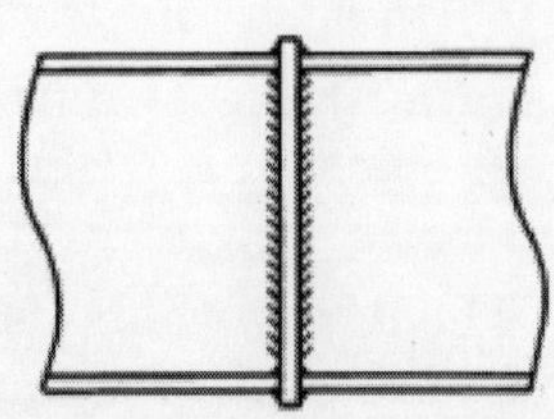

beam with division plate

Figure: *Typical welded beam splices*

Welded shop splices are often employed to join shorter lengths delivered from the mills or stockists. In these circumstances the welds are invariably made 'full strength' by butt welding the flanges and the web. Small cope holes may be formed in the web to facilitate welding of the flange.

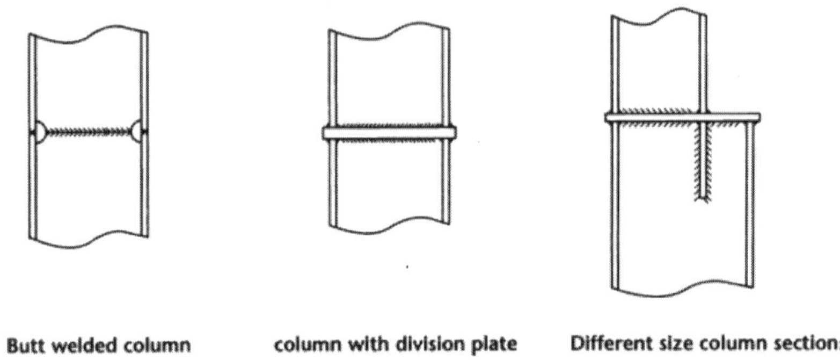

Figure: *Typical welded column splices*

Where the sections being joined are not from the same 'rolling' and consequently vary slightly in size because of rolling tolerances, a division plate is commonly provided between the two sections. When joining components of a different serial size by this method, a web stiffener is needed in the larger section (aligned to the flange of the smaller section), or a haunch may be provided to match the depth of the larger size.

A site splice can be made with fillet welded cover plates, as an alternative to a butt welded detail. Bolts may be provided in the web cover plates for temporary connection during erection.

Design Basis

For welded splices the general design basis is:

- For statically indeterminate frames, whether designed plastically or elastically, full strength welds should be provided to the flanges and the web
- In statically determinate frames, splices may be designed to resist a design moment that is less than the member moment resistance, in which case:
    - o The flange welds should be designed to resist a force equal to the design moment divided by the distance between flange centroids.
    - o The web welds should be designed to resist the design shear.

- o If there is an axial force it should be shared between the flanges and the welds designed for this force in addition to that due to the design moment.

The full strength requirement for indeterminate is needed to ensure that a splice is strong enough to accommodate any inaccuracy in the design moment, arising for example, from frame imperfections, modelling approximations or settlement of supports.

Column Bases

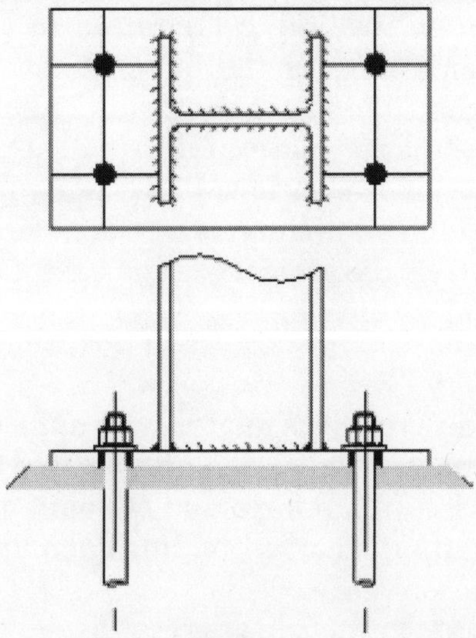

Figure: *A typical unstiffened column base plate*

An example of a column base which is able to transmit moment and axial force between steel members and concrete substructures at the base of columns is shown in the figure on the left. The example shows a column base with an unstiffened base plate. Stiffened base plate connections and column bases cast in pockets are other options available. However, rigid base connections are not used because of the associated foundation costs.

Design Basis

In terms of design, a column base connection is essentially a bolted end plate connection with certain special features:

- Axial forces are more important than is generally the case in end plate connections.
- In compression, the design force is distributed over an area of steel-to-concrete contact that is determined by the strength of the concrete and the packing mortar or grout.

- In tension, the force is transmitted by holding down bolts that are anchored in the concrete substructure.

As a consequence, an unstiffened base plate tends to be very thick, by comparison with end plates of beam-to-column connections.

More often than not, the moment may act in either direction and symmetrical details are chosen. However, there may be circumstances, e.g. some portal frames, in which asymmetrical details may be appropriate.

The connection will usually be required to transmit horizontal shear, either by friction or via the bolts. It is not reasonable that horizontal shear is distributed evenly to all the bolts passing through clearance holes in the base plate, unless washer plates are welded over the bolts in the final position. If the horizontal shear is large, a shear stub welded to the underside of the base plate may be more appropriate. In all cases, the grouting of the base is a critical operation, and demands special attention.

Design Method

The design process requires an iterative approach in which a trial base plate size and bolt configuration are selected and the resistances to the range of combined axial force and moment are then evaluation. The relevant steps and the corresponding calculations are described in SCI P398 Section 5.5.

Classification of Column Base Connections

The rigidity of the base connection has generally greater significance on the performance of the frame than other connections in the structure. Most unstiffened base plates are substantially stiffer than a typical end plate detail. The thickness of the base plate and pre-compression from the column contribute to this. However, no base connection is stiffer than the foundation and, in turn, the soil to which its moment is transmitted. Much can depend on the characteristics of these other components, which include propensity to creep under sustained loading. The base connection cannot be regarded as 'rigid' unless the concrete base it joins is itself relatively stiff. Often this will be evident by inspection.

Connections in Trusses

Truss connections are generally welded, except for site splices.

The connections between the members of a truss were traditionally bolted but welded connections are now preferred, especially in tubular construction. Nevertheless it may be necessary to introduce splices in the chord members if the complete trusses are too long for transportation. These splices provide for bolting individual lengths of trusses on site and should be located and detailed carefully if they are architecturally important.

Connections in open section trusses are lapped and welded.

Traditional truss construction based on open sections (angles and tees) used bolts and gusset plates. The position of the bolts was usually detailed to minimise eccentricities. In welded construction the internal members are typically attached directly to the top and bottom chords. These chords are usually continuous except at changes of direction or where splices are necessary to reduce the length of the assembly for transportation. Internal members lap onto the chord members and are simply connected by fillet welding.

Connections in trusses using heavier sections are welded but may require more careful preparation.

Long span or heavily loaded trusses often use I, H or channel sections. Internal members are generally cut to fit directly against the flanges of the chord members and are profile welded to form the connection.

Connections in tubular trusses are usually welded.

Figure: *Intersecting angles should not be too small.*

Tube to tube joints, such as those in tubular trusses, are usually welded, because full profile welded joints not only look better but are also cheaper than creating elaborate bolted joints. It is usually cheaper to make truss connections if the chords are made from rectangular tubes, rather than circular. This is because the ends of the diagonals can be straight cut, rather than cut to the more complex intersection profile if circular chords are used. In either case the end of the diagonal may be cut in the same plane as the chords if a fillet weld is adequate. This will normally be adequate, but if a butt weld is needed the end of the diagonal must be properly prepared. A number of fabricators use profiling machines which automatically cut the diagonal to the correct line and end preparation for welding.

It is important that the truss geometry does not produce too small an intersection angle between members. The limit is about 30 degrees for rectangular hollow sections and 20 degrees for circular hollow sections, and below this value it is difficult to get the welding electrode in to make a weld.

Connections between Steel and Concrete

Connections between steel and concrete must account for differences in strength and accuracy between the two materials.

Connections between steel and concrete are fundamentally different from normal steelwork connections for a number of reasons:

- Safe working stresses in concrete are typically in the order 1/10 to 1/15 of those for steel;
- The accuracy with which in-situ concrete can be formed is much less than for factory fabricated steel;
- Concrete has negligible tensile strength, and so any tensile forces applied to a concrete element must usually be carried back into the reinforcement.

The connection between a column base and concrete foundation is detailed to distribute the stresses in the concrete and provide adjustment on site for inevitable inaccuracies.

A very common column base connection involves casting holding down bolts into the concrete foundation. The concrete base is poured under typical site conditions, and a high degree of precision is unrealistic. After the concrete has hardened, the top level of the base can be checked, and shims placed so that when the column is erected, it will be placed at the correct level. The bolt boxes allow the bolts to be adjusted sideways to fit into the holes in the baseplate. The bolts serve to locate the column and provide some stability until the frame is braced. They may also be required to resist uplift or bending of the column base, if the building design requires this. Once the column position has been checked, the space under the baseplate can be grouted, and the baseplate can then distribute the high stresses in the steel section over a large area of concrete in bending.

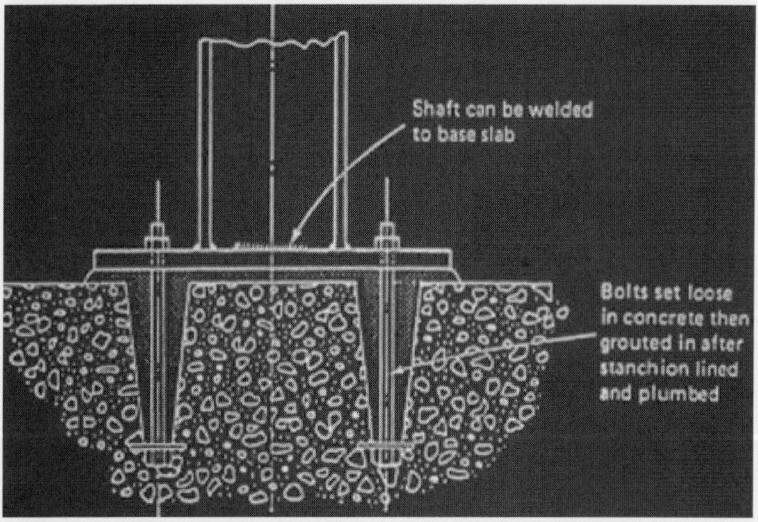

Figure: *Tensile or shear forces in steel-concrete connections can be accommodated by using expanding bolts or resin anchor bolts.*

A different type of connection occurs when tension or shear is to be applied to a concrete slab or wall. Expanding bolts can be used to transfer loads into concrete. These bolts have to work at lower stresses and be at much larger centres and edge distances than bolts for connecting steel. Such connections can therefore become very complicated. If fire resistance is not a problem, resin anchors can be used as an alternative to expanding bolts.

Beam-Beam Connections

The principles of beam-beam connections are similar to those for beam-to-column connection, although some additional preparation may be necessary if beams are to be connected with the top flanges level as is normal.

Beam-beam connections are normally detailed so that the top flanges of the two beams are level. The end of the secondary (supported) beam is therefore often notched to enable attachment to the web of the main (supporting) beam.

The simplest concept for a beam-beam connection is to support one beam directly on the top flange of the other. This is often used with the parallel beam system, providing an efficient means of distributing services horizontally and vertically above and between the main beams respectively. For other framing arrangements for floors this is not a common detail since it leads to deep construction zones. It is, however, commonly used in roof construction where purlins, acting as secondary structural beams, are supported directly on the main beam or rafter.

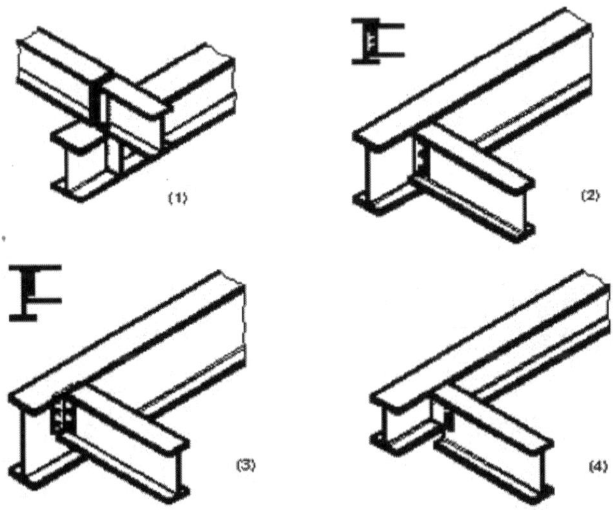

Other Functions of Connections

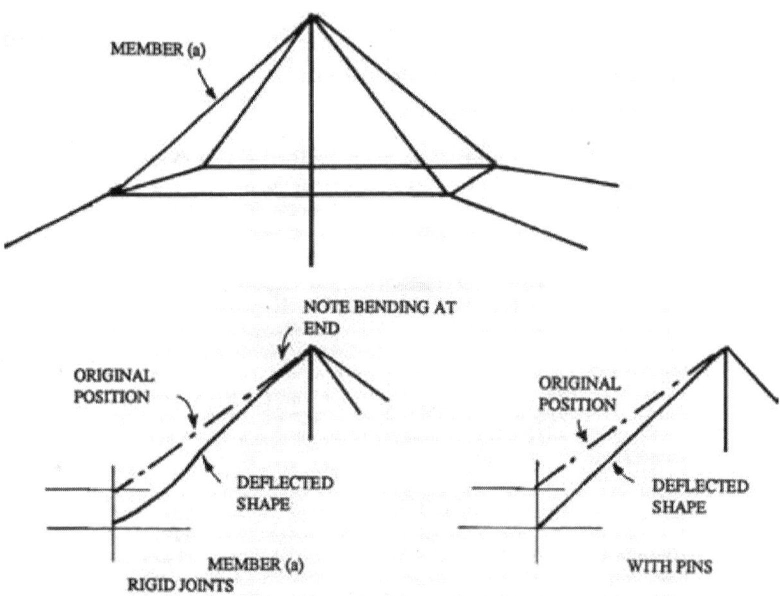

Figure: *At movement joints, special connection details are used.*

Connections can influence structural behaviour. Connections can even be used to control the structural behaviour. The figure below shows a connection for a leisure pool roof at Windsor. The four rectangular tubes (a) are working mainly as a set of ties, carrying equal loads. The overall system is fairly flexible, and parasitic bending moments could develop in the tubes because of the deflection of the roof. By introducing pin joints in the members it was possible to eliminate these bending moments and minimise the size of the tubes.

At movement joints in buildings, special connection details are needed. Movement joints may be used for several reasons, such as to allow for thermal expansion or relative vertical movements of structure supported on different foundations. These can be accommodated by providing sliding or rotational connections.

Connections can also be Used for Decoration.

Connections which are visible can have a significant impact on the overall appearance. In such cases the details require careful design, and the decorative effect may be the principal design generator.

Cast Iron in Moulds

Cast iron components were manufactured by casting the molten metal in sand moulds that allowed fairly complex shapes to be produced.

Cast iron was formed into structural elements by pouring the molten and extremely fluid material into moulds formed in damp sand. These moulds were normally in two halves which were brought together initially against a timber 'pattern' that left an impression in the sand of the desired shape of the iron component. The two halves of the mould were separated, the timber pattern removed and the mould re-assembled ready to receive the molten iron. When the metal had cooled and solidified, the sand was simply brushed away and then used again to form another mould. This meant that each individual cast iron element was specially made and that the shape could be as complex as the two-part sand mould would allow.

The ends of structural elements were shaped during the casting process to facilitate jointing.

By the end of the 18th century, the craft of sand moulding was highly developed so that intricate and complicated shapes could be formed quickly and accurately. Much of this skill went into stylistic enterprises like mass producing Corinthian columns, but it did mean that the problem of connecting one element to another could be tackled in a diverse range of ways.

In the Iron Bridge at Coalbrookdale, which was the first major cast iron structure, the joint configurations were similar to those used in carpentry.

Not unnaturally, at Iron Bridge the designers of this unique structure borrowed ideas from established materials. They opted for arrangements that were based on carpentry joints. The moulds were shaped so that a tongue was cast into the end of one piece and a groove was cast into the end of another. The pieces could then be slotted into each other and a cast iron wedge locked them together. In this way, the lightweight framework of the bridge could be assembled with no special jointing pieces other than the wedges. Looking at the bridge today, many extra bolted clamps and fasteners are visible. These are later attempts to strengthen the bridge, but they must not be taken to mean that the original connections were basically unsound. The defects in the bridge structure were due to movement of the abutments and the interlocking joints were always a satisfactory if somewhat elaborate means of connection.

Figure: *Structural frameworks with cast iron beams and columns and brick jack arch floors were developed in the late eighteenth century to provide fireproof construction for industrial buildings.*

By the end of the eighteenth century cast iron mill structures were being erected and these had to be justified on a purely functional basis. A significant requirement was for a textile mill that would be resistant to fire. In the 1790s and the 1800s a number of mills were built which were all variations on the theme of masonry jack floors, timber beams and cast iron columns. Later, cast iron was used for the beams as well. These structures were much less combustible than their all-timber predecessors, and the way the different materials connected together to give mutual structural support was a triumph for the adaptability of the shaped cast iron end pieces. Not only did the columns lock together vertically over one another, but the thrust from the jack arches was absorbed by the angled seating and transferred into the horizontal beam. In some cases tie rods were threaded through the connectors to prevent outward thrust being applied to the external masonry walls.

This drawing shows the superb way in which the extremely complex junction between cast iron column, tie rod, timber beam and brick jack arch floor including sand filling was resolved in the earliest cast iron mills of the late eighteenth century. Cast iron enabled such complex junctions to be achieved because it allowed skilled craftsmen to make geometrically elaborate shapes.

As the 19th century progressed, designers became more and more confident with cast iron, and spurred on by rapid industrialisation and particularly the railway network, they produced more and more ambitious structures. Cast iron, however, has one major structural inadequacy: it is very weak in tension. This is no problem with compression arches for bridges and roofs, but any structure with bending or tension forces was vulnerable. By using large sections with big connections, tension stresses could be kept quite low. As confidence grew, however, sections became more highly stressed, and longer span railway bridges required complex and elaborate special junctions to keep the tension stresses within bounds. When the Dee Bridge near Chester collapsed with the loss of five lives in 1847 and the greatest engineer of his day, Robert Stephenson, only narrowly avoided major censure, it was realised that a material that was both brittle and of low tensile strength could have only limited use for structures. Subsequently, wrought iron and steel became the predominant structural alloys of iron and carbon.

This complicated device was used by Robert Stephenson to make a bridge beam that utilised the tensile strength of expensive wrought iron with cheap cast iron which was only reliably strong in compression.

This jointing system was used for the railway bridge over the River Dee but failed in 1847 soon after it had been opened. Such designs where disparate parts are struggling to compensate for the inherent weakness of each other are an example of basically poor engineering practice. Curiously Stephenson's contemporary design for the High Level Bridge over the River Tyne demonstrated how good design could overcome the same problem.

Dowels Rivets and Bolts

Rivets and bolts transmit load by a combination of bearing pressure, shear and friction on the interface between the components being joined.

The structural principle of the rivet and the bolt is just the same as that of the traditional dowel or peg. In essence, whenever it is desired to transfer load from one plate to another, the method is to drill a hole in each plate and then insert a circular bar through these holes.

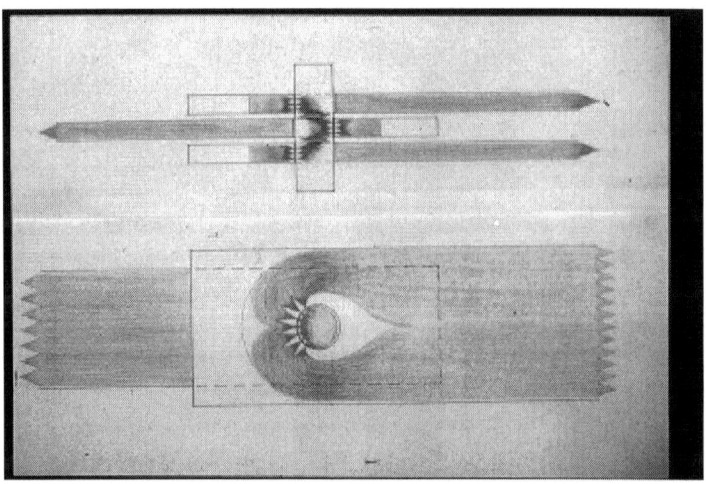

The system relies on the pressure at the interface of the plate and the bar to transfer load and also on the shear stresses in the cross section of the bar. Iron and steel have adequate shear strength so that simply increasing the cross sectional area increases the capacity of the bar. To get adequate bearing between the bar and the plate, the ductile character of wrought iron and steel helps the action of the joint. The holes drilled in the plates have in practice to be slightly larger than the bar. If the circle of the bar and the circle of the hole remained perfect, they would only touch at a point. With wrought iron and steel connections, both bar and hole will distort significantly to give a larger area of contact. This principle is used for the dowel, the bolt and the rivet, but it was the rivet which was the predominant means of connecting wrought iron and steel from the mid-19th to the mid-20th century.

This curious diagram is an attempt to illustrate the flow of forces between steel plates connected by a bar. The principle is the same whether the bar is a dowel, a rivet or a bolt. The lower diagram shows the forces flowing through the plate going around the hole cut in the plate and concentrating behind the bar. It should be apparent that at the cross section of the plate that goes through the hole, the stresses are most heavily concentrated. Also that for a reasonable contact area between the plate and the bar, a degree of ductile deformation is required. Were the hole in the plate and the bar to remain perfectly circular they would touch only at a point.

The positive virtue of a ductile material like steel or wrought iron is that by deforming by quite small amounts it spreads the loading over a reasonable area. The upper diagram is a section through the bar showing the transfer of load between the plates and the sections of the bar that have to transfer the load by shear stresses. In order to make the diagram clearer and to give room to indicate the shear planes the plates are shown slightly apart. In practice of course, rivets and bolts clamp the plates together. Indeed the friction between the plates gave traditional joints added strength and it is utilised by modern high strength friction grip bolts.

Rivets make very effective connections between components in both wrought iron and steel.

In essence, the rivet is simply a bar with its ends hammered over to prevent it slipping out of the hole. This is easy to do with soft metals like gold, silver, or copper and has been a traditional method of jointing since prehistory. It was impossible to achieve with cast iron because cast iron cannot be squashed, it simply fractures. It is a very effective way of joining ductile metals such as wrought iron and steel. Rivets were heated in portable braziers and held in position with tongs while the protruding end was bashed over with a big hammer to form strong reliable and positive fixings.

In fact, the connection had more strength than simply the shear across the cross section of the rivet since the hammering of the head gripped the plates together and load could be transferred through friction between the plates. Also, the heat and pressure would often induce some pressure welding if the conditions were favourable.

The Eiffel Tower used 2.5 million rivets. On site they were fitted by this type of four person team. The rivet was heated to red heat in a portable brazier and transferred from brazier to pre-drilled hole by the person holding it in position with the tongs. Finally while restrained at the far side it was beaten with a hammer to form the fixing head. Look at any large scale nineteenth century structure and consider the man power and organisation required.

In terms of the visual expression of the joint, the change from cast to wrought iron was a step backwards. The relation between the parts was limited by the overlapping of flat surfaces. While this was ideal for the fabrications of steam boilers, it made the connection of columns and beams, struts and ties more difficult. Initially, sections were small in size and restricted in profile to flats, tees and angles. Larger sections would be made up from combinations of these basic profiles, and here there was scope for richer and more varied joint design. As the 19th century progressed, the introduction of steel and the development of the industry meant that the rolled sections got larger and the jointing tended to be cruder.

This column head is a particularly interesting juxtaposition of a timber roof purlin with localised increase in depth, a wrought iron edge beam made up from riveted angles and flats, cast iron decorative brackets with both opposing curves fixed by bolts and a slender cast iron column.

Figure: *Rivet patterns at joints contributed to the visual interest of large structures.*

The most visually interesting structures of this period tended to be the very large showpiece structures built for the great exhibitions.

Eiffel's great tower for the Paris exhibition of 1888 was a tour de force of joint design. As the scale of the structures got larger, it was not sensible to increase the size of the rivets, there just had to be more of them. At close quarters, the pattern of the uncountable numbers of rivets arranged in consciously designed patterns gives human scale to the vast structure. The care with which the geometry of these patterns was organised showed that engineers like Eiffel were as alive to the potential of rewarding detailed design as any architect. The Gallery of Machines for the same exhibition was not only a gigantic enclosure of space, but its structure was built up from riveted sections in a way that demonstrated a controlled richness of detail that can still be admired in the famous photograph a hundred years later.

7

Structural Framing

Framing (Construction)

Framing, in construction is the fitting together of pieces to give a structure support and shape and sometimes is used as a noun such as "the framing" or "framing members". Framing materials are usually wood, engineered wood, or structural steel. Building framing is divided into two broad categories, heavy-frame construction (heavy framing) if the vertical supports are few and heavy such as in timber framing, pole building framing, or steel framing or many and smaller called light-frame construction (light framing) including balloon, platform and light-steel framing. Light-frame construction using standardized dimensional lumber has become the dominant construction method in North America and Australia because of its economy. Use of minimal structural materials allows builders to enclose a large area with minimal cost, while achieving a wide variety of architectural styles. Historically mankind fitted naturally shaped wooden poles together as framework and then began using joints to connect the timbers, a method today called traditional timber framing.

Timber framing was superseded by balloon framing beginning in the 1830s in America which is made up of many light-weight wall members called studs rather than using fewer, heavier supports called posts, and was nailed together rather than using joinery. The studs in a balloon frame extend two stories from sill to plate. Platform framing superseded balloon framing and is the standard wooden framing method today. The name comes from each floor level being framed as a separate unit or platform. Modern light-frame structures usually gain strength from rigid panels (plywood and other plywood-like composites such as oriented strand board (OSB) used to form all or part of wall sections) but until recently carpenters employed various forms of diagonal bracing to stabilize walls. Diagonal bracing remains a vital interior part of many roof systems, and in-wall wind braces are required by building codes in many municipalities or by individual state laws in the United States. Special framed shear walls are becoming more common to help buildings meet the requirements of earthquake engineering and wind engineering.

The alternative to framed construction is generally called *mass wall* construction which is made from horizontal layers of stacked materials such as log building, masonry, rammed earth, adobe, etc.

Walls

Wall framing in house construction includes the vertical and horizontal members of exterior walls and interior partitions, both of bearing walls and non-bearing walls. These *stick* members, referred to as studs, wall plates and lintels (*headers*), serve as a nailing base for all covering material and support the upper floor platforms, which provide the lateral strength along a wall. The platforms may be the boxed structure of a ceiling and roof, or the ceiling and floorjoists of the story above. The technique is variously referred to colloquially in the building trades as *stick and frame*, *stick and platform*, or *stick and box* as the sticks (studs) give the structure its vertical support, and the box-shaped floor sections with joists contained within length-long post and lintels (more commonly called *headers*), support the weight of whatever is above, including the next wall up and the roof above the top story. The platform also provides the lateral support against wind and holds the stick walls true and square. Any lower platform supports the weight of the platforms and walls above the level of its component headers and joists.

Framing lumber should be grade-stamped, and have a moisture content not exceeding 19%.

There are three historically common methods of framing a house.

- Post and beam, which is now used predominantly in barn construction.
- Balloon framing using a technique suspending floors from the walls was common until the late 1940s, but since that time, platform framing has become the predominant form of house construction.
- Platform framing often forms wall sections horizontally on the sub-floor prior to erection, easing positioning of studs and increasing accuracy while cutting the necessary manpower.

The top and bottom plates are end-nailed to each stud with two nails at least 3.25 in (83 mm) in length (***16d*** or ***16 penny*** nails). Studs are at least doubled (creating posts) at openings, the jack stud being cut to receive the lintels (headers) that are placed and end-nailed through the outer studs.

Wall sheathing, usually a plywood or other laminate, is usually applied to the framing prior to erection, thus eliminating the need to scaffold, and again increasing speed and cutting manpower needs and expenses. Some types of exterior sheathing, such as asphalt-impregnated fibreboard, plywood, oriented strand board and waferboard, will provide adequate bracing to resist lateral loads and keep the wall square. (Construction codes in most jurisdictions require a stiff plywood sheathing.) Others, such as rigid glass-fibre, asphalt-coated fibreboard, polystyrene or polyurethane board, will not. In this latter case, the wall should be reinforced with a diagonal wood or metal bracing inset into the studs. In jurisdictions subject to strong wind storms (hurricane countries, tornado alleys) local codes or state law will generally require both the diagonal wind braces and the stiff exterior sheathing regardless of the type and kind of outer weather resistant coverings.

Corners

A multiple-stud post made up of at least three studs, or the equivalent, is generally used at exterior corners and intersections to secure a good tie between adjoining walls and to provide nailing support for the interior finish and exterior sheathing. Corners and intersections, however, must be framed with at least two studs.

Nailing support for the edges of the ceiling is required at the junction of the wall and ceiling where partitions run parallel to the ceiling joists. This material is commonly referred to as 'dead wood' or backing.

Exterior Wall Studs

Wall framing in house construction includes the vertical and horizontal members of exterior walls and interior partitions. These members, referred to as studs, wall plates and lintels, serve as a nailing base for all covering material and support the upper floors, ceiling and roof.

Exterior wall studs are the vertical members to which the wall sheathing and cladding are attached. They are supported on a bottom plate or foundation sill and in turn support the top plate. Studs usually consist of 1.5 in × 3.5 in (38 mm × 89 mm) or 1.5 in × 5.5 in

(38 mm × 140 mm) lumber and are commonly spaced at 16 in (410 mm) on centre. This spacing may be changed to 12 in (300 mm) or 24 in (610 mm) on centre depending on the load and the limitations imposed by the type and thickness of the wall covering used. Wider 1.5 in × 5.5 in (38 mm × 140 mm) studs may be used to provide space for more insulation. Insulation beyond that which can be accommodated within a 3.5 in (89 mm) stud space can also be provided by other means, such as rigid or semi-rigid insulation or batts between 1.5 in × 1.5 in (38 mm × 38 mm) horizontal furring strips, or rigid or semi-rigid insulation sheathing to the outside of the studs. The studs are attached to horizontal top and bottom wall plates of 1.5 in (38 mm) lumber that are the same width as the studs.

Interior Partitions

Interior partitions supporting floor, ceiling or roof loads are called loadbearing walls; others are called non-loadbearing or simply partitions. Interior loadbearing walls are framed in the same way as exterior walls. Studs are usually 1.5 in × 3.5 in (38 mm × 89 mm) lumber spaced at 16 in (410 mm) on centre. This spacing may be changed to 12 in (300 mm) or 24 in (610 mm) depending on the loads supported and the type and thickness of the wall finish used.

Partitions can be built with 1.5 in × 2.5 in (38 mm × 64 mm) or 1.5 in × 3.5 in (38 mm × 89 mm) studs spaced at 16 or 24 in (400 or 600 mm) on centre depending on the type and thickness of the wall finish used. Where a partition does not contain a swinging door, 1.5 in × 3.5 in (38 mm × 89 mm) studs at 16 in (410 mm) on centre are sometimes used with the wide face of the stud parallel to the wall. This is usually done only for partitions enclosing clothes closets or cupboards to save space. Since there is no vertical load to be supported by partitions, single studs may be used at door openings. The top of the opening may be bridged with a single piece of 1.5 in (38 mm) lumber the same width as the studs. These members provide a nailing support for wall finish, door frames and trim.

Lintels (Headers)

Lintels (or, headers) are the horizontal members placed over window, door and other openings to carry loads to the adjoining studs. Lintels are usually constructed of two pieces of 2 in (nominal) (38 mm) lumber separated with spacers to the width of the studs and nailed together to form a single unit. The preferable spacer material is rigid insulation. The depth of a lintel is determined by the width of the opening and vertical loads supported.

Wall Sections

The complete wall sections are then raised and put in place, temporary braces added and the bottom plates nailed through the subfloor to the floor framing members. The braces should have their larger dimension on the vertical and should permit adjustment of the vertical position of the wall.

Once the assembled sections are plumbed, they are nailed together at the corners and intersections. A strip of polyethylene is often placed between the interior walls and the exterior wall, and above the first top plate of interior walls before the second top plate is applied to attain continuity of the air barrier when polyethylene is serving this function.

A second top plate, with joints offset at least one stud space away from the joints in the plate beneath, is then added. This second top plate usually laps the first plate at the corners and partition intersections and, when nailed in place, provides an additional tie to the framed walls. Where the second top plate does not lap the plate immediately underneath at corner and partition intersections, these may be tied with 0.036 in (0.91 mm) galvanized steel plates at least 3 in (76 mm) wide and 6 in (150 mm) long, nailed with at least three 2.5 in (64 mm) nails to each wall.

Balloon Framing

Balloon framing is a method of wood construction – also known as "Chicago construction" in the 19th century – used primarily in Scandinavia, Canada and the United States (up until the mid-1950s). It utilises long continuous framing members (studs) that run from the sill plate to the top plate, with intermediate floor structures let into and nailed to them. Here the heights of window sills, headers and next floor height would be marked out on the studs with a storey pole. Once popular when long lumber was plentiful, balloon framing has been largely replaced by *platform framing.*

It is not certain who introduced balloon framing in the United States. However, the first building using balloon framing was possibly a warehouse constructed in 1832 in Chicago, Illinois, by George Washington Snow. Architectural critic Sigfried Giedion cited Chicago architect J. M. Van Osdel's 1880s attribution, as well as A. T. Andreas' 1885 *History of Chicago,* to credit Snow as 'inventor of the balloon frame method.' In 1833, Augustine Taylor (1796–1891) constructed St. Mary's Catholic Church in Chicago using the balloon framing method.

In the 1830s, Hoosier Solon Robinson published articles about a revolutionary new framing system, called "balloon framing" by later builders. Robinson's system called for standard 2x4 lumber, nailed together to form a sturdy, light skeleton. Builders were reluctant to adopt the new technology, however, by the 1880s, some form of 2x4 framing was standard.

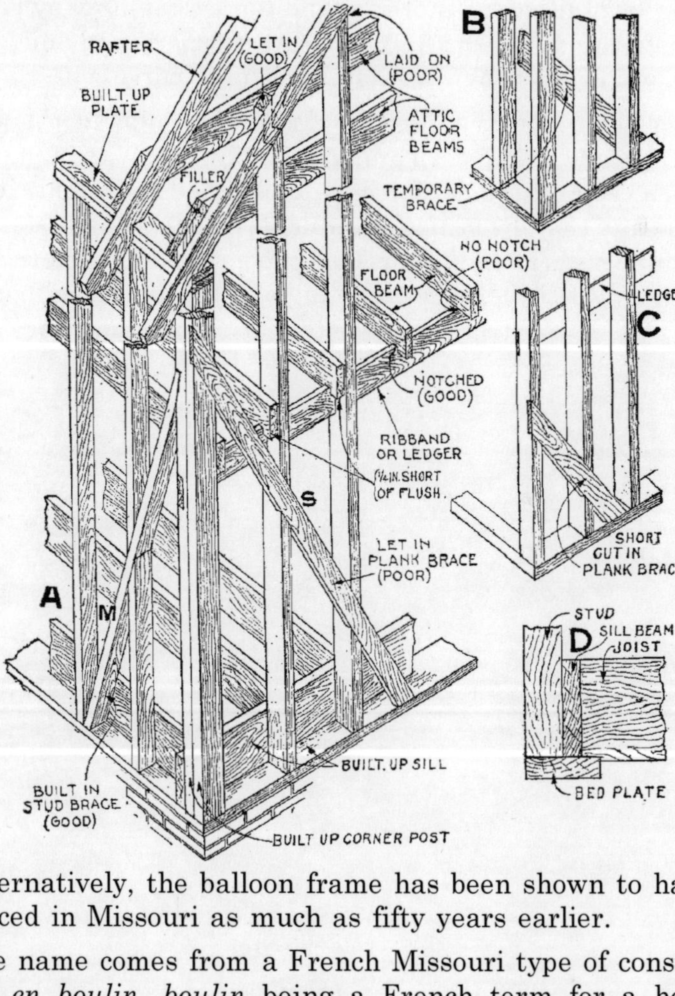

Alternatively, the balloon frame has been shown to have been introduced in Missouri as much as fifty years earlier.

The name comes from a French Missouri type of construction, *maison en boulin, boulin* being a French term for a horizontal scaffolding support. Historians have also fabricated the following story: As Taylor was constructing his first such building, St. Mary's Church, in 1833, skilled carpenters looked on at the comparatively thin framing members, all held together with nails, and declared this method of construction to be no more substantial than a balloon. It would surely blow over in the next wind! Though the criticism proved baseless, the name stuck.

Although lumber was plentiful in 19th-century America, skilled labour was not. The advent of cheap machine-made nails, along with water-powered sawmills in the early 19th century made balloon framing highly attractive, because it did not require highly skilled carpenters, as did the dovetail joints, mortises and tenons required by post-and-beam construction. For the first time, any farmer could build his own buildings without a time-consuming learning curve.

It has been said that balloon framing populated the western United States and the western provinces of Canada. Without it, western boomtowns certainly could not have blossomed overnight. It is also a fair certainty that, by radically reducing construction costs, balloon framing improved the shelter options of poorer North Americans. For example, many 19th-century New England working neighbourhoods consist of balloon-constructed three-story apartment buildings referred to as triple deckers.

Figure: *A very unusual example of balloon framing: The Jim Kaney Round Barn, Adeline, Illinois, U.S.A.*

The main difference between platform and balloon framing is at the floor lines. The balloon wall studs extend from the sill of the first story all the way to the top plate or end rafter of the second story. The platform-framed wall, on the other hand, is independent for each floor.

Balloon framing has several disadvantages as a construction method:

1. The creation of a path for fire to readily travel from floor to floor. This is mitigated with the use of firestops at each floor level.
2. The lack of a working platform for work on upper floors. Whereas workers can readily reach the top of the walls being erected with platform framing, balloon construction requires scaffolding to reach the tops of the walls (which are often two or three stories above the working platform).
3. The requirement for long framing members.
4. In certain larger buildings, a noticeable down-slope of floors towards central walls, caused by the differential shrinkage of the wood-framing members at the perimeter versus central walls. Larger balloon-framed buildings will have central bearing walls which are actually platform framed and thus will have horizontal sill and top plates at each floor level, plus the intervening floor joists, at these central walls. Wood will shrink much more across its grain than along the grain. Therefore, the cumulative shrinkage in the centre of such a building is considerably more than the shrinkage at the perimeter where there are many fewer horizontal members. This problem, unlike the first three, takes time to develop and become noticeable.
5. Present-day balloon framing buildings often have higher heating costs, due to the lack of insulation separating a room from its exterior walls. However, this can be remedied through the addition of insulation, as with any other framed building.

Since steel is generally more fire-resistant than wood, and steel framing members can be made to arbitrary lengths, balloon framing is growing in popularity again in light gauge steel stud construction. Balloon framing provides a more direct load path down to the foundation. Additionally, balloon framing allows more flexibility for tradesmen in that it is significantly easier to pull wire, piping and ducting without having to bore through or work around framing members.

Steel Frame

Steel frame usually refers to a building technique with a "skeleton frame" of vertical steelcolumns and horizontal I-beams, constructed in a rectangular grid to support the floors, roof and walls of a building which are all attached to the frame. The development of this technique made the construction of the skyscraper possible.

Concept

The rolled steel "profile" or cross section of steel columns takes the shape of the letter "I". The two wide flanges of a column are thicker and wider than the flanges on a beam, to better withstand compressive stress in the structure. Square and round tubular sections of steel can also be used, often filled with concrete. Steel beams are connected to the columns with bolts and threaded fasteners, and historically connected by rivets. The central "web" of the steel I-beams is often wider than a column web to resist the higher bending moments that occur in beams.

Wide sheets of steel deck can be used to cover the top of the steel frame as a "form" or corrugated mold, below a thick layer of concrete and steel reinforcing bars. Another popular alternative is a floor of precast concrete flooring units with some form of concrete topping. Often in office buildings the final floor surface is provided by some form of raised flooring system with the void between the walking surface and the structural floor being used for cables and air handling ducts.

The frame needs to be protected from fire because steel softens at high temperature and this can cause the building to partially collapse. In the case of the columns this is usually done by encasing it in some form of fire resistant structure such as masonry, concrete or plasterboard. The beams may be cased in concrete, plasterboard or sprayed with a coating to insulate it from the heat of the fire or it can be protected by a fire resistant ceiling construction.

The exterior "skin" of the building is anchored to the frame using a variety of construction techniques and following a huge variety of architectural styles. Bricks, stone, reinforced concrete, architectural glass, sheet metal and simply paint have been used to cover the frame to protect the steel from the weather.

Light Steel Framing

Thin sheets of galvanized steel can be formed into steel studs for use as a building material for rough-framing in commercial or residential construction (pictured), and many other applications. The dimension of the room is established with horizontal track that is anchored to the floor and ceiling to outline each room. The vertical studs are arranged in the tracks, usually spaced 16" apart, and fastened at the top and bottom.

The primary shapes used in residential construction are the C-shape stud and the U-shaped track, and a variety of other shapes.

Framing members are generally produced in a thickness of 12 to 25 gauge. The wall finish is anchored to the two flange sides of the stud, which varies from 1-1/4" to 3" thick, and the width of web ranges from 1-5/8" to 14". Rectangular sections are removed from the web to provide access for electrical wiring.

Steel mills produce galvanized sheet steel, the base material for light-gauge steel. Sheet steel is then roll-formed into the final profiles used for framing.

The sheets are zinc coated (galvanized) to prevent oxidation and corrosion. Steel framing provides excellent design flexibility due to the inherent strength of steel, which allows it to span over a longer distance than wood, and also resist wind and earthquake loads. Steel framing is surprisingly more energy efficient than wood since you can put insulation into the studs, which you cannot do with wood studs, causing a 2 inch non-insulated space every 16 inch. Construction with steel framing contributes to thermal bridges between the outside environment and interior conditioned space.

The spacing between studs is 16 inches on centre for homes exterior and interior walls. In office suites the spacing is 24 inches on centre for all walls except for elevator and staircasewells.

Evaluating Structural Framing

Evaluating structural framing can be difficult for several reasons:

- the quality of work varies widely;
- building standards and practices have changed over the years; and
- different methods have been used and considered acceptable in different parts of North America.

The following is an overview of what you can do to improve your skills in evaluating structural framing, as well as a discussion of the limitations of this type of inspection.

In evaluating homes, at one end of the spectrum are great architectural designs and engineering, quality workmanship, and top-quality materials. At the other end of the spectrum, all those things are poor and failing. Homes that are either in great shape or terrible shape are easy to inspect because they require very little judgment.

In most of the homes you'll inspect, the quality of design, workmanship and material will lie somewhere in the middle. You'll often have to decide whether something is a defect or just poor-quality work, whether it's dangerous and needs immediate correction, or whether it will last and be safe, even though it's ugly? Making these decisions correctly is really the hard part.

Knowing the Fundamentals

As a home inspector, you'll be looking at structural work that may be currently under construction, or several hundred years old, and all types of homes in between. What was acceptable 80 years ago may not be acceptable now. While this is especially true for parts of the home like the electrical system, basic framing methods haven't changed as radically. Most of what you'll see will be variations on western platform framing, balloon framing, or post-and-beam. Once you understand the basic framing methods, you'll at least have an idea of what structural configuration is probably behind the drywall.

Inspecting in Context

You'll need to evaluate homes in the context of the time in which they were originally built, taking into consideration locally accepted framing practices, and the general quality of the home. Don't expect a 300-year-old cottage in Boston, Massachusetts, to be built like a brand new mansion in Pacific Palisades, California.

Speaking of mansions... Don't assume that high-end homes have been built using good quality materials or workmanship. Those assumptions can cause you to make major misses, and a major miss in a multi-million dollar home can make you cry all the way to your lawyer's office.

Three-hundred years ago, there were highly skilled carpenters, other carpenters whose skills were terrible, and carpenters whose skills lay between the extremes. That hasn't changed, and the same is true for other occupations connected with the building of homes.

Load Path

Homes are built to transfer the structural load down through the framing to the foundation and soil. This is called the "load path" and it's where most of the forces in a home are concentrated.

Understanding the different types of basic framing methods will help you to identify load paths in homes built using different framing methods. Being able to locate key point loads will also help you know where to look for structural problems.

Point loads are loads that are concentrated in a small area, such as where a post supporting a ridge beam rests on a header above a window. Point loads are high-stress areas. Load paths don't always transfer the load downward. Uplift can be a strong force, too. Forces may move down, up, and laterally during an earthquake, so if you live in an area subject to seismic events, such as Northridge, California, you'll need to learn about the framing methods that have been developed to deal with shakers.

Framing Support

Since the structural framing is supported by the foundation and soil, any movement of those components will be reflected in the framing. Keep in mind that there can be a number of reasons for soil movement, such as:

- changes in soil volume with the gain or loss of moisture content. Moisture can cause certain clays to expand as they gain moisture, and some types may contract with loss of moisture. These are called "expansive soils." Excessive soil moisture can also affect the ability of the soil to bear a load. When expansive soils are the cause, the problem may continue; and/or
- inadequate compaction during original construction. When soil is disturbed during excavation, it must be re-compacted by mechanical means before the foundation is installed on it. If soil is not re-compacted, the weight of the structure will cause soil to consolidate as soil particles are pressed together. The result is settling of the soil and anything resting on it. Once the soil reaches a compacted state, settling will stop. This type of problem usually takes place within the first few years after original construction, then stops.

Foundation undermining from erosion can be caused by:

- erosion caused by water flowing from underground springs or poorly routed roof drainage. This can undermine the foundation (bad roof drainage caused the undermining in the photo above): and/or
- nearby excavation. Soil supporting the foundation is supported by the surrounding soil. Digging too near the foundation can interrupt the cone of support beneath load-bearing soil.

Knowing what to look for will require some knowledge about the soils and construction practices common to the areas in which you work. It takes time to build that library of knowledge, but you can speed the process by talking to local contractors and building officials.

Where to Look

Much of the structure is hidden, so look for evidence of failure in the components you can see, such as the materials covering the floor, walls and roof.

When you're faced with difficult decisions, remember what you do as a home inspector. You perform a visual General Home Inspection designed to identify safety issues, and system and component defects.

Design and Engineering

Structural design and engineering are important because the size and strength of structural framing members and the ways in which they're connected should be adequate. You don't have to perform calculations in order to determine adequacy. Instead, you'll be looking for signs of failure.

Bibliography

Anbuvelan, K. : *Management Concepts for Civil Engineers*, Laxmi Publications, Delhi, 2005.

Arthur Kennelly: *Recent Types of Dynamo-Electric Machinery*, P.F. Collier and Sons New York, 1902.

Banks R. F., and M. L. Kennedy: *The Technology of Cement and Concrete*, John Wiley & Sons, New York, 1955.

Bansal, R.K. : *Basic Civil Engineering and Engineering Mechanics*, Laxmi Publications, Delhi, 2011.

Barnes, P. and Bensted, J.: *Structure and Performance of Cements*, Spon Press, New York, 2001.

Benjamin Henry: *The Engineering Drawings of Benjamin Henry Latrobe*, Yale University Press, NY, 1980.

Brownell, W. E.: *Structural Clay Products*, Springer-Verlag Publisher, N.Y., 1976.

China, T.: *Building Industry, Pride of the Pacific*, Honolulu, Hawaii, 1991.

Davey, N.: *History of Building Materials*, Phoenix House, London, 1961.

David Fisher: *Rules of Thumb for Engineers and Scientists*, Gulf Pub, Houston, 1991.

Dimes, Francis B.: *Conservation of Building and Decorative Stone*, Heineman, Butterworth, 1990.

Eglinton, M. S.: *Concrete and Its Chemical Behavior*, Thomas Telford, London, 1987.

Fersht, A.: *Structure and Mechanism in Protein Science: A Guide to Enzyme Catalysis and Protein Folding*, W.H. Freeman, New York,1999.

Fitchen, John: *Building Construction Before Mechanization*. MIT Press, 1986.

Fred W.: *Sheet Metal Machine Processes*, Delmar, N.P., 1975.

Ghosh S. N., and A. K. Chatterjee: *Advances in Cement Technology*, Pergamon Press, N.Y., 1983.

Gille, B.: *Engineers of the Renaissance*, MIT Press, Cambridge, 1966.

Hasan, Syed Danish : *Civil Engineering Materials and Their Testing*, Narosa, Delhi, 2011.

Johnston, Jr., E.R.: *Mechanics of Materials*, McGraw-Hill, Inc., New York, 1981.

Komendant, A. E.: *Contemporary Concrete Structures,* McGraw-Hill Book Co., New York, 1972.

MacCullough, G.H.: *Elements of Strength of Materials,* Van Nostrand Co., New York, 1957.

Malinowski, R.: *Concrete and Mortars in Ancient Aqueducts,* Concrete International: Design and Construction. Detroit, Michigan, 1979.

McCormick, T. : *Metal Technics*, Brooks Waterloo-Jacaranda Press, Milton, Qld, 1989.

Mielenz, R. C.: *Concrete as a Modern Material, Modern Materials,* Academic Press, New York, 1965.

Orchard, D. F.: *Concrete Technology,* Applied Science Publishers, London, 1973.

Peckworth, H. F.: *Concrete Pipe Handbook,* American Concrete Pipe Assoc., Chicago, 1959.

Popovics, S.: *Concrete-Making Materials,* McGraw-Hill Book Co., New York, 1979.

Pradeep P. Phule: *The Science & Engineering of Materials*, Thomson-Engineering, NY, 2005.

Taylor, W. H.: *Concrete Technology and Practice,* Angus and Robertson Publisher, 1965.

Utkin, V.I.: *Sliding Modes in Control and Optimization*, Springer-Verlag, Berlin, 1992.

Wayne, Beaty, H. : *Standard Handbook for Electrical Engineers,* McGraw-Hill, New York, 1978.

Wilby, C. B.: *Concrete for Structural Engineers,* Newnes Butterworths, London, 1977.

William D.: *Materials Science and Engineering – An Introduction*, John Wiley and Sons, UK, 2000.

William M.: *Metal Alloys and Patinas for Castings for Metalsmiths, Jewelers and Sculptors*, University Press, Kent State, 1976.

Index